普通高等教育“十二五”计算机类规划教材

Android 程序设计教程

丁 山 编

机 械 工 业 出 版 社

本书介绍了基于 Android 系统的程序设计技术，内容涵盖了 Android 相关的各个领域，本书大致可以分成两个部分，第一部分为理论篇，主要介绍了 Android 操作系统概况、Android 生命周期与组件通信、Android 用户界面设计、Android 数据存储与交互、Android 后台服务与事件广播、媒介与网络、Android NDK 等内容。第二部分为实践篇，主要介绍了 Android 通信应用、定位与 Google 地图开发、语音与短信服务、Android 传感器应用等内容。

本书内容丰富，叙述浅显易懂，程序实例具有典型性，随书光盘中收录了本书所有例题的源代码，可供读者参考和学习。本书配有免费的教学课件，欢迎选用本书作为教材的教师登录机械工业出版社教材服务网 www.cmpedu.com 下载或发邮件到 wangkang_maizi9@126.com 索取。

本书可作为高等院校电子信息类、计算机科学与技术等专业高年级学生和研究生的教材，也可以作为学习 Android 系统程序设计的工程技术人员的参考用书。

图书在版编目（CIP）数据

Android 程序设计教程/丁山编.—北京：机械工业出版社，2014.12 (2016.8 重印)

普通高等教育“十二五”计算机类规划教材

ISBN 978-7-111-48823-1

Ⅰ.①A… Ⅱ.①丁… Ⅲ.①移动终端—应用程序—程序设计—高等学校—教材 Ⅳ.①TN929.53

中国版本图书馆 CIP 数据核字（2014）第 290190 号

机械工业出版社（北京市百万庄大街 22 号 邮政编码 100037）

策划编辑：王 康 责任编辑：王 康 刘丽敏

版式设计：常天培 责任校对：王 欣

封面设计：张 静 责任印制：常天培

涿州市京南印刷厂印刷

2016 年 8 月第 1 版第 2 次印刷

184mm×260mm · 20.5 印张 · 496 千字

标准书号：ISBN 978-7-111-48823-1

ISBN 978-7-89405-589-7（光盘）

定价：42.00 元（含 1DVD）

凡购本书，如有缺页、倒页、脱页，由本社发行部调换

电话服务 网络服务

服务咨询热线：010-88379833 机 工 官 网：www.cmpbook.com

读者购书热线：010-88379649 机 工 官 博：weibo.com/cmp1952

教育服务网：www.cmpedu.com

 金 书 网：www.golden-book.com

前　言

Android 是 Google 于 2007 年 11 月推出的一款开放的嵌入式操作系统平台，由于其完全开源的特性，广泛应用于手机、平板电脑、家电及其他嵌入式系统设计中，车载设备、智能电视、VoIP 电话和医疗设备等厂商纷纷推出 Android 系统产品。Android 正以空前的速度吸引着大批开发者的加入，尤其是应用开发工程师。本书以 Android 系统的程序设计开发为主体，并结合真实的案例向读者介绍 Android 的基本组件的使用及程序开发的整个流程，内容涵盖了所有和 Android 相关的领域。

本书力求全面、实用，对例题做到详细分析和解释，既可以帮助读者学习理解知识和概念，降低学习难度，又具有启发性，帮助读者更加轻松、迅速地理解和掌握本书内容。

本书在内容的组织上分为理论篇和实践篇共 11 章，其中第 1 ~ 7 章为理论篇，第 8 ~ 11 章为实践篇。各章的具体内容如下：

第 1 章主要对 Android 的发展、特点、环境搭建和体系结构进行简要介绍。并且讲解了 JDK、Eclipse、Android SDK 软件的下载及安装的基本知识。对 Android 应用程序进行解析，提高读者对程序的创建、目录的结构、资源的管理以及对程序权限的理解。最后讲解如何调试 Android 程序。

第 2 章主要讲述了 Android 生命周期与组件之间的通信。生命周期主要讲述了 Android 四大组件之一的 Activity 生命周期，包括生命周期函数、栈结构和基本状态三方面。组件的通信靠 Intent 实现，讲述了 Intent 基本构成、形成和过滤器。

第 3 章主要从 Android 用户界面开发出发，讲述了开发过程中经常使用到的控件，包括菜单、常用基础控件、对话框与消息框。界面中控件的结构及位置等需要通过有效的界面布局控制，Android 中提供了 5 种界面布局格式，即线性布局、相对布局、表格布局、绝对布局和框架布局。界面中还有一种必要的操作处理——外部操作的响应，通过有效的事件处理机制完成。

第 4 章主要讲述了 Android 数据存储与交互方面的内容，系统中数据交互主要通过五种方式实现，共享优先数据机制、SQLite 数据库、File 文件机制、内容提供器控件和网络存储。其中在应用程序中最常用也是最有效的数据交互方式是使用 SQLite 数据库。

第 5 章主要讲述了 Android 后台服务、事件广播和常驻程序。后台服务由系统提供的 Service 组件实现，可分为本地服务和远程服务。事件广播机制主要依靠 BroadCast Reciver 组件实现。常驻程序 AppWidget 又称为窗口小部件，是在 HomeScreen 上显示的小部件，开发时常用 AppWidgetProvider 和 AppWidgetProviderInfo 类实现。

第 6 章介绍了 Android 平台下通过程序实现音频、视频播放等操作。分别介绍了从源文件播放、文件系统播放和流媒体播放等方式。并且介绍了 Android 图形绘制与特效，包括图形的平移、旋转及缩放等操作，保存指定格式图形文件。编写专业的绘图或控制图形动画的应用程序。以及如何使用 Android 手机中内置的高性能 WebKit 内核浏览器浏览网页，使用 HTTP 和 URL 获得网络资源等内容。

第 7 章介绍了 Android NDK 的相关知识，从 NDK 的简单介绍到开发环境的配置，以及开发流程。使用 NDK 实现一些对代码性能要求较高的模块并将这些模块嵌入到 Android 应用程序中会大大地提高程序效率，比如用 NDK 开发 OpenGL。此外，如果项目中包含了大量的逻辑计算或者是3D 特效，这时 Android NDK 便会显示出它超强的功能。

第 8 章介绍了 Android 平台下的几种通信方式，即 Socket 通信、蓝牙及 WiFi。其中对它们的通信方式、通信中所需的各种 API 及其使用方法进行了介绍。在 Socket 通信中主要介绍了它的通信模型以及通信各部分的实现，并通过实例展示了其具体的通信过程。在蓝牙中主要介绍了蓝牙系统的基本构成，在 Android 下的各种 API 及通信方式。最后对 WiFi 的操作做了详细的介绍。

第 9 章介绍了 GPS 的概念、系统架构以及底层驱动的编写，并通过例子讲解了 GPS 在 Android 上的应用。

第 10 章介绍了 Android 中对语音及短消息的访问。其中重点介绍了利用 Telephony 类来监听来电与去电信息；利用 SMSManager 来发送和接收短消息，并用 PendingIntent 对发送消息进行跟踪。

第 11 章介绍了 Android 系统所支持的传感器类型，如何使用传感器 API 来获取传感器数据，如何通过 SensorManager 来注册传感器监听器，如何在 SensorEventListener 中对传感器进行监听，如何使用几种常用的传感器等。最后通过使用最常用的加速度传感器开发有趣的应用来进一步介绍传感器开发的流程。

本书内容充实，系统全面，重点突出。阐述循序渐进，由浅入深。书中所有例题均在 Eclipse + ADT 环境下运行通过。本书配有免费的电子课件，欢迎选用本书作为教材的老师登录 www. cmpedu. com 下载或发邮件到 wangkang_maizi9@ 126. com 索取。

参加本书编写、校对及程序测试工作的还有研究生刘海瑞与马文翼，在此表示感谢。

由于作者水平有限，书中难免有错误和不足之处，恳请各位专家和读者批评指正。

编　者

目 录

实 践 篇

理 论 篇

第1章 Android 操作系统

作为首个真正完全开放的手机移动平台，Android 以其开源性及强劲的功能被称为目前世界上最为流行的手机操作系统之一，本章主要对 Android 的发展、特点、环境搭建和体系结构进行简要介绍。

1.1 Android 简介

1.1.1 Android 的起源

Android 系统最初由安迪·鲁宾（Andy Rubin）等人开发研制，最初是为创建一个数码相机的先进操作系统，但是后来发现市场需求不够大，加上智能手机市场的快速成长，于是 Android 被改造成一款面向智能手机的操作系统，2005 年 8 月 Android 被 Google 收购。

Android 一词英文本意为“人形机器人”，2007 年 11 月 5 日，Google 公司正式对外展示了基于 Linux 内核的开放源代码移动设备操作系统并宣布将其命名为 Android。自此 Android 便以最具开放性的手机平台开发系统在操作系统中崭露头角。为了更好地开发与推广 Android 手机操作系统，2007 年 11 月 Google 公司与 34 家手机相关企业携手建立了开放手机联盟（Open Handset Alliance，OHA）。开放手机联盟的组成成员涵盖了手机和其他终端制造商、移动运营商、半导体厂商、软件厂商，例如 HTC（宏达电子）、SAMSUNG（三星）、Motorola（摩托罗拉）、中国移动等知名品牌公司。然而苹果、微软、诺基亚、RIM、黑莓等公司并没有在联盟成员之列。开放手机联盟建立的目的是支持谷歌可能发布的手机开发系统和应用软件，并共同开发名为 Android 的开放源代码移动系统，开发多种技术，以大幅削减移动设备和服务的开发和推广成本。随后，Google 以 Apache 免费开放源代码许可证的授权方式，发布了 Android 的源代码。让生产商推出搭载 Android 操作系统的智能手机，随着 Android 操作系统的逐步完善和发展，Android 操作系统后来逐渐拓展到平板电脑及其他领域。

1.1.2 Android 的发展史

Android 的发展与 Google 密不可分。2007 年 Google（谷歌）在收购了 Android 公司后，将其特有的谷歌服务与 Android 手机操作系统相连，赋予了 Android 操作系统全新的灵魂。

谷歌移动服务（Google Mobile Service，GMS）是谷歌的一项服务，该服务致力于让用户在利用移动电话或其他移动设备时能够使用谷歌搜索、谷歌地图、Gmail、YouTube、Android Market 等谷歌服务产品。谷歌公司将 GMS 内嵌到 Android 手机系统中，并同时给予 Android 手机生产商不同程度的授权。由于 GMS 的强大功能，GMS 成为 Android 操作系统的灵魂，大部分用户使用 Android 手机，其实就是为了使用谷歌所提供的各项服务。

Android 1. 5 版本发布后，Android 手机操作系统便开始以甜点命名系统版本。继 2008 年 9 月 Android 1. 1 发布后又陆续发布了 Android 1. 5 Cupcake（纸杯蛋糕）、Android 1. 6 Donut（甜甜圈）、Android 2. 0/2. 1 Éclair（法式奶油夹心甜点，也译为松饼）、Android 2. 2 Froyo（冻酸奶）、Android 2. 3 Gingerbread（姜饼）、Android 3. 0/3. 1/3. 2 Honeycomb（蜂巢）、Android 4. 0 Ice Cream Sandwich（冰激凌三明治）、Android 4. 1/4. 2/4. 3 Jelly Bean（果冻豆）。到目前为止，最新的 Android 版本是 Android 4. 4 KitKat（奇巧）。

Android 操作系统是基于 Linux 的开源操作系统。除了应用于手机移动平台外，它在平板电脑方面的应用也日趋广泛，并且有报道称 Android 开始应用于军事方面。

1. 1. 3　Android 的特点

Android 操作系统为何会脱颖而出，这与市场的选择和 Android 的优越性紧密相关。

手机从其出现到目前市场上流行的形形色色的功能样式，最受人们欢迎并且被人们普遍使用的是功能手机和智能手机。功能手机包含了基本的手机应用并集成了一定的手机扩展功能，然而功能手机和智能手机并不可混为一谈。功能手机所具有的功能是手机生产厂商利用 Java 软件集成于手机之中的，而智能手机内置与计算机类似的操作系统，可以根据需要下载和安装不同的软件来扩展手机的功能，而不受手机生产厂商的限制。所谓智能手机，即“像个人电脑一样，具有独立的操作系统，可以由用户自行安装软件、游戏等第三方服务商提供的程序，通过此类程序来不断对手机的功能进行扩充，并可以通过移动通信网络来实现无线网络接入的这样一类手机的总称”。

智能手机的特点如下：

（1）拥有一般手机的全部功能，如能够进行正常的手机通话、收发短信等手机通信功能，并且具备智能手机的典型功能之一，即多任务功能和复制粘贴功能。

（2）具备手机无线网络接入能力和掌上电脑的功能。

（3）具备手机功能扩展能力。使用者可以根据自身的需要下载安装相应功能软件，从而达到扩展手机功能的目的。

目前市场上的智能手机操作系统主要有开放手机联盟的 Android 操作系统、诺基亚的 Symbian 操作系统、苹果的 iOS 操作系统、RIM 的 BlackBerry 操作系统等。自 iOS 操作系统和 Android 操作系统推进市场以后，Symbian 操作系统的市场份额已被占去大多半。

和 Symbian、iOS 系统比较起来，Android 系统的特点及优势如下：

（1）可自动切换无线网络，节省上网费用。

（2）操作界面更简洁、个性，与实际使用联系紧密；更易上手，操作方便。

（3）互联网连接使用简单便捷，可称之为最佳的互联网移动终端。

（4）支持多任务运行，切换简单快捷，流畅无阻。

（5）支持与微软 Exchange 的同步，办公娱乐两不误。

（6）全新开源系统，软件数量和增长速度远超过 Windows Mobile。无“证书”限制，安装软件更自由，系统发展更具前景。软件安装卸载更方便，无需第三方平台软件。

（7）强大的 Linux 内核，内存管理更优秀，不容易死机。

1.2 Android 体系结构

Android 是基于 Linux 内核并开放源代码的移动平台操作系统，它采用了分层式架构，主要分为四层，从低到高依次为 Linux 核心层（LINUX KERNEL）、系统运行层（LIBRARIES）、应用程序框架层（APPLICATION FRAMEWORK）及应用程序层（APPLICATIONS）。其系统架构如图 1-1 所示。

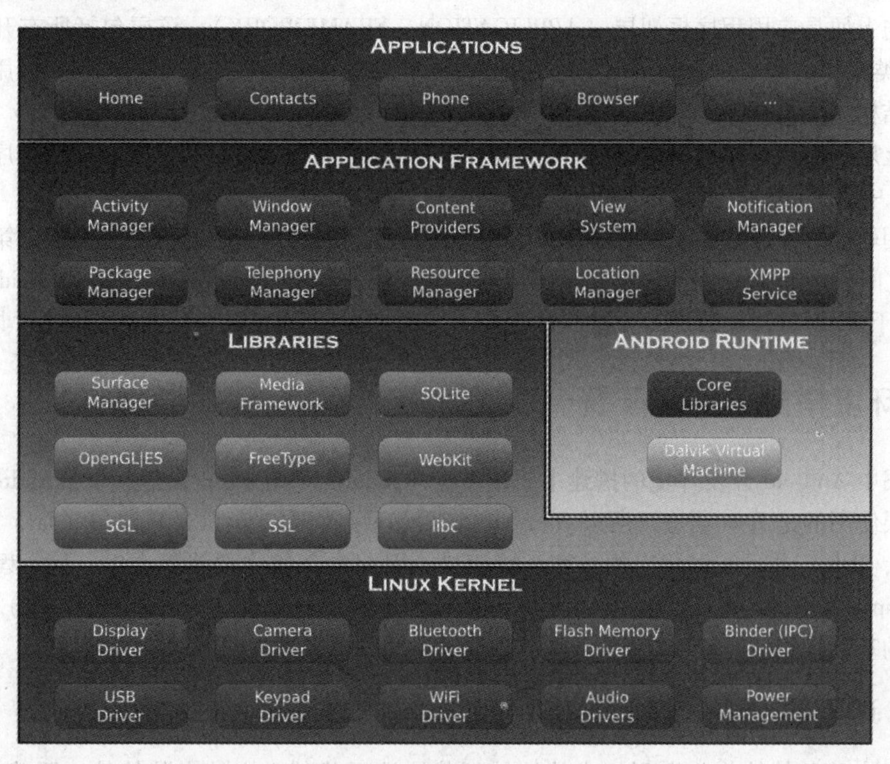

图 1-1　Android 的系统架构

Android 的分层结构特点是低层为高层提供统一的服务，高层使用低层所提供的服务，并屏蔽低层和本层的差异。当低层结构发生变化时并不会影响高层的结构及功能的实现。下面将从底层 Linux 核心层到高层应用程序层，依次讲解各层的功能及特点。

由于 Android 是以 Linux 为内核开发的。对于 Linux 核心层（LINUX KERNEL），其主要功能是隐藏硬件特性并为高层提供一系列统一的系统服务及驱动程序，例如内存管理、进程管理、网络协议栈等系统服务及 WiFi 驱动、蓝牙驱动等相关功能的驱动程序。核心层结构如图 1-1 中 LINUX KERNEL 部分所示。

Linux 核心层是用户空间和内核空间的分界线。并且仅 Linux 核心层属于内核空间，其余三层均属于用户空间。Linux 核心层是硬件和软件之间的抽象层，通过 C 语言开发并提供其基本功能。

再向上一层主要是由系统运行层（LIBRARIES）构成，还包括虚拟机（Virtual Machine）。系统运行层的基本结构如图 1-1 中 LIBRARIES 部分所示。

由于 Android 系统需要支持 Java 代码的运行，系统运行层主要包括系统运行库 LIB 和 Android 运行环境，通常称之为中间层。与移动设备的平台应用相关，并通过 Java 调用接口函数实现与上层之间的通信。该层由 C/C ++ 开发，主要包括媒体库、Web 浏览引擎、关系数据库引擎、图形库等功能。Android 运行环境以虚拟机 Dalvik Virtual Machine 为主，虚拟机以编码格式运行。其中 Dalvik 虚拟机和标准的 Java 虚拟机之间的区别主要体现在，Dalvik 是基于寄存器的，而标准的 Java 虚拟机是基于堆栈的，并且 Dalvik 经过优化可在有限的内存中同时运行多个虚拟机程序实例，每一个 Dalvik 应用可作为一个独立的 Linux 进程执行。

再向上便是应用程序框架层（APPLICATION FRAMEWORK）。该层包括所有开发所用的 SDK 类库和某些未公开的接口类库，是 Android 移动平台的核心机制的体现。应用程序框架层的结构如图 1-1 中 APPLICATION FRAMEWORK 部分所示。

最上层是应用程序层，包括各种应用软件，如通话程序、短信程序等。应用程序层结构如图 1-1 中 APPLICATIONS 部分所示。

应用软件由各公司自行开发，以 Java 作为编写程序的一部分。从低层到高层，第二层和第三层之间，是本地代码层和 Java 代码层的接口。第三层和第四层之间，是 Android 的系统 API。对于 Android 应用程序的开发，第三层以下的内容是不可见的，仅考虑系统 API 即可。

1.3 环境搭建及环境配置

本书中 Android 开发环境的搭建主要是基于 Windows 32 位操作系统，由于 Android 的应用框架层使用的是 Java 语法，所以 Android 的开发环境需要安装 Java 开发包（Java Development Kit，JDK）并且配置相应的开发环境变量。因此本节将主要讲解 JDK、Java 程序开发环境 Eclipse 和 Android 开发包（Android Software Development Kit，即 Android SDK）的下载及安装的基本知识。

1.3.1 JDK 的下载和安装

在安装 JDK 软件前首先到相应的软件下载地址下载其对应的安装软件。下载地址为：http://Java. sun. com/Javase/downloads/index. jsp。下载 JDK 方法如下。打开对应链接，在弹出的页面中将会看到很多下载版本，选择对应下载版本，此处选择 Java SE 7。接着将会进入 Java SE Downloads 的下载页面如图 1-2 所示。

此页面还包括其他与 Java 安装相关的下载项，可根据自身需要进行下载，主要使用的是 JDK（其中 JDK 中包含 JRE），单击 Download 进行下载。

将会出现如图 1-3 所示下载页面。

在注册登录后便可以下载和自己计算机所安装的系统对应的 JDK 版本，在此处使用的是 Windows x86 下所对应的安装版本。下载完成后单击应用程序文件进行安装（注意在此处程序的安装路径可在安装过程中自行设定）。

本书附有已下载好的 JDK 的安装包，此处以安装包内应用程序进行安装，双击安装包中的 JDK 安装应用程序 jdk-7u51-windows-i586. exe，将会弹出如图 1-4 所示的安装对话框。选择“接受”选项，然后在弹出的对话框中依次单击“下一步”。安装过程如图 1-5 ~ 图 1-8 所示。本软件默认安装在 C:\Program Files\Java\jdk1. 7. 0_51\中。

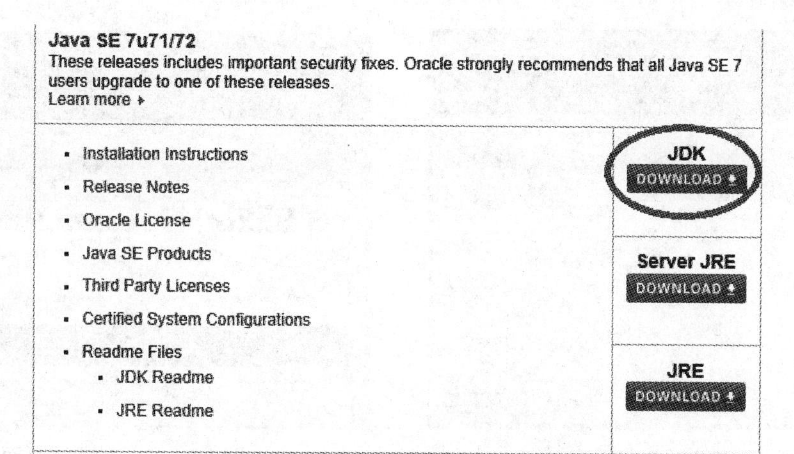

图 1-2　JDK 的下载页面 1

Java SE Development Kit 7u71

You must accept the Oracle Binary Code License Agreement for Java SE to download this software.

Thank you for accepting the Oracle Binary Code License Agreement for Java SE; you may now download this software.

Product / File Description	File Size	Download
Linux x86	119.44 MB	jdk-7u71-linux-i586.rpm
Linux x86	136.76 MB	jdk-7u71-linux-i586.tar.gz
Linux x64	120.81 MB	jdk-7u71-linux-x64.rpm
Linux x64	135.63 MB	jdk-7u71-linux-x64.tar.gz
Mac OS X x64	185.84 MB	jdk-7u71-macosx-x64.dmg
Solaris x86 (SVR4 package)	139.36 MB	jdk-7u71-solaris-i586.tar.Z
Solaris x86	95.48 MB	jdk-7u71-solaris-i586.tar.gz
Solaris x64 (SVR4 package)	24.68 MB	jdk-7u71-solaris-x64.tar.Z
Solaris x64	16.36 MB	jdk-7u71-solaris-x64.tar.gz
Solaris SPARC (SVR4 package)	138.74 MB	jdk-7u71-solaris-sparc.tar.Z
Solaris SPARC	98.62 MB	jdk-7u71-solaris-sparc.tar.gz
Solaris SPARC 64-bit (SVR4 package)	23.94 MB	jdk-7u71-solaris-sparcv9.tar.Z
Solaris SPARC 64-bit	18.35 MB	jdk-7u71-solaris-sparcv9.tar.gz
Windows x86	127.78 MB	jdk-7u71-windows-i586.exe
Windows x64	129.52 MB	jdk-7u71-windows-x64.exe

图 1-3　JDK 的下载页面 2

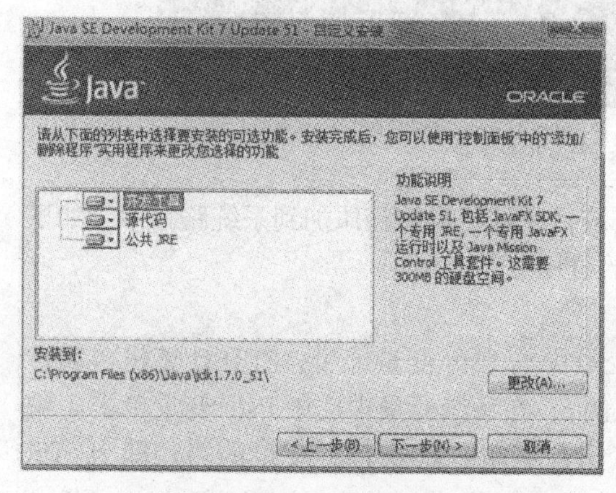

图 1-4　JDK 的安装页面 1

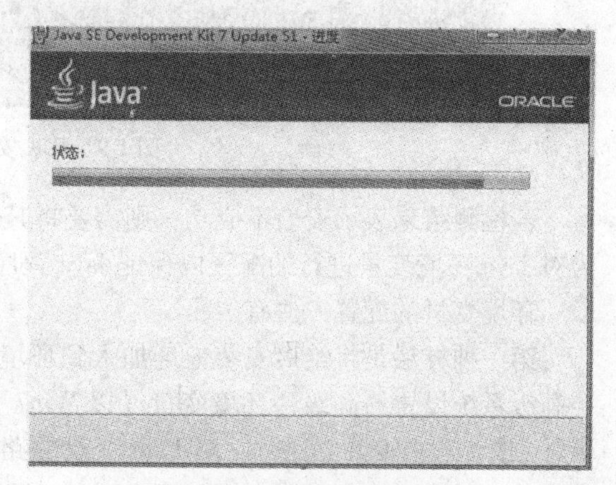

图 1-5　JDK 的安装页面 2

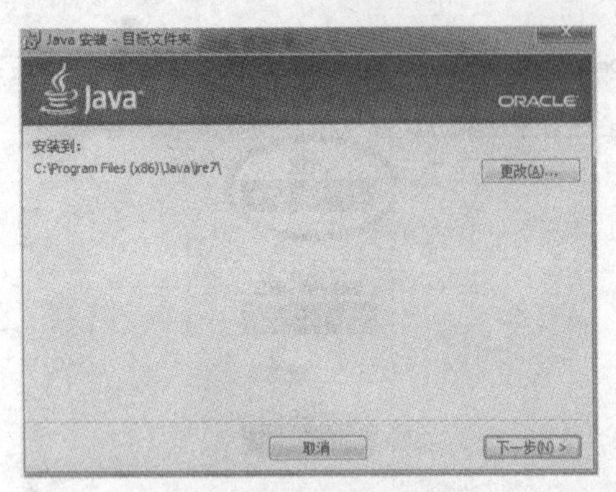

图 1-6　JDK 的安装页面 3

图 1-7　JDK 的安装页面 4

注意在安装过程中，将最后一步中显示自述文件前边的勾去掉，然后安装完成。

在 JDK 安装完成后需要对其进行进一步的检测以确定软件成功安装。具体的检测方法如下所述。

单击“开始”→“运行”（对于 Win7 可直接通过组合键 < Win + R > 实现），在弹出的对话框中输入“cmd”并按 < Enter > 键确定后，将会弹出 cmd 命令器窗口。在弹出的 cmd 窗口中输入 java -version（注意 java 后面有一个空格）并按 Enter 确定。若显示如下信息则表示安装成功，如图 1-9 所示。

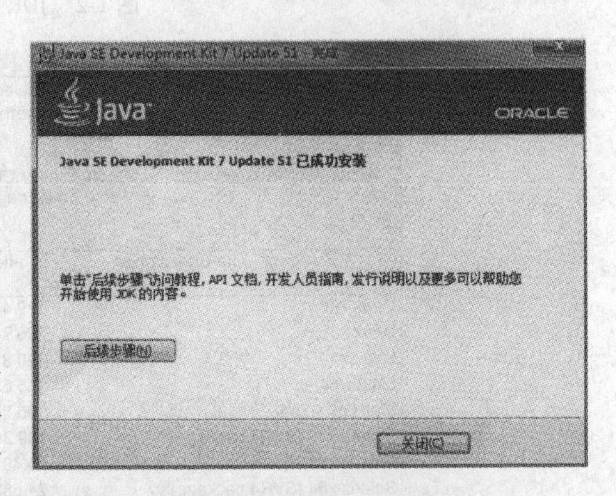

图 1-8　JDK 的安装页面 5

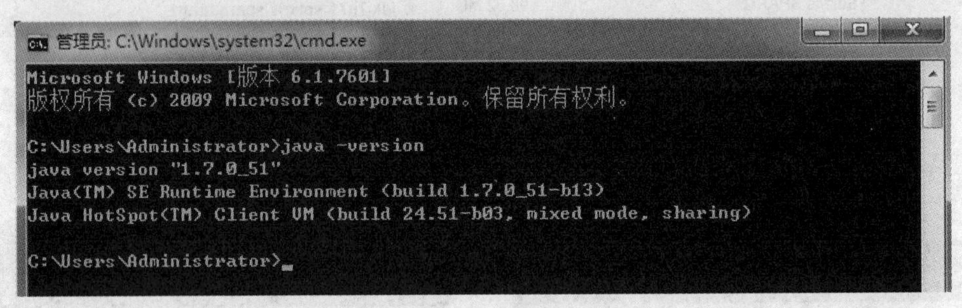

图 1-9　JDK 安装测试页面 1

若检测结果表示安装不成功，则需要将其安装目录的绝对路径添加到系统路径中，如下为对 Java 环境变量进行的配置以保证 Java 程序的可用性。

环境变量的配置方法有二：

第一种方法是将绝对安装地址加入到环境变量中的 Path 变量即可：打开计算机→属性→高级系统设置→高级→环境变量（以 Win7 为例），在系统变量中选中 Path 变量单击“编辑”（或者双击 Path 变量），对 Path 变量的路径进行编辑，此处 Path 变量的路径即为 SDK 的安装地址及 Java 的安装地址。例如地址(;F:\Android-sdk-windows\tools;C:\Program Files\

Java\jdk1.7.0_51\bin;)。

第二种方法是设置 JAVA_HOME、Path、ClassPath 这三个系统变量的路径（方法同上）。在系统变量处选择新建，在新建变量对话框中变量名输入 JAVA_HOME，变量值处输入 Java 的安装地址，此处为 C:\Program Files\Java\jdk1.7.0_51；新建系统变量 ClassPath 并将变量值赋为 Win7→Win7；编辑 Path 变量将其赋值为.;%JAVA_HOME%\bin;。

此处 JAVA_HOME 表示 Java 的安装路径，即将变量值写为其安装路径即可；Path 变量的含义是系统在任何路径下都可以识别 Java 命令；ClassPath 为 Java 加载类（class or lib）路径，只有类在 ClassPath 中，Java 命令才能识别。

在配置好环境变量的前提下，再次检测环境变量是否配置成功。打开 cmd 命令器，在命令器中输入 Java -version 若出现如图 1-10 所示状态则表示配置成功且 Java 程序安装成功并可使用。

图 1-10 JDK 安装测试页面 2

1.3.2 下载安装 Eclipse

安装并测试成功 JDK 软件后，下一步要安装 Eclipse 软件。在安装 Eclipse 软件前首先到相应的软件下载地址下载软件安装包。Eclipse 的下载地址：http://www.eclipse.org/downloads/。输入下载地址后，进入下载页面，如图 1-11 所示。

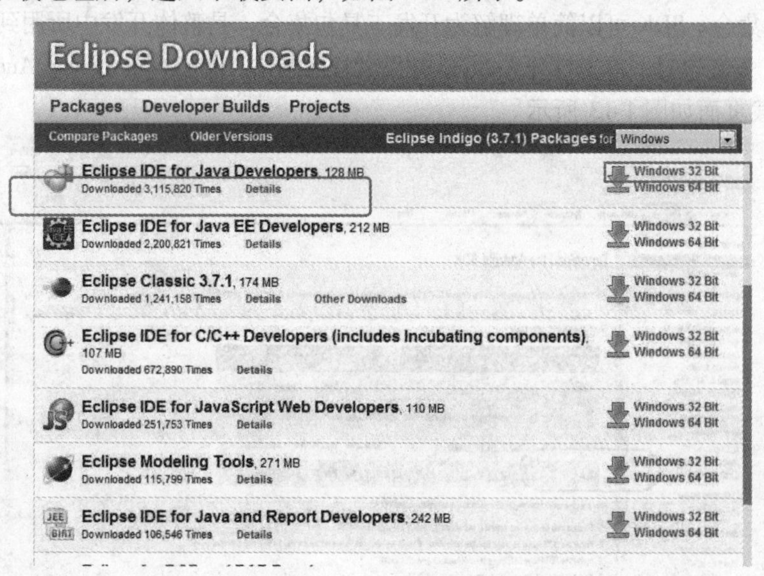

图 1-11 Eclipse 的下载页面 1

选择 Eclipse IDE for Java Developers 右侧与你的处理器位数相对应的项进行下载（此处以 32 位处理器为例）。选择完成后将进入如图 1-12 所示下载页面。

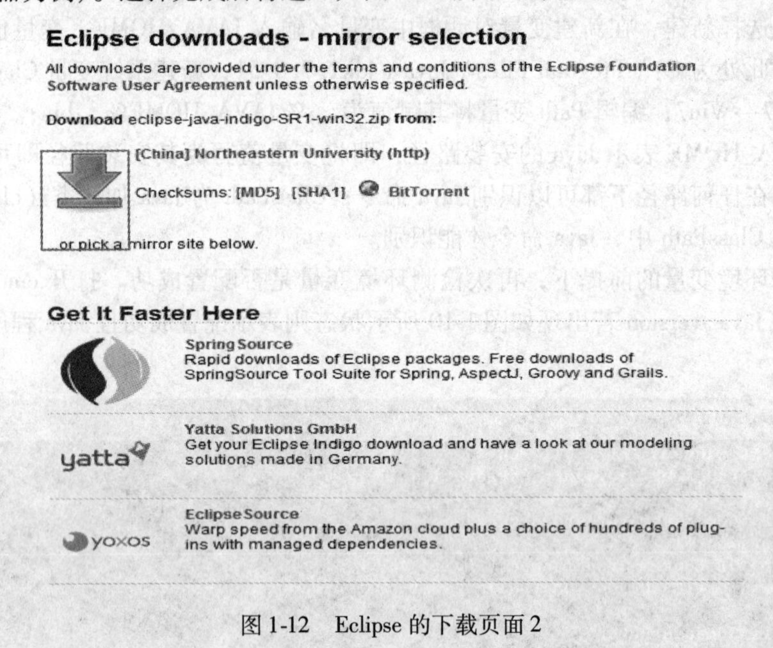

图 1-12　Eclipse 的下载页面 2

下载完成后解压即可使用，Eclipse 下载包在随书光盘中，读者可以直接从光盘中获取，解压便可使用（在附录中下载其压缩包解压即可使用）。

1.3.3　SDK 的下载和安装

SDK 一般是指一些被软件工程师用于作为特定的软件包、软件框架、硬件平台、操作系统等建立应用软件的开发工具的集合。在 Android 中，它为开发者提供了库文件以及其他开发所用到的工具的集合。SDK 可以简单理解为开发工具包集合，是整体开发中所用到的工具包。

下载 SDK 安装包时，首先打开 Android 开发者社区网站：http://developer.Android.com/，并转入 SDK 下载页面如图 1-13 所示。

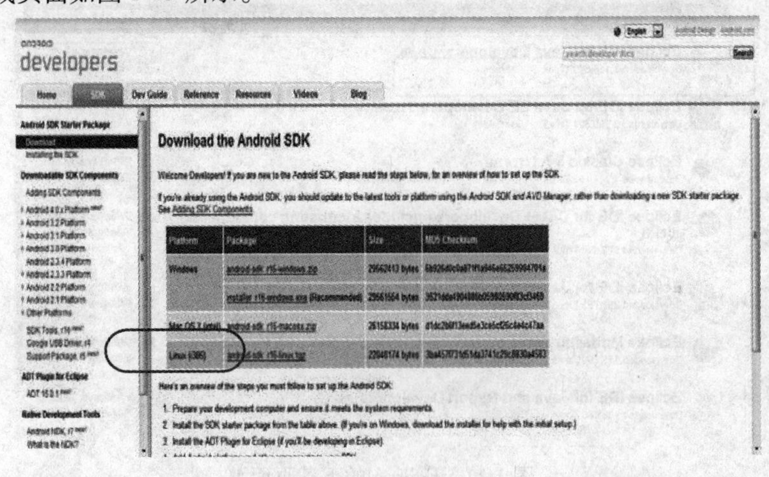

图 1-13　SDK 的下载

选择与所安装的系统一致的下载项，此处我们选择用于 Windows 平台的 Android-SDK r16-windows. zip，单击进入下载页面，下载并保存。下载完成后解压下载的文件，注意解压的路径在环境变量的设置中会涉及使用，此处假设将下载的压缩包解压在 F:\Android\ 目录下，配置环境变量方法同上，若使用第二种方法配置，此处只需更改 SDK 文件中的 tools 的绝对路径即可，具体方法如下：

计算机→属性→高级系统设置→高级→环境变量，在打开的环境变量对话框中，对系统变量进行设置，单击“新建”，在变量名处输入 SDK_HOME，变量值处输入前面提到的解压后 SDK 的路径，此处即为 F:\android-sdk-windows，设置完成后单击“确定”，如图 1-14 所示。

在此基础上，还需编辑路径的变量，即同样在打开的环境变量对话框中完成此操作，在环境变量对话框中对系统变量进行操作，选中变量 Path 对其进行编辑，将变量名改为% SDK_HOME% \tools；即可，如图 1-15 所示。

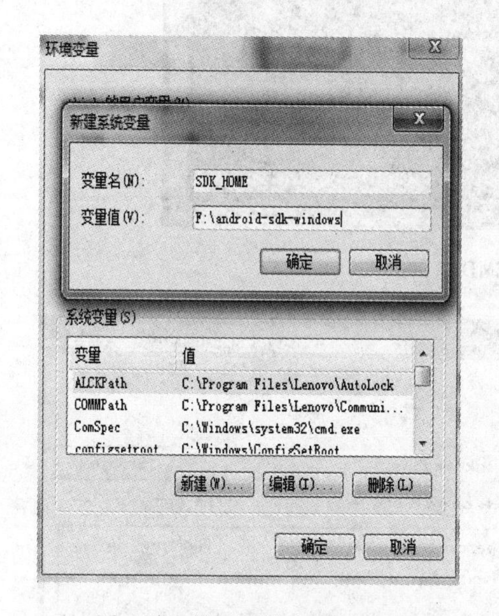

图 1-14 设置环境变量页面 1

图 1-15 设置环境变量页面 2

变量值有问题也可以直接将 Path 的变量值设置为；F:\Android 光盘内容\ DISK-ANDRIOD\ Android-sdk-windows \ tools；C:\Program Files \ Java \ jdk1. 7. 0_51 \ bin；免去以上两步的繁琐。

利用 cmd 命令测试窗口进行 SDK 软件测试：打开 cmd 命令窗口，输入测试命令 android -h，若显示如图 1-16 所示信息则表示 SDK 软件安装成功。

完成 SDK 的安装后，配置 Android SDK 的大概方法如下。打开 Eclipse 开发环境，依次选择 window→Preferences→Android，将会出现 SDK 的配置窗口，如图 1-17 所示。

在 Android 中，为开发者提供了库文件以及其他开发所用到的工具。简单理解为开发工具包集合，是整体开发中所用到的工具包，如果不用 Eclipse 作为开发工具，就不需要下载 ADT，只下载 SDK 即可开发。

```
Microsoft Windows [版本 6.1.7601]
版权所有 (c) 2009 Microsoft Corporation。保留所有权利。

C:\Users\Administrator>android -h

      Usage:
      android [global options] action [action options]
      Global options:
  -h --help       : Help on a specific command.
  -v --verbose    : Verbose mode, shows errors, warnings and all messages.
     --clear-cache: Clear the SDK Manager repository manifest cache.
  -s --silent     : Silent mode, shows errors only.
                                                          Valid
                                                          actions
                                                          are
                                                          composed
                                                          of a verb
                                                          and an
                                                          optional
                                                          direct
                                                          object:
-   sdk              : Displays the SDK Manager window.
-   avd              : Displays the AVD Manager window.
-  list              : Lists existing targets or virtual devices.
-  list avd          : Lists existing Android Virtual Devices.
-  list target       : Lists existing targets.
-  list sdk          : Lists remote SDK repository.
- create avd         : Creates a new Android Virtual Device.
-  move avd          : Moves or renames an Android Virtual Device.
- delete avd         : Deletes an Android Virtual Device.
- update avd         : Updates an Android Virtual Device to match the folders
                       of a new SDK.
- create project     : Creates a new Android project.
- update project     : Updates an Android project (must already have an
                       AndroidManifest.xml).
- create test-project : Creates a new Android project for a test package.
- update test-project : Updates the Android project for a test package (must
                       already have an AndroidManifest.xml).
- create lib-project : Creates a new Android library project.
- update lib-project : Updates an Android library project (must already have
                       an AndroidManifest.xml).
```

图 1-16　CMD 窗口

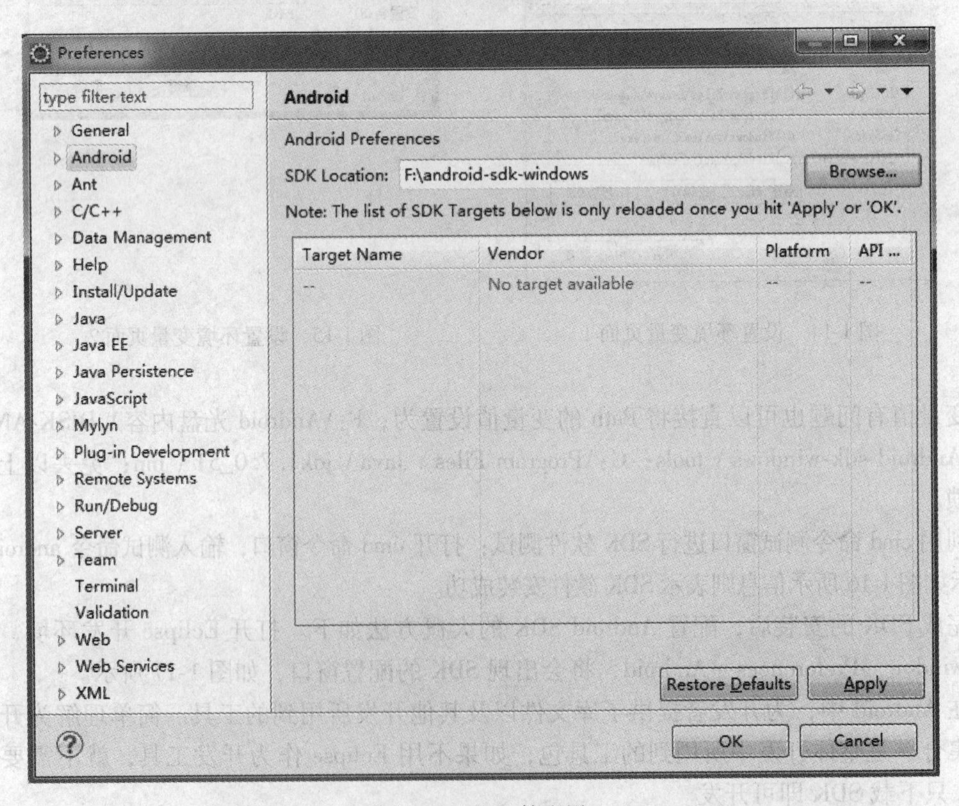

图 1-17　SDK 的配置

1.3.4 创建 AVD

Dalvik 虚拟机是 Android 操作系统的核心基础，并且 Dalvik 虚拟机的特性会直接影响 Android 操作平台的性能。Dalvik 虚拟机的主要功能是使用 . dex 文件格式保存已经编译的程序。

每一个 Dalvik 应用作为一个独立的 Linux 进程执行。下面将解释如何使用 cmd 命令器查询和建立模拟器。使用组合键 < Win + R > 打开运行窗口，输入 cmd 进入 cmd 窗口，查询模拟器信息输入 android list target 并按 < Enter > 键，将会出现如图 1-18 所示模拟器信息。

```
管理员: C:\Windows\system32\cmd.exe
     Type: Platform
     API level: 19
     Revision: 1
     Skins: HVGA, QVGA, WQVGA400, WQVGA432, WSVGA, WVGA800 (default), WVGA854, W
XGA720, WXGA800, WXGA800-7in
     ABIs : armeabi-v7a
----------
id: 13 or "Google Inc.:Google APIs:19"
     Name: Google APIs
     Type: Add-On
     Vendor: Google Inc.
     Revision: 1
     Description: Android + Google APIs
     Based on Android 4.4 (API level 19)
     Libraries:
      * com.google.android.media.effects (effects.jar)
          Collection of video effects
      * com.android.future.usb.accessory (usb.jar)
          API for USB Accessories
      * com.google.android.maps (maps.jar)
          API for Google Maps
     Skins: WVGA854, WQVGA400, WSVGA, WXGA800-7in, WXGA720, HVGA, WQVGA432, WVGA
800 (default), QVGA, WXGA800
     ABIs : armeabi-v7a
```

图 1-18 对 AVD 进行查询

输出结果表示系统共含有 6 个模拟器，各自的 ID 分别为 1、2、3、4、5、6。若要建立模拟器其方法是在 cmd 窗口输入 android create avd-- <新建 AVD 的名称> -- <新建 AVD 的 ID 号>，确定后在 cmd 窗口将会显示如图 1-19 所示信息。

```
管理员: C:\Windows\system32\cmd.exe
C:\Users\think>android create avd --ff --target 9
Error: Flag '--ff' is not valid for 'create avd'.

Usage:
  android [global options] action [action options]

Global options:
  -v --verbose   Verbose mode: errors, warnings and informational messages are pr
inted.
  -h --help      Help on a specific command.
  -s --silent    Silent mode: only errors are printed out.

Action "create avd":
  Creates a new Android Virtual Device.
Options:
  -c --sdcard    Path to a shared SD card image, or size of a new sdcard for the
new AVD
  -t --target    Target id of the new AVD [required]
  -n --name      Name of the new AVD [required]
  -p --path      Location path of the directory where the new AVD will be created

  -f --force     Force creation (override an existing AVD)
  -s --skin      Skin of the new AVD

C:\Users\think>
```

图 1-19 建立 AVD

如果不利用 cmd 命令器编辑建立 AVD，也可以在创建 Android 工程项目时在 Eclipse 中创建 AVD 模拟器。方法如下：

打开 Eclipse 界面后，在打开的界面选择 window 再选择 Android SDK and AVD Manager 项出现如图 1-20 所示对话框。

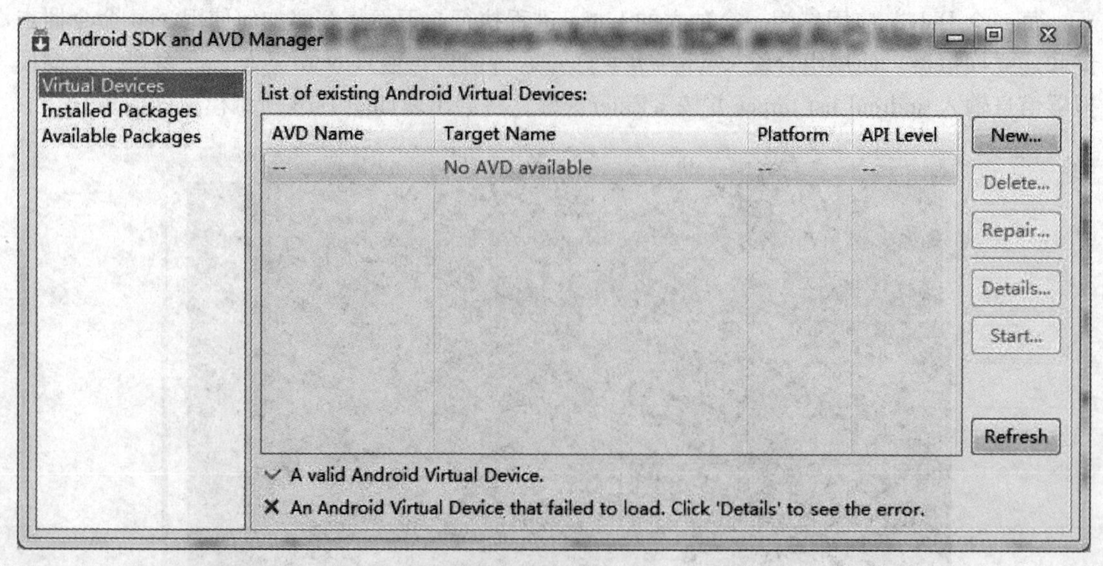

图 1-20　新建 AVD1

选择对话框右上角的 New 选项出现新建 AVD 对话框，按图 1-21 所示进行设置。界面各项所代表的意思如下：

Name：新建的 AVD 的名称，这里我们输入 AVD1。

Target：选择 Android 2. 1-update1-API Level 7。

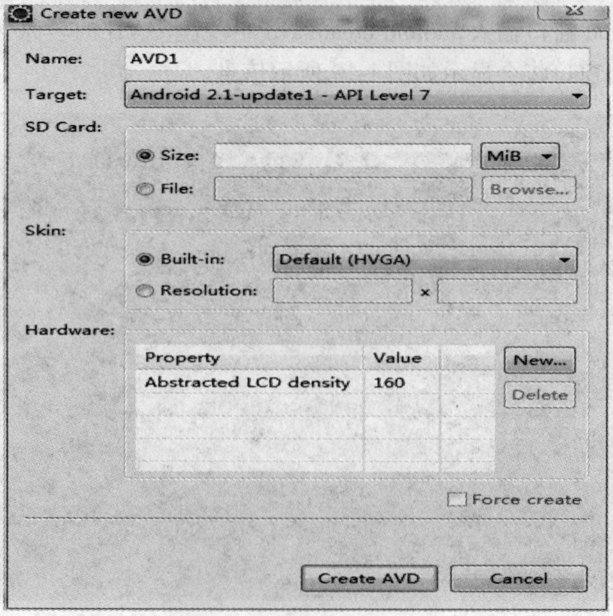

图 1-21　新建 AVD2

单击 Create AVD 后在弹出的窗口中直接选择 OK，在原有窗口的对应项将会出现 AVD1 这个模拟器。

1.3.5 安装 ADT

Android 所用的开发工具是 Eclipse，在 Eclipse 的 IDE 下安装 ADT，为 Android 开发提供相应开发工具的升级或者变更。ADT 即为 Android 的开发插件，可理解为在 Eclipse 下开发工具的升级下载工具。

安装插件 ADT 的步骤为：首先打开 Eclipse 开发环境，依次选择 Help→Install New Software，将会弹出如图 1-22 所示界面，选择右侧 Add 选项。

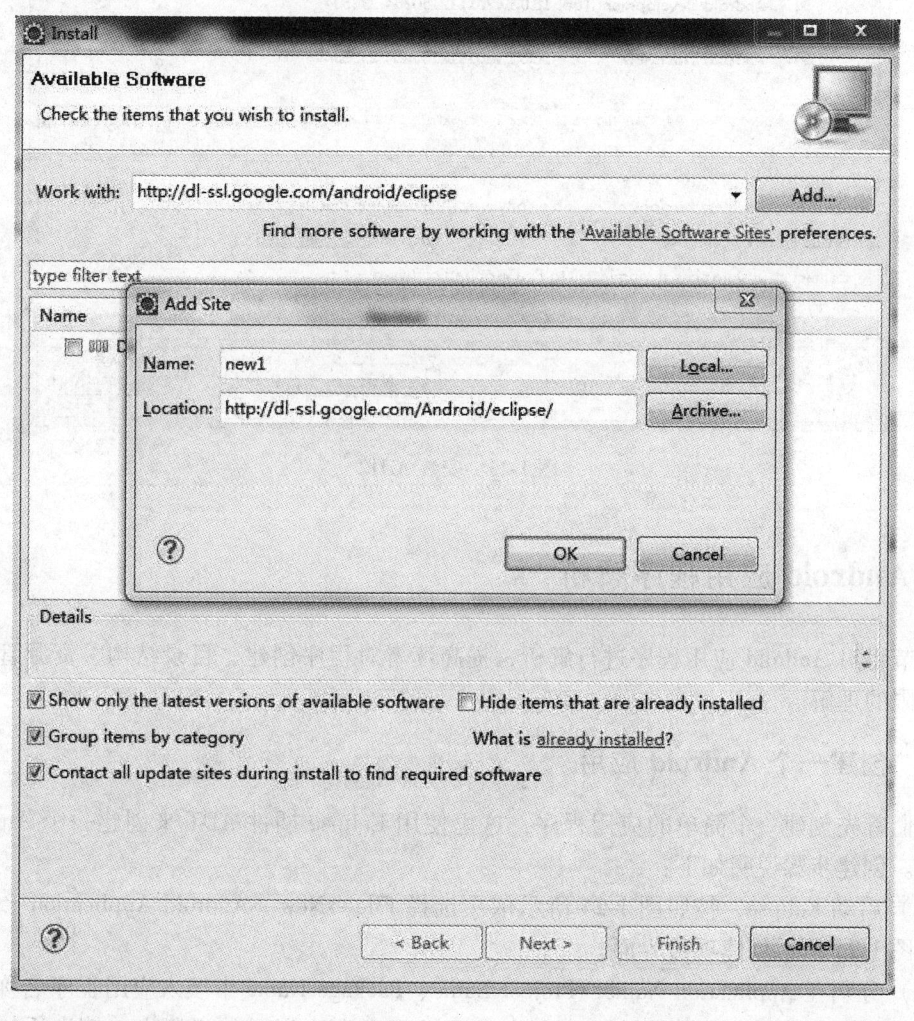

图 1-22 安装 ADT1

在弹出的小菜单栏中分别填入插件的名称及插件的下载地址，输入完毕后单击 OK。则在原窗口将会出现如图 1-23 所示信息。

在接下来的界面按照一般的软件安装顺序即可完成对 ADT 的安装。到此为止基本完成了对 Android 开发环境的搭建及配置。

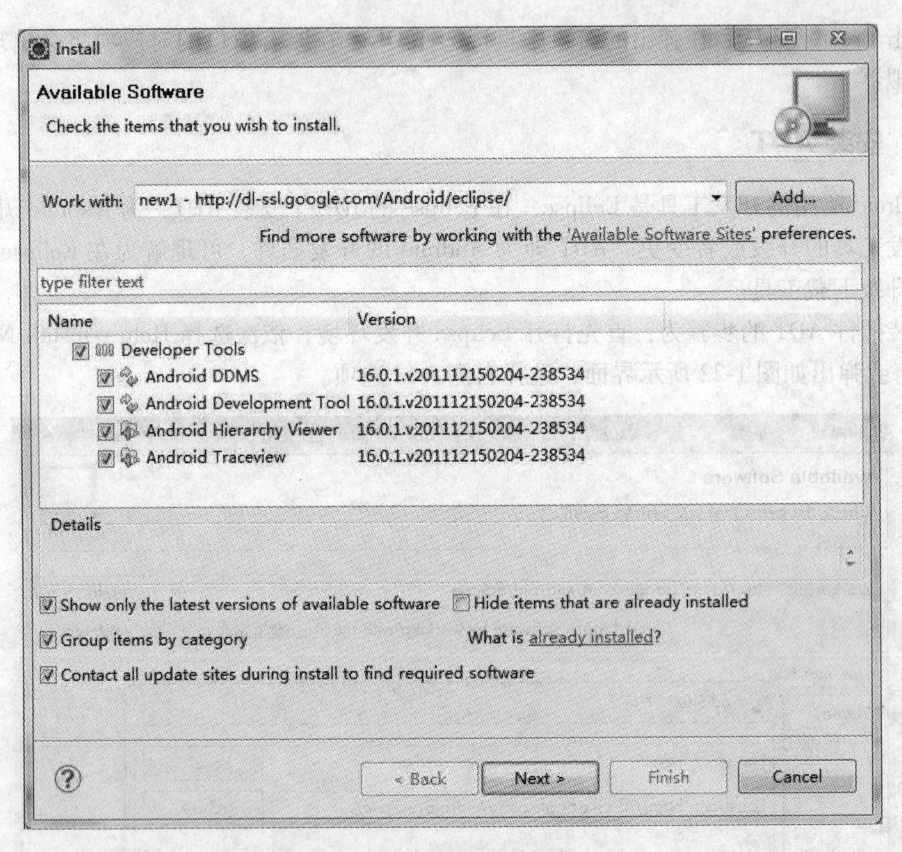

图 1-23 安装 ADT2

1.4 Android 应用程序解析

本节将对 Android 应用程序进行解析，提高读者对程序创建、目录结构、资源管理以及程序权限的理解。

1.4.1 创建一个 Android 应用

我们首先创建一个简单的应用程序，这里使用 Eclipse 插件 ADT 来创建一个 Android 应用程序。创建步骤说明如下：

（1）启动 Eclipse，按照图 1-24 所示依次选择 File→New→Android Application Project 将弹出如图 1-25 所示创建项目界面。

（2）分别在 Application Name、Project Name、Package Name 中填入应用程序名称、项目名称、应用程序包名并选择最低 SDK 版本、目标 SDK 版本以及编译版本，填入后如图 1-26 所示并单击 Next。

（3）在弹出的界面（图 1-27）中分别选择 Create custom launcher icon（创建本地图标）、Create activity（创建活动）、Create Project in Workplace（在工作区创建项目）并单击 Next。

（4）此时进入到图标属性界面一般默认即可，单击 Next 进入 Activity 选项界面，分别填入 Activity Name、Layout Name 单击 Finish 即可创建一个应用，如图 1-28 所示。

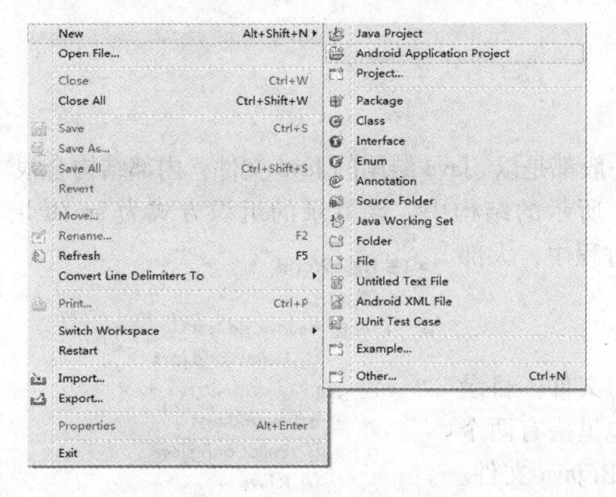

图 1-24　创建 Android 项目界面 1

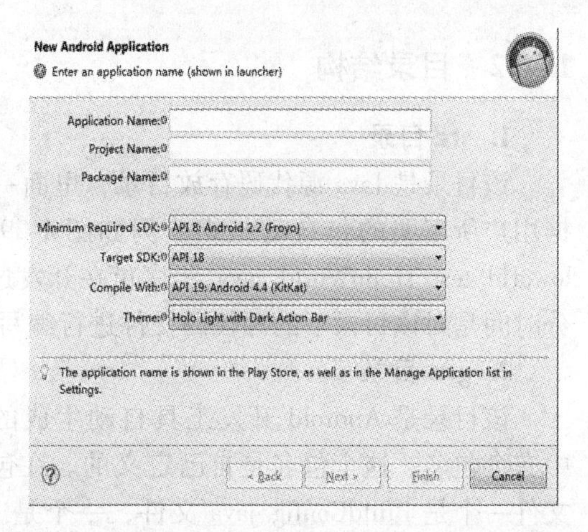

图 1-25　创建 Android 项目界面 2

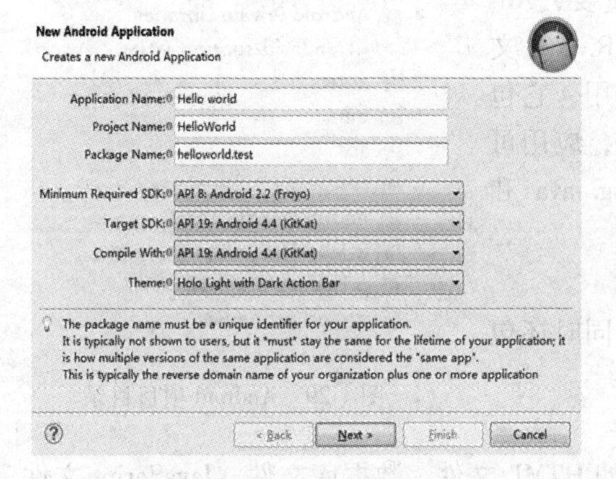

图 1-26　创建 Android 项目界面 3

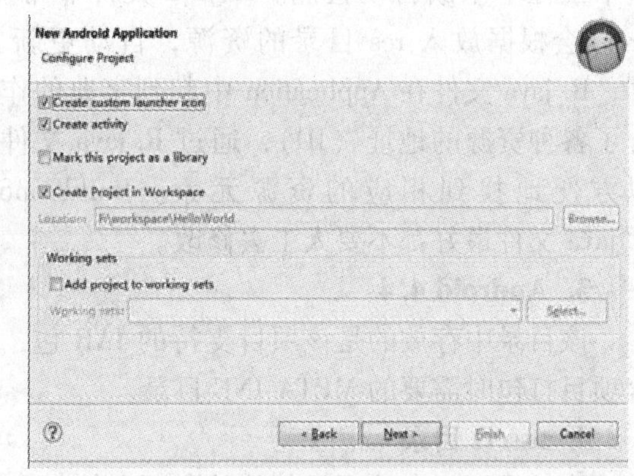

图 1-27　创建 Android 项目界面 4

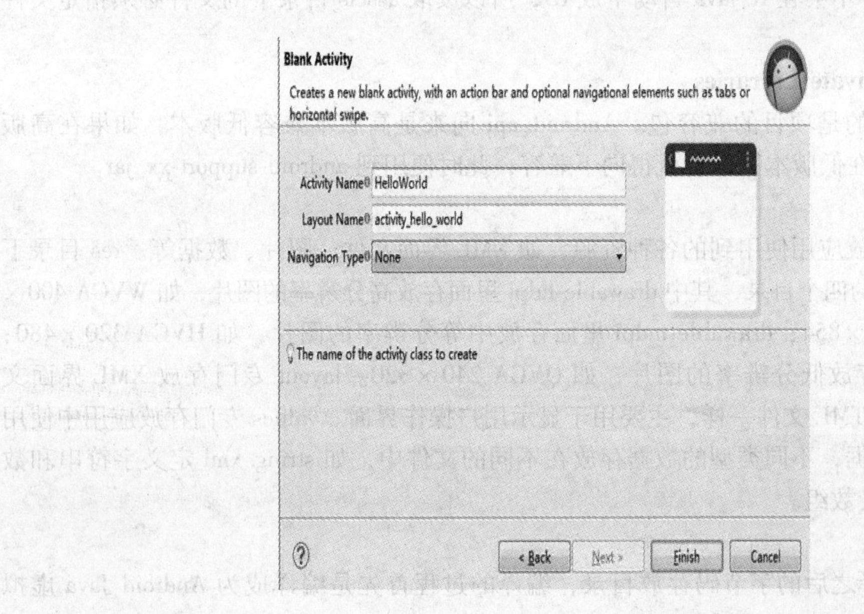

图 1-28　创建 Android 项目界面 5

1.4.2　目录结构

1. src 目录

该目录是 Java 源代码存放目录，里面一般都是以 . Java 结尾的 Java 文件，内部结构会根据用户所声明的包自动组织，例如图 1-29 所示的结构中，该目录的组织方式为 src/helloworld. test/HelloWorld. java，程序员在开发过程中，大部分时间是对该目录下的源代码文件进行编写。

2. gen 目录

该目录是 Android 开发工具自动生成的文件。目录中有个包名，这个包名是自己定义的。在包里头有两个文件一个是 BuildConfig. java 文件，一个是 R. java 文件。BuildConfig. java 文件用于 Android 代码的辅助检查，在整个工程中不断自动检测。R. java 文件非常重要，Android 会根据放入 res 目录的资源，自动更新 R. java 文件，R. java 文件在 Application 中起到字典的作用，它包含了各种资源的地址（ID），通过 R. java 文件，应用可以方便地找到相应的资源元素。BuildConfig. java 和 R. java 文件最好都不要人工去修改。

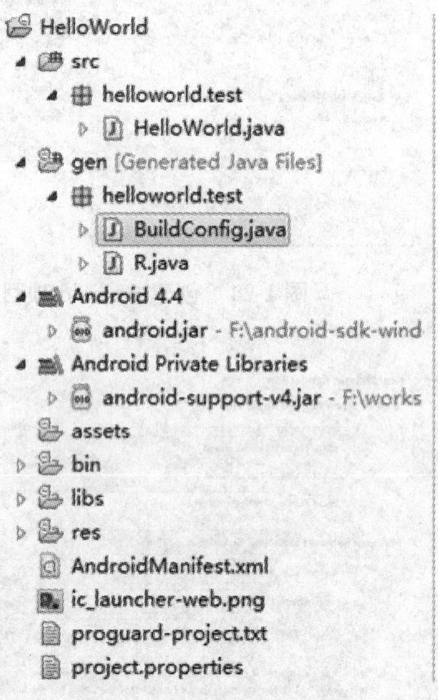

图 1-29　Android 项目目录

3. Android 4.4

该目录中存放的是该项目支持的 JAR 包，同时还包含项目打包时需要的 META-INF 目录。

4. assets 目录

该目录用于存放项目相关的资源文件，比如 HTML 文件、数据库文件、JavaScript 文件，assets 目录下的文件不会在 R. java 自动生成 ID，所以读取 assets 目录下的文件必须指定文件的路径。

5. Android Private Libraries

该目录中存放的是项目的兼容包。Android. api 向来是高版本兼容低版本，如果在高版本上开发的 apk 想在低版本上运行就得向下兼容，此时便用到 android-support-xx. jar。

6. res 目录

该目录用于存放应用使用到的各种资源，如 XML 界面文件、图片、数据等。res 目录下有以 drawable 开头的四个目录，其中 drawable-hdpi 里面存放高分辨率的图片，如 WVGA 400 × 800 和 FWVGA 480 × 854；drawable-mdpi 里面存放中等分辨率的图片，如 HVGA 320 × 480；drawable-ldpi 里面存放低分辨率的图片，如 QVGA 240 × 320。layout 专门存放 XML 界面文件，XML 文件同 HTML 文件一样，主要用于显示用户操作界面。values 专门存放应用中使用到的各种类型的数据，不同类型的数据存放在不同的文件中，如 string. xml 定义字符串和数值，arrays. xml 定义数组。

7. bin 目录

bin 目录是编译之后的字节码存放目录，编译的过程首先是编译成为 Android Java 虚拟机（Dalvik Virtual Machine）文件 classes. dex，再把该 classes. dex 文件打包成为 apk，apk 是

Android 平台上安装的应用程序包，类似于 Windows 应用程序 setup. exe 安装文件。

8. AndroidManifest. xml 文件

AndroidManifest. xml 是每个 Android 程序中必需的文件。它位于整个项目的根目录，描述了 package 中暴露的组件（activities、services 等）、它们各自的实现类、各种能被处理的数据和启动位置。除了能声明程序中的 Activities、ContentProviders、Services 和 Intent Receivers，还能指定 permissions（安全控制）和 instrumentation（测试）。

1.4.3 Android 中的资源访问

Android 中的资源是在代码中使用的外部文件。资源在 Android 架构中扮演了重要角色。Android 中的资源是与可运行应用所绑定的文件（类似于音乐文件或者描述窗口布局的文件）或者值（例如对话框的标题）。这些文件与值的绑定方式可以允许在改变时不必重新编译应用。

资源的一些常见例子包括字符串（Strings）、颜色（Colors）、位图（Bitmaps）和布局（Layouts）。资源中的字符串允许使用其资源 ID，而不是在代码中进行硬编码。这种间接使用资源的方法可以只改变资源而不用改变代码。

Android 中有相当多的资源。我们就以最常见的资源 String 为例进行介绍。这些文件作为应用程序的一部分被编译到应用程序当中。Android 中支持大量的资源文件，接下来将对各种资源的使用进行介绍。

1. 字符串资源

Android 允许在一个或者多个 XML 资源文件中定义字符串。这些包含字符串的资源文件位于 res/values 目录下。虽然常见的是名字为 strings. xml 的文件，但是事实上其名称可以是任意的。例 1-1 给出了一个字符串资源的例子。

【例 1-1】 String. xml

```
1  <? xml version = "1.0" encoding = "utf-8"? >
2  <resources>
3      <string name = "hello" >hello </string>
4      <string name = "app_name" >hello appname </string>
5      </resorces>
```

注意在一些 Eclipse 版本中，<resources>需要 xmlns 进行限制。只要有 xmlns 描述就可以，而不必管 xmlns 到底指向哪里。例如下面的两个例子都可以。

```
<resources xmlns = "http://schmas.android.com/apk/res/android">
<resources xmlns = "default namespaces">
```

甚至连第一行（指明这是一个 XML 文件，其编码格式是什么）都可以忽略，Android 依然可以很好地运行。

当该文件创建或者进行更新时，Android 的 ADT 会自动在根目录的包中生成一个 R. java 文件，其中不同的字符串有着不同的 ID 号（注：无论有多少资源文件，只要一个 R. java 文件生成）。

对于例 1-1 中的资源，在 R. java 中对应的结构如下：

【例 1-2】 R. java

```
1   package helloworld.test;
2   public final class R {
3   …other entries denpend on your project and applicaton
4     public static final class string {
5         public static final int app_name = 0x7f050000;
6         public static final int hello_world = 0x7f050002;
7         }
8     }
```

请注意 Android 是如何定义顶层类的：public final class R。其内部类为 public static final 类型，名为 string。R. java 文件通过创建这个内部类作为 string 的命名空间，来统一管理 string 资源。

其中的两个 static final 的 int 数据是与之相对应的字符串资源的 ID。可以通过下面的形式在任何地方使用资源 ID：R. string. hello_world。

这些自动生成的 ID 指向的是 int 类型，而不是 string 类型。大多数方法使用字符串时，以这些 int 型的 ID 作为输入，android 会在需要时将这些 int 数据转换为 string。

将所有的字符串定义在 strings. xml 里仅仅是一个约定俗成的规范。任何文件，只要其内部结构类似于例 1-1，且位于 res/values 目录下，Android 都可以进行处理。这个文件的结构有一个根节点 <resources>，里面有一系列 <string> 子元素。每一个 <string> 子元素里都有一个 name 属性，与 R. java 里的 ID 相对应。

2. 布局资源

在 Android 中，屏幕上显示的图像往往从 XML 文件中加载而来，而 XML 文件则视为资源。这与 HTML 文件描述网页的内容与布局十分相似。这些 XML 文件称为布局资源。布局文件是 Android UI 编程中的关键性文件。以例 1-3 展示的代码为例。

【例 1-3】 使用布局文件的 Java 代码

```
1   public class MainActivity extends Activity {
2   @ Override
3   protected void onCreate(Bundle savedInstanceState) {
4      super.onCreate(savedInstanceState);
5      setContentView(R.layout.activity_main);
6      }
```

setContentView（R. layout _ activity _ main）这一行可以看出，存在着一个静态的类 R. layout，其内部有一个常量 activity_main（int 类型）指向了一个由布局资源文件所定义的视图。该布局文件的名称为 activity_main. xml，位于 res/layout 目录下。换句话说，这个表达式需要程序员创建一个文件 res/layout/main. xml，并在文件中定义必要的布局。activity_main. xml 的内容可以如例 1-4 所示。

【例 1-4】 布局文件的声明

```
1    <?xml version = "1.0" encoding = "utf-8"? >
2    <LinearLayout xmlns:android = "http://schemas.android.com/apk/res/android"
3       android:orientation = "vertical"
```

```
4        android:layout_width = "fill_parent"
5        android:layout_height = "fill_parent"
6        >
7        <TextView
8            android:layout_width = "fill_parent"
9            android:layout_height = "wrap_content"
10           android:text = "@ string/hello_world" />
11    </LinearLayout>
```

文中第1行声明了XML的版本及编码方式，第2行定义了一个名为LinearLayout的根节点以及XML的命名空间，第3行定义了以垂直方向排列子布局（也可以是水平布局），第4、5行定义了子布局的宽度和高度为填满整个父空间，第7行定义了一个文本控件，并在其中定义了文本控件的排列方式。第9行的wrap_content表示强制性地使视图扩展以显示全部内容。

需要为每个屏幕（或activity）定义一个单独的布局文件。更确切地说，每一个布局需要对应一个专门的文件。如果打算画两个屏幕，就可能需要两个布局文件：res/layout/screen1_layout. xml和res/layout/screen2_layout. xml。（注：res/layout目录下的每一个文件都会根据其文件名（不包括扩展名）生成一个独一无二的常量。对于layout而言，布局文件的数量是关键，而对于string资源，单独的string资源的数量才是关键所在。）

资源文件中的具体视图，例如TextView等可以通过R. java中自动生成的资源ID进行引用，如：

```
TextView tv = (TextView) findViewById(R.id.text1);
tv.setText("Try this text instead");
```

本例中，通过Activity类中的findViewById方法来定位到TextView。其中常量R. id. text1与TextView中定义的ID相对应。如：

```
<TextView id = "@ +id/text1" >
        …
</TextView>
```

其中，属性ID的值，也就是常量text1被用来唯一标识该TextView和Activity中的其他View区别开来。“+”表示如果id. text1不存在的话则创建一个。下面将介绍更多关于ID属性的内容。

先不考虑各种资源的类型（目前为止我们已经接触到了string和layout两种），Android的Java代码中是通过其ID来引用资源的。在XML文件中，使用的将ID与资源项相联系的符号被称作资源引用符（Resources Reference Syntax）。它不仅通过ID来进行资源定位，而且是定义string、layout和image等资源的方法。这种通用的方法是如何实现资源定位的呢？通过其用法可以看出，ID其实就是一个能追踪类似string等资源的数字。假设工程中有一组ID数字，那么可以选择其中一个与某个控件相关联。这种资源引用有如下结构：

```
@ [package:]type/name
```

其中，type与R. java中的资源类型的命名空间相对应。例如：

R. drawable

R. id

R. layout

R. string

R. attr

R. plural

R. array

而 name 则是这些资源的名称，与之相对应的资源引用符号如下：

drawable

id

layout

string

attr

plurals

string-array

同时，在 R. java 中它也代表着一个 int 型常量。

如果不指定 package 名字，则 type/name 会基于本地资源以及本地生成的 R. java 文件。

如果其形式为 android：type/name，则引用会基于 Android 包以及 android. R. java。可以使用任何 Java 的包名来作为 package，这样与之相对应的 R. java 文件会处理相关引用。

3. 字符串数组

可以设置一个字符串数组作为资源，将其置于 res/values 目录下的任意文件中。需要使用 string-array 作为 XML 的节点。这个节点是 XML 节点的一个子节点，类似于 string 节点。例 1-5 是一个在资源文件中定义字符串数组的例子。

【例 1-5】 字符串数组的含义

```
1    <resources... >
2      .......Other resources
3        <string-array name="test_array">
4              <item>one</item>
5              <item>two</item>
6              <item>three</item>
7        </string-array>
8     ......other resources
9    </resources>
```

一旦定义了这样一个字符串数组，就可以在 Java 文件中如例 1-6 这样使用。

【例 1-6】 字符串数组的使用

```
1    //从 Activity 中获取资源对象
2    Resources res = your-acitivity.getResources();
3    String[] strings = res.getStringArray(R.array.test_array);
4    //打印字符串
5    for(String s : strings) {
6        log.d("example", s);}
```

4. 复数

Plurals 资源是一个字符串集合。这些字符串根据某个数量的不同有不同的表示形式。

例 1-7 将告诉我们如何使用 plurals 来实现字符串根据数量不同而发生改变。

【例 1-7】 复数资源的定义

```
1    <resources>
2    <plurals name="eggs_in_a_nest_test">
3    <item quantity="one">There is 1 egg</item>
4    <item quantity="other">There are %d eggs</item>
5    </plurals>
6    </resoureces>
```

两种不同表达式在同一个 plurals 里。现在可以在 Java 代码里通过赋予一个数量值来使用这个 plurals 资源。如例 1-8 所示，getQuantityString( )的第一个参数是 plurals 资源的 ID。第二个参数选择使用哪个字符串。当其值为 1 时，使用的是该字符串本身。当字符串不是 1 的时候，必须制定第三个参数来替换%d，如果在 plurals 里使用格式化字符串，则应该至少传入 3 个参数。第二个参数可能造成混淆，其实其关键在于该值是否等于 1。

【例 1-8】 复数资源的使用

```
1    Resources res = your-acitivity.getResoureces();
2    String s1 = res.getQuantityString(R.plurals.eggs_in_a_nest_test, 0, 0);
3    String s2 = res.getQuantityStirng(R.plurals.eggs_in_a_nest_test, 1, 1);
4    String s3 = res.getQuantityStirng(R.plurals.eggs_in_a_nest_test, 10, 10);
```

通过上述代码，会根据传入的值不同，而返回合适的字符串。

下面介绍如何在 XML 资源文件中定义普通字符串、带引号字符串、HTML 字符串和可替换字符串。

【例 1-9】 利用 XML 语法定义字符串

```
1    <resources>
2    <string name="simple_string>simple string</string>
3    <string name="quoted_string>"quoted 'xyz' string"</string>
4    <string name="double_quoted_string">\"double quotes\"</string>
5    <string name="java_format_string"> hello %2$s Java format string. %1$s again </string>
6    </resources>
```

XML 字符串资源文件需要放在 res/values 目录下，其文件名可以任意。带引号字符串需要通过转义字符输出双引号。字符串还允许以 Java 格式化字符串的形式进行定义。

Android 还允许在 string 节点内部使用 XML 的子节点，如 <b> <i> 及其他简单的 HTML 字体标记。可以通过复合的 HTML 字符串来定义字符的样式以便在 textview 中输出。

例 1-10 表示如何在 Java 代码中使用这些字符串。

【例 1-10】 在 Java 文件中使用字符串

```
1    //使用简单的字符串,并在 textview 中显示
2    String simpleString = activity.getString(R.stirng.simple_string);
3    textView.setText(simpleString);
4    //读取一个带引号的字符串,并在 textview 中显示
5    String quotedString = acitivity.getString(R.string.quoted_string);
```

```
6    textView.setText(quotedString);
7    //读取一个带双引号的字符串,并在 textview 中显示
8    String doubleQuotedString = activity.getString(R.string.double_quoted_string);
9    textView.setText(doubleQuotedString);
10   //读取一个 Java 格式化字符串
11   String javaFormatString = acitivity.getString(R.string.java_format_string);
12   //通过传入参数格式化字符串
13   String substitutedString = String.format(javaFormatString, "hello", "Android");
14   textView.setText(substitutedString);
15   //读取一个 HTML 字符串,并显示
16   String htmlString = activity.getString(R.string.tagged_string);
```

5. 颜色资源

和字符串资源一样，也可以使用资源引用符来间接地引用颜色资源。这使得 Android 可以将颜色资源本地化，并且可以提供主题。一旦在资源文件中定义了颜色资源标识符，就可以在 Java 代码中通过其 ID 引用颜色资源。对比字符串资源，其 ID 位于 <包名>. R. string 命名空间中，颜色资源位于 <包名>. R. color 命名空间中。

Android 在其自带的资源文件中也定义了一套基本的颜色。这些 ID 位于 android. R. color 命名空间中。例 1-11 列出了一些在 XML 资源文件中定义 color 资源的例子。

【例 1-11】 颜色资源的定义

```
1    <resources>
2    <color name="red">#f00</color>
3    <color name="blue">#0000ff</color>
4    <color name="green">#f0f0</color>
5    <color name="main_back_ground_color">#ffffff00</color>
6    </resources>
```

例 1-11 的内容需要位于 res/values 目录下，其文件名可以任意选取。Android 会扫描所有文件，从中选取 <resources> 和 <color> 节点来计算出其 ID。例 1-12 列出如何在 Java 代码中使用 color 资源。

【例 1-12】 颜色资源在 Java 文件中的使用

```
Int mainBackGroundColor = activity.getResources().getColor(R.color.main_back_ground_color);
```

例 1-13 列出如何在定义 View 时使用 color 资源。

【例 1-13】 颜色资源在布局文件中的使用

```
1    <TextView android:layout_widht="fill_parent"
2            android:layout_height="wrap_content"
3            android:textColor="@color/red"
4                android:text="Sample text to show red color"/>
```

6. 尺寸资源

像素、英寸、点都是尺寸的一种，均可以在 XML 文件和 Java 文件中使用。可以用这些尺寸资源样式化和本地化 Android 的 UI 而不必修改 Java 代码。例 1-14 列出如何在 XML 文件中使用尺寸资源。

【例 1-14】 尺寸资源的定义

```
1    <resources>
2    <dimen name="mysize_in_pixels">1px</dimen>
3    <dimen name="mysize_in_dp">1dp</dimen>
4    <dimen name="medium_size">100sp</dimen>
5    </resources>
```

可以用以下任意单位定义尺寸：Px（像素）、in（英寸）、mm（毫米）、pt（点）、dp（无关像素，基于 dpi（每英寸像素）为 160 的屏幕而定义，尺寸根据屏幕密度而动态变化）、sp（缩放无关像素，允许用户调整大小的尺寸，主要用于字符）。

在 Java 中，需要实例化一个资源对象来获取尺寸。可以在 activity 中调用 getResourece( ) 方法，然后根据该资源实例和尺寸资源 ID 来获取实际尺寸值。

【例 1-15】 在 Java 文件中使用尺寸资源

```
float dimen = activity.getResources().getDimension(R.dimen.mysize_in_pixels);
```

注：在 Java 中需要调用 Dimension 的全称，而在 R. java 命名空间中选取的是其缩写 dimen，相对于 Java 代码，在 XML 中使用尺寸资源，需要使用简写 dimen，见例 1-16。

【例 1-16】 在布局文件中使用尺寸资源

```
1    <TextView android:layout_width="fill_parent"
2            android:layout_height="wrap_content"
3            android:textSize="@dimen/medium_size"/>
```

7. 图像资源

Android 会为存储在 res/values 目录下的图像资源生成资源 ID 号。支持的图像类型为 gif、png 和 jpg 格式。每个图像资源会根据其名称生成独特的 ID 号。如果一个图像的文件名称为 sample_image. jpg，则生成的 ID 号为 R. drawable. sample_image。

注：如果有两个文件的名称相同会得到错误提示，res/drawable 下面的子目录下的图像会被忽略。该目录下的任何文件都不会读取，可以在其他的 XML 文件中引用 res/drawable 的图像资源，如例 1-17 所示。

【例 1-17】 在布局文件中使用图像资源

```
1    <Button
2        android:id="@+id/button1"
3        android:layout_width="fill_parent"
4        android:layout_height="wrap_content"
5        android:text="fail"
6        android:background="@drawable/simple_image"
```

可以通过 Java 代码来使用图像资源，将其设置到一个 UI 对象里，如 Button。

【例 1-18】 在 Java 文件中使用图像资源

```
1    //调用 getDrawable 来获取图像
2    BitmapDrawable d = activity.getResource().getDrawable(R.drawable.sample_image);
3    //然后使用 drawable 设置背景
4    button.setBackgroundDrawable(d);
5    //或者直接通过资源 ID 使用
6    button.setBackgroundResource(R.drawable.sample_image);
```

注：这些背景相关的办法属于 View 类。因此绝大多数 UI 空间都可以使用。

8. 颜色图片资源

在 Android 中，图像是 drawable 资源的一种。Android 还支持其他的 drawable 资源，比如 color-drawable 资源。它本质上是一个带颜色的矩形。要想定义一个带颜色的矩形，需要在 res/values 目录下的 XML 文件里通过名为 drawable 的节点来标识。例 1-19 是一些 Color-Drawable 的例子。

【例 1-19】 Color-Drawable 的定义

```
1    <resources>
2    <drawable name="red_rectangle">#f00</drawable>
3    <drawable name="blue_rectangle">#0000ff</drawable>
4    <drawable name="green_rectangle>#f0f0</drawable>
5    </resources>
```

例 1-20 和例 1-21 分别演示了如何在 Java 代码和 XML 文件中使用 color-drawable 资源。

【例 1-20】 在 Java 文件中使用 color-drawable 资源

```
1    ColorDrawableredDrawable=(ColorDrawable)activity.getResource().getDrawable(R.drawable.red_
2    rectangle);
3    textView.setBackgroundDrawable(redDrawable);
```

【例 1-21】 在布局文件中使用 color-drawable 资源

```
1    <TextView android:layout_width="fill_parent"
2           android:layout_height="wrap_content"
3           android:textAlign="center"
4           android:background="@drawable/red_rectangle"/>
```

9. 使用任意 XML 资源文件

除了之前介绍的结构化的资源，Android 还允许将任意的 XML 文件作为资源。这种方法将对资源的使用扩展到了对任意 XML 文件的使用。首先，这种方法通过使用以这些文件为基础而生成的资源 ID 来很方便地使用它们。第二，这种方法能本地化这些 XML 资源文件。第三，可以有效地编译和在设备上存储这些 XML 文件。

允许通过这种方式读取的 XML 文件存储在 res/xml 目录下。例 1-22 列出了一个名为 res/xml/test.xml 的例子。

【例 1-22】 XML 文件的使用

```
1    <rootelem1>
2       <subelem1>
3            Hello World from a xml sub element
4       </subelem1>
5      </rootelem1>
```

正如其他 XML 资源文件一样，AAPT（Android Asset Packaging Tool）先将该文件编译，然后再放到应用包里。下面给出 XmlPullParser 实例来解析这些文件。可以使用例 1-23 中的代码获取一个 XmlPullParser 的实例（可以在任意 Context 下使用）。

【例 1-23】 读取 XML 资源

```
1    Resources res = activity.getResources();
2    XmlResourceParser xpp = res.getXml(R.xml.test);
```

返回的 XmlResourceParser 是 XmlPullParser 的一个实例，它还实现了 java.util.AttributeSet 接口。

10. Raw 资源

除了 XML 文件外，Android 还允许使用 raw 文件。这些资源位于 res/raw 目录下。这些 raw 文件可以是音频、视频或 txt 文档。它们都可以通过资源 ID 来进行本地化或者引用。与在 res/xml 目录下的 XML 文件不同，这些 raw 文件不会被编译，但是会打包到应用包里对应的位置。然而，每个文件都会在 R.java 文件中生成相应的标识符。如果创建了一个 res/raw/test.txt 的文件，那么可以使用例 1-24 中的代码读取该文件。

【例 1-24】 读取 RAW 资源

```
1    String getStringFromRawFile(Activity activity) throws IOException{
2    Resources r = activity.getResources();
3    InputStream is = r.openRawResource(R.raw.test);
4    String myText = convertStreamToString(is);
5    is.close();
6    return myText;}
7    String convertStreamToString(InputStream is) throws IOException {
8    ByteArrayOutputStream baos = new ByteArrayOutputSteam();
9    int i = is.read();
10   while(i ! = -1) {
11   baos.write(i);
12   i = is.read();}
13   return baos.toString();}
```

注：如果文件名称是通过复制自动产生的，则 ADT 会产生一个编译错误。这是所有基于文件名称产生资源 ID 的情况共有的问题。

11. Assets

Android 还提供了其他的一些目录用来存放文件，如 assets。assets 与 res 属于同一级目录。也就是说 assets 并不是 res 下的子目录。在 assets 目录下的文件不会在 R.java 中生成资源 ID。必须指定路径来读取这些文件。文件路径是基于 assets 的相对路径。如例 1-25 所示，可以使用 AssetManager 来读取文件。

【例 1-25】 读取 assets 文件

```
1    String getStringFromAssetFile(Activity activity) {
2    AssetManager am = activity.getAssets();
3    InputStream is = am.open("test.txt");
4    String s = convertStreamToString(is);
5    is.close;
6    return s;}
```

12. 应用程序的权限

一个 Android 应用可能需要权限才能调用 Android 系统的功能，一个 Android 应用也可能被其他应用调用，因此它也需要声明调用自身所需要的权限。

（1）声明该应用自身所拥有的权限

通过为 <manifest.../>元素添加 <uses-permission.../>子元素即可为自身声明权限。例如在 <manifest.../>元素里添加如下代码：

```
<!--声明该应用本身需要打电话的权限-->
<uses-permission android:name="android.permission.CALL_PHONE"/>
```

（2）声明调用该应用自身所需的权限

通过为应用的各组件元素，如 <activity.../>元素添加 <uses-permission.../>子元素即可声明调用该程序所需的权限。例如在 <activity.../>元素里添加如下代码：

```
<!--声明调用本身需要发送短信的权限-->
<uses-permission android:name="android.permission.SEND_SMS"/>
```

通过上面的介绍可以看出，<uses-permission.../>元素的用法并不难。而且 Android 提供了大量的权限，这些权限位于 Manifest.permission 类中。Android 常用权限及说明如表 1-1 所示。

表 1-1　Android 常用权限及说明

权　限	说　明
ACCESS_NETWORK_STATE	允许应用程序获取网络状态信息的权限
ACCESS_WIFI_STATE	允许应用程序获取 WiFi 网络状态信息的权限
BATTERY_STATE	允许应用程序获取电池状态信息的权限
BLUETOOTH	允许应用程序链接匹配的蓝牙设备的权限
BLUETOOTH_ADMIN	允许应用程序发现匹配的蓝牙设备的权限
BROADCAST_SMS	允许应用程序广播收到短信提醒的权限
CALL_PHONE	允许应用程序拨打电话的权限
CAMERA	允许应用程序使用照相机的权限
CHANGE_NETWORK_STATE	允许应用程序改变网络连接状态的权限
CHANG_WIFI_STATE	允许应用程序改变 WiFi 网络连接状态的权限
DELETE_CACHE_FILES	允许应用程序的删除缓存文件权限
DELETE_PACKAGES	允许应用程序删除安装包的权限
FLASHLIGHT	允许应用程序访问闪光灯的权限
INTERNET	允许应用程序打开网络 Socket 的权限
MODIFY_AUDIO_SETTINGS	允许应用程序修改全局声音设置的权限
PROCESS_OUTGOING_CALLS	允许应用程序监听、控制、取消呼出电话的权限
READ_CONTACTS	允许应用程序读取用户的联系人数据的权限
READ_HISTORY_BOOKMARKS	允许应用程序读取历史书签的权限
READ_OWNER_DATA	允许应用程序读取用户数据的权限
READ_PHONE_STATE	允许应用程序读取电话状态的权限
READ_PHONE_SMS	允许应用程序读取短信的权限
REBOOT	允许应用程序重启系统的权限

（续）

权　限	说　明
RECEIVE_SMS	允许应用程序接收、监控、处理短信的权限
RECEIVE_MMS	允许应用程序接收、监控、处理彩信的权限
RECORD_AUDIO	允许应用程序录音的权限
SEND_SMS	允许应用程序发送短信的权限
SET_ORIENTATION	允许应用程序旋转屏幕的权限
SET_TIME	允许应用程序设置时间的权限
SET_TIME_ZONE	允许应用程序设置时区的权限
SET_WALLPAPER	允许应用程序设置桌面壁纸的权限
VIBRATE	允许应用程序访问振动器的权限
WRITE_CONTACTS	允许应用程序写入用户联系人的权限
WRITE_HISTORY_BOOKMARKS	允许应用程序写历史书签的权限
WRITE_OWNER_DATA	允许应用程序写用户数据的权限
WRITE_SMS	允许应用程序写短信的权限

1.5 Android 系统的调试与下载

1. 直接调试

首先，应该将手机设置为可调试模式，即找到手机的开发者选项，将 USB 调试处的勾打上，如图 1-30 所示。

图 1-30　将手机设置为可调试模式

其次，在 Eclipse 选中要进行调试的项目，单击右键依次选择 Run As→Run Configurations→Android Application，在 Android 选项卡下单击 Browse 会弹出如图 1-31 所示界面，选择要调试的工程，如本例的 MyVideo，选后单击 OK。在 Target 选项卡下选择 Always prompt to pick device 会弹出如图 1-32 所示界面，单击 Run，弹出如图 1-33 所示界面。

最后，在 Choose a running android device 中选择手机即可，本文为 htc_d516t。

2. 利用生成的 apk 调试

首先需要找到要调试工程的 apk 文件。常用的有以下两种方式。

（1）使用 Eclipse 在 Package Explorer 中，选择 AndroidManifest. xml 文件，在 Manifest 选项卡中找到 Export an unsigned APK，单击即可生成 apk。如图 1-34 所示。

（2）在 Eclipse 的 Package Explorer 中，选择工程的 bin 目录，在其中就有 apk 文件，在其上单击右键选择复制即可（或者是进入工程文件夹，比如我们的是 F:\workspace\MyVideo\bin\MyVideo. apk，复制相应的 apk）。然后，将得到的 apk 文件下载到手机里，单击该 apk，根据提示即可安装到手机里。

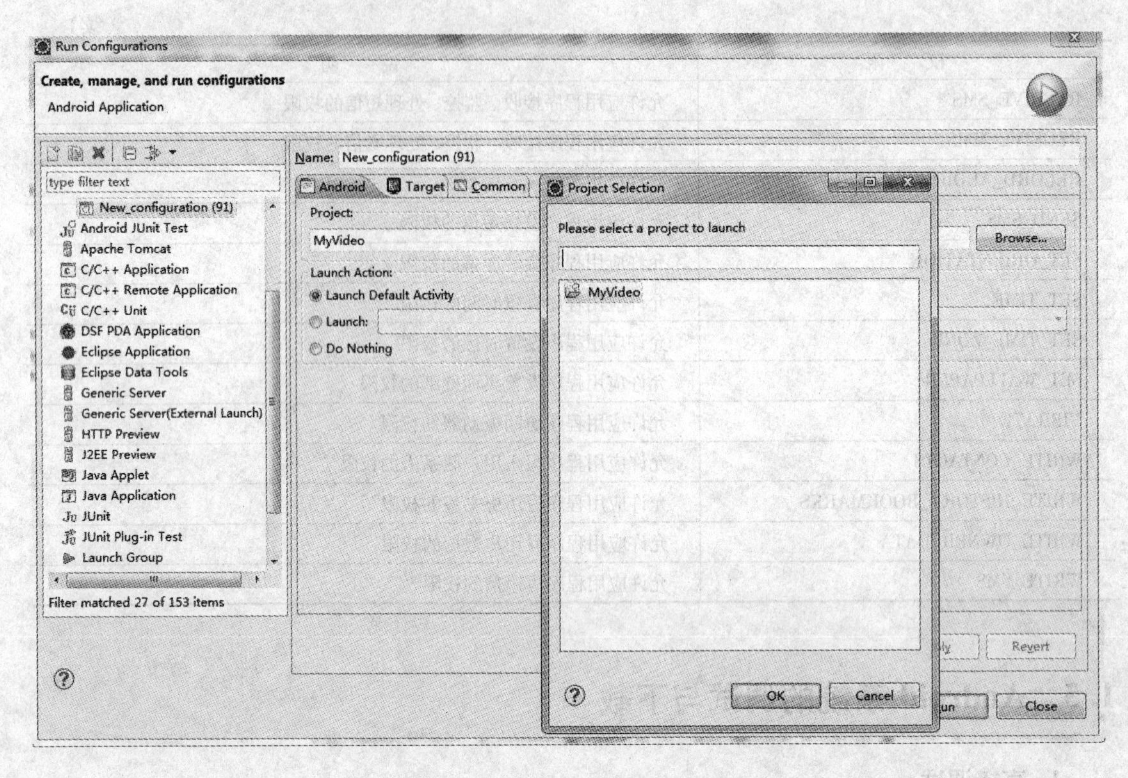

图 1-31　在 Eclipse 中设置调试模式

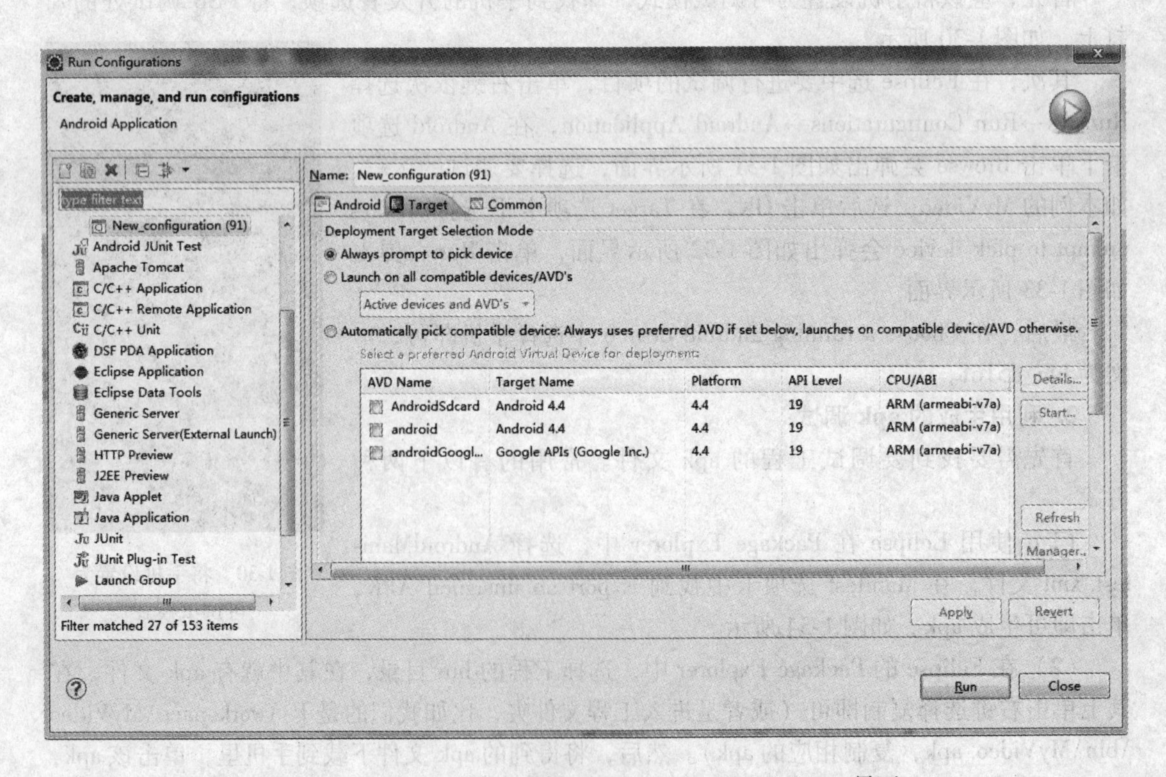

图 1-32　Target 选项卡下选择 Always prompt to pick device 界面

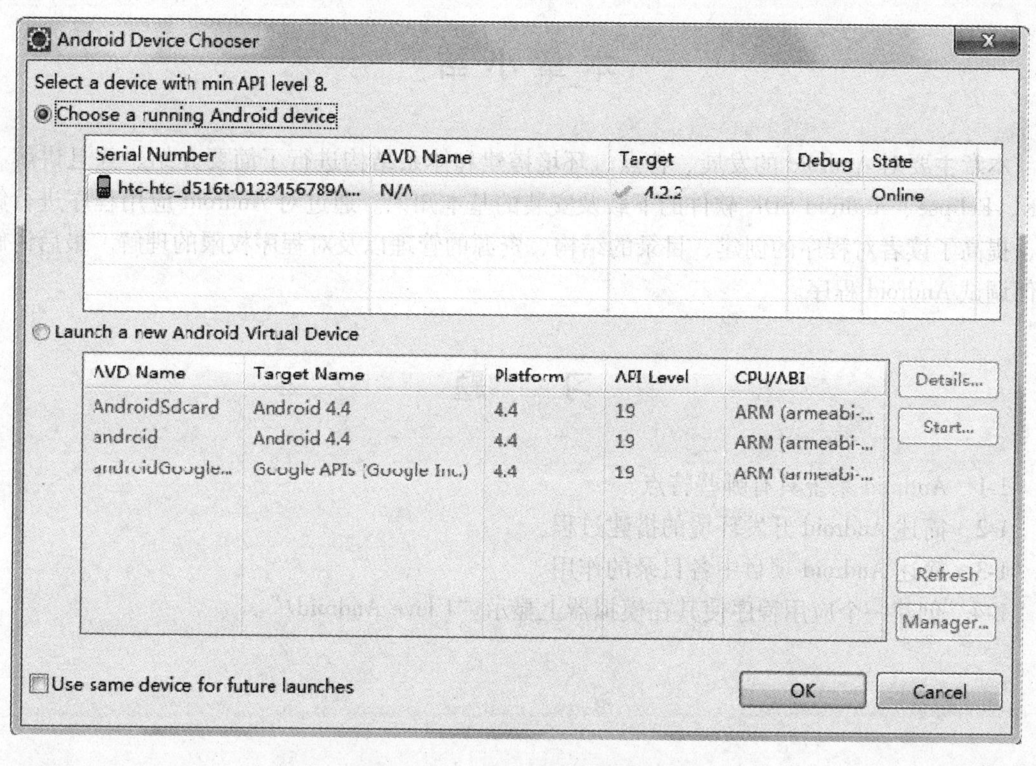

图 1-33 单击 Run 弹出界面

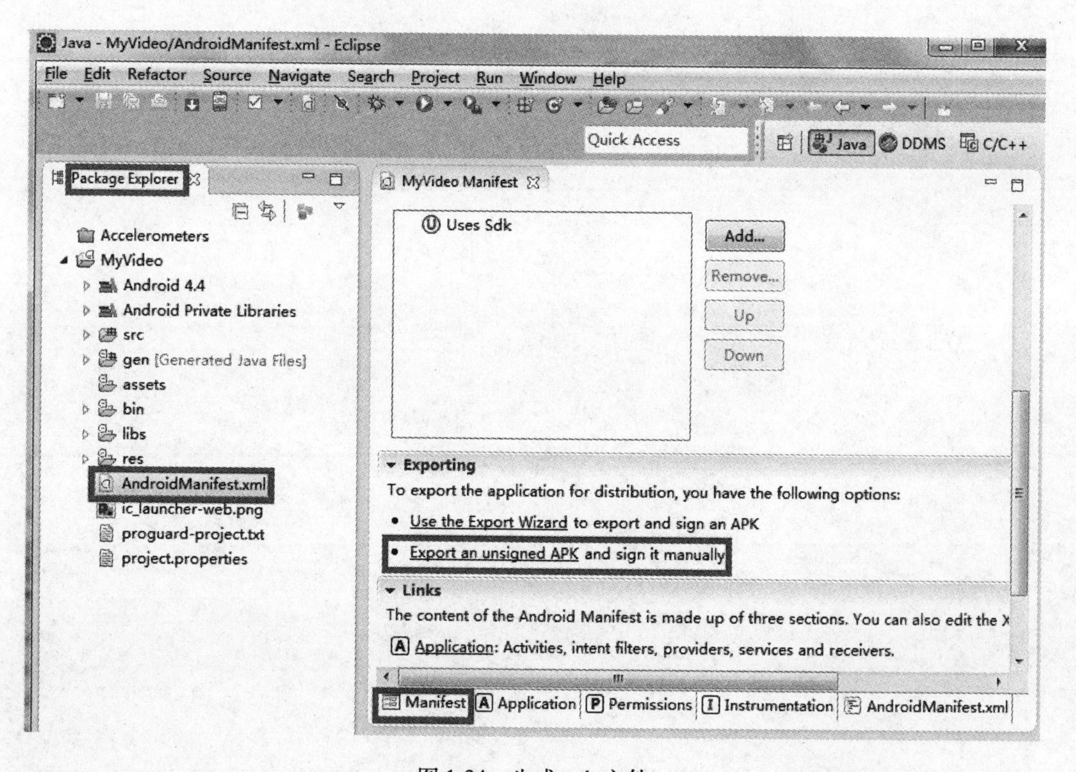

图 1-34 生成 apk 文件

本章小结

本章主要对 Android 的发展、特点、环境搭建和体系结构进行了简要介绍。并且讲解了 JDK、Eclipse、Android SDK 软件的下载及安装的基本知识。通过对 Android 应用程序进行解析，提高了读者对程序的创建、目录的结构、资源的管理以及对程序权限的理解。最后讲解如何调试 Android 程序。

习题

1-1 Android 系统具有哪些特点？

1-2 简述 Android 开发环境的搭建过程。

1-3 简述 Android 项目中各目录的作用。

1-4 创建一个应用程序使其在模拟器上显示“I love Android!”。

第 2 章　Android 生命周期与组件通信

本章首先讲述了 Android 程序的生命周期，并以 Activity 为例讲述了 Android 组件生命周期的生命周期函数，栈结构和基本状态转换等三方面内容。接着讲述了组件的通信，介绍了 Android 系统的组件通信机制，其中包括使用 Intent 启动组件的原理和方法，Intent 过滤器的原理与匹配机制，广播消息的接收和发送方法等内容。

2.1　Android 生命周期

生命周期（Life Cycle）是指使用过程以及报废或处置等废弃回到自然的过程，这个过程构成了一个完整的人工产物的生命周期。而软件生命周期（Systems Development Life Cycle，SDLC）又称为软件生存周期或系统开发生命周期，是软件的产生直到报废的生命周期。通常情况下由软件定义、开发和维护三个阶段组成（见图 2-1）。软件定义包括问题定义、可行性研究和需求分析三个阶段；软件开发包括总体设计、详细设计、编码和单元测试、综合测试四个阶段；软件维护分为四类：改正性维护、适应性维护、完善性维护和预防性维护。软件生命周期中软件维护既可以指向软件的再定义也可以完全废弃软件而结束周期循环。这种按时间划分阶段的方法是软件工程中的一种基本思想原则，即逐步推进原则。但随着新的面向对象的设计方法和技术的成熟，软件生命周期设计方法的指导意义正在逐步减少。

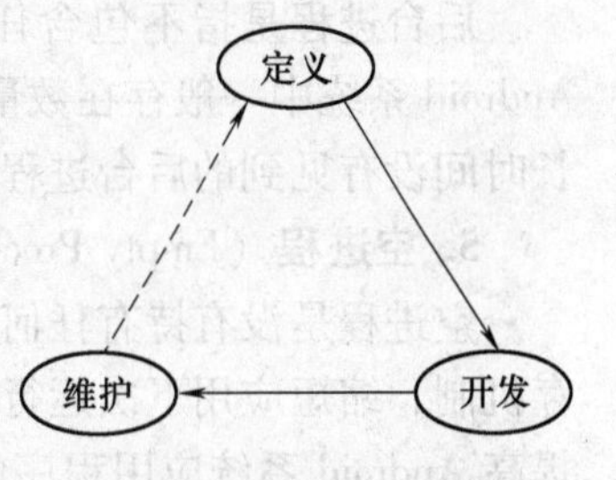

图 2-1　软件生命周期

Android 系统中，每个 Android 应用都将运行在自己的 Linux 进程当中。当一个程序或其某些部分被请求时，它的进程就被创建了；当这个程序没有必要再运行下去且系统需要回收这个进程的内存用于其他程序时，这个进程就“死亡”了。Android 系统主动管理资源，为了保证高优先级程序的正常运行或者为了减轻系统内存负载，Android 系统会主动终止低优先级的程序，可见程序的生命周期是由 Android 系统控制而非程序自身。Android 系统中的进程优先级如图 2-2 所示，按照优先级从低到高的顺序分别为空进程、后台进程、服务进程、可见进程和前台进程。

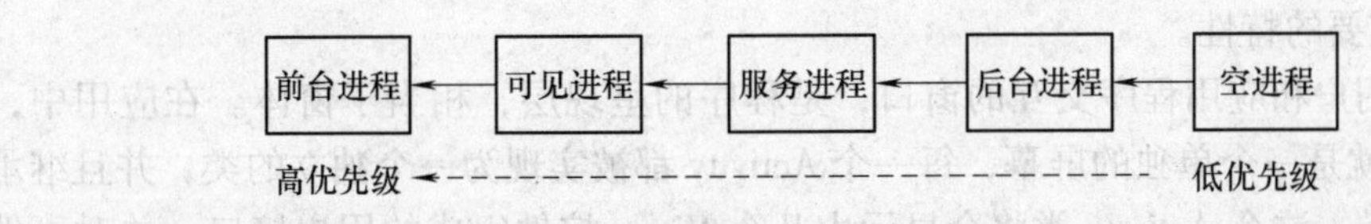

图 2-2　进程优先级

1. 前台进程（Foreground Process）

前台进程亦称为活动进程，是指当前正和用户进行交互的承载应用程序的进程，即正在前台运行的进程，说明用户正在通过该进程与系统进行交互。前台进程是 Android 系统中最重要的进程，优先级最高，在 Android 系统中包含以下四种情形：

（1）Activity 正在与用户进行交互。

（2）进程被 Activity 调用，而且这个进程正在与用户进行交互。

（3）进程服务正在执行声明中的回调函数，如 OnCreate( )、OnStart( )或 OnDestroy( )。

（4）进程的 BroadCastReceiver 在执行 OnReceive( )函数。

2. 可见进程（Visible Process）

可见进程是指可见但是非活动的进程，是指部分程序界面能够被用户看见，却不在前台与用户交互，不影响界面事件的进程。例如，当一个 Activity 被部分遮挡时就可视为可见进程，如果一个进程包含服务，且这个服务正被用户可见的 Activity 调用，此进程同样被视为可见进程。Android 系统一般存在少量的可见进程，只有在特殊情况下，Android 系统才会为保证前台进程的资源而清除可见进程。

3. 服务进程（Service Process）

服务进程是指包含已启动服务的进程。服务进程没有用户界面，在后台长期为用户服务运行，例如后台的任务管理等。一般情况下，Android 系统除非不能保证前台进程或可见进程所必要的资源，否则不强行清除服务进程。

4. 后台进程（Background Process）

后台进程是指不包含任何已经启动的服务而且没有任何用户可见的 Activity 的进程。Android 系统中一般存在数量较多的后台进程，在系统资源紧张时，系统将优先清除用户较长时间没有见到的后台进程。

5. 空进程（Empty Process）

空进程是没有持有任何活动应用组件的进程。保留空进程的唯一理由是为了提供一种缓存机制，缩短应用下次运行时的启动时间。空进程在系统资源紧张时会被首先清除，但为了提高 Android 系统应用程序的启动速度，Android 系统会将空进程保存在系统内存中，在用户重新启动该程序时，空进程会被重新使用。

2.2 Activity 组件

2.2.1 Android 组件简介

Android 应用程序中主要包括四种类型的组件：Activity、Service、Broadcast Receiver 和 Content Provider。在 Android 中，一个应用程序可以使用其他应用程序的组件，这是 Android 系统一个非常重要的特性。

Activity 是用户和应用程序交互的窗口，是程序的呈现层，相当于窗体。在应用中，一个 Activity 通常就是一个单独的屏幕，每一个 Activity 都被实现为一个独立的类，并且继承于 Activity 这个基类。这个 Activity 类将会显示由几个 Views 控件组成的用户接口，并对事件做出响应。大部分的应用都会包含多个 Activity。例如，一个短消息应用程序将会有一个屏幕用于显示联系人列表，第二个屏幕用于写短消息，同时还会有用于浏览旧短消息及进行系统设置的屏幕。每一个这样的屏幕，就是一个 Activity，当打开一个屏幕时，之前的那一个屏幕会被置为暂停状态，并且压入历史堆栈中，用户可以通过回退操作返回到以前打开过的屏幕。虽然很多 Activity 一起工作共同组成了一个应用程序，但每一个 Activity 都是相对独立的。

Service 服务是一种程序，没有可视化的用户界面，但它会在后台一直运行。它可以运行很长时间，相当于后台的一个服务。例如，一个服务可以在用户做其他事情的时候在后台播放背景音乐、从网络上获取一些数据或者计算一些东西并提供给需要这个运算结果的 Activity 使用。每个服务都继承自 Service 基类，通过 startService（Intent service）可以启动一个 Service，通过 Context. bindService( )可以绑定一个 Service。

Broadcast Receiver 被称为广播接收器，是一个专注于接收广播通知信息，并做出对应处理的组件。应用程序可以拥有任意数量的广播接收器以对所有感兴趣的通知信息予以响应，所有的接收器均继承自 BroadcastReceiver 基类。广播接收器没有用户界面，然而可以启动一个 Activity 来响应它们收到的信息，或者用 NotificationManager 来通知用户。BroadcastReceiver 注册的有两种方式，一种是可以在 AndroidManifest. xml 中注册，另一种可以在运行时的代码中使用 Context. registerReceiver( )进行注册。用户还可以通过 Context. sendBroadcast( )将它们自己的 Intent Broadcasts 广播给其他的应用程序。

Content Provider 内容提供者将一些特定的应用程序数据供给其他应用程序使用。数据可以存储于文件系统、SQLite 数据库或其他方式。内容提供者继承于 ContentProvider 基类，为其他应用程序取用和存储管理的数据实现了一套标准方法。然而，应用程序并不直接调用这些方法，而是使用一个 ContentResolver 对象，调用它的方法作为替代。ContentResolver 可以与任意内容提供者进行会话，与其合作来对所有相关的交互通信进行管理。

本节主要讲述 Activity 组件，其他的组件将在后续的章节中详细讲述。

2.2.2 Activity 生命周期

前面提到 Activity 是用户和应用程序交互的窗口，是呈现给手机用户的界面窗体。下面从 Activity 的创建出发探究 Activity 组件如何用于应用开发，下面代码创建了一个 Activity。

【例 2-1】 创建一个生命周期的 NewActivity. Java 代码。

```
1   package com.NewActivity;
2
3   import android.app.Activity;
4   import android.os.Bundle;
5   import android.widget.Button;
6   import android.widget.TextView;
7
8   public class NewActivity extends Activity {
9       private TextView m_txtNewAct = null;
10  private Button m_btnNewAct = null;
11  /* * Called when the activity is first created. * /
12  @Override
13  public void onCreate(Bundle savedInstanceState) {
14  super.onCreate(savedInstanceState);
15  setContentView(R.layout.main);
16  this.m_btnNewAct = (Button)findViewById(R.id.btnNewAct);
```

```
17  this.m_txtNewAct = (TextView)findViewById(R.id.txtNewAct);
18  this.m_txtNewAct.setText("文本控件");
19  this.m_btnNewAct.setText("按钮控件");
20      }
21      }
```

从代码第 8 行可以看出，一个 Activity 就是一个 Java 类，而且必须继承自 Activity 基类。代码第 13 行复写了基类的 onCreate( )方法，并且在 14 行调用了父类的 onCreate( )方法，这在创建 Activity 时同样是必须的，在 Activity 第一次运行时总会首先执行 onCreate( )方法，此方法的作用类似于其他面向对象语言中的构造函数。代码 15 行采用 setContentView( )方法设置布局文件，代码 16 和 17 行的 findViewById( )方法功能是通过控件 ID 属性获得所需的控件对象，此方法返回 View 类，View 类是 Android 系统的控件基类，可以通过强制类型转换转化成所需的控件对象。代码 18 行和 19 行完成相应控件显示文本的设置。创建 Activity 时，除了需要继承 Activity 基类和复写 onCreate( )方法外，每个 Activity 还必须在 AndroidManifest.xml 中注册。

AndroidManifest.xml 文件中声明 Activity 的代码如下：

```
……
1   <application android:icon = "@drawable/icon" android:label = "@string/app_name" >
2   <activity android:name = ".NewActivity"
3   android:label = "@string/app_name" >
4   <intent-filter >
5   <action android:name = "android.intent.action.MAIN" />
6   <category android:name = "android.intent.category.LAUNCHER" />
7   </intent-filter >
8   </activity >
9   </application >
……
```

在应用程序 application 元素中声明了前面所定义的 Activity，在 application 元素中同样可以声明 Service、BroadcastReceiver 和 ContentProvider。代码第 2 行属性 android：name 定义了实现 Activity 类的名称，其值有两种实现形式，一种是使用全称 com.NewActivity，另外一种是使用简化后的类名称 .NewActivity，其中的“.”不可省略。第 3 行属性 android：label 定义了 Activity 的标签名称，此名称将在 Activity 界面上面以标题形式显示，@string/app_name 是一种资源引用方式，其真实值是 res/values/string.xml 文件中 app_name 元素代表的字符串值。intent-filter 元素中包含了两个子元素 action 和 category，这些元素的意义将在后面详细介绍，在这里 intent-filter 的功能就是设置程序的启动主窗体为包含它的 Activity。

至此，在创建的 Activity 上可以按照需要布局必要的界面控件，如上面的 Activity 上添加了一个 Button 控件和一个 TextView 控件，其声明的代码如下所示：

```
……
1    <TextView
2    android:layout_width = "fill_parent"
3    android:layout_height = "wrap_content"
```

```
4      android:id="@ +id/txtNewAct"
5      />
6      <Button
7      android:layout_width="fill_parent"
8      android:layout_height="wrap_content"
9      android:id="@ +id/btnNewAct"
10     />
……
```

属性 android：layout_width 定义控件横向宽度，属性 android：layout_height 定义控件纵向高度，属性 android：layout_width 和属性 android：layout_height 的可选值均为 fill_parent 与 wrap_content。fill_parent 值代表填充父控件，wrap_content 值代表按照内容填充。属性 android：id 定义控件唯一标识名称，“@ +id”告诉系统在 R. java 文件中生成相应的值。

Activity 的创建预示着整个组件生命周期的开始，Activity 生命周期是指 Activity 从启动到销毁的全过程，在生命周期中起重要作用的是它的事件回调函数。Activity 提供了七个生命周期的事件回调函数，在这些事件回调函数中添加相应的功能代码可以实现或者完成相应的功能。

系统中 Activity 类实现的事件回调函数：

```
1   public class Activity extends ApplicationContext
2   {
3       protected void onCreate(Bundle savedInstanceState);
4       protected void onStart();
5       protected void onRestart();
6       protected void onResume();
7       protected void onPause();
8       protected void onStop();
9       protected void onDestroy();
10  }
```

各个事件回调函数的用法可参考表 2-1，其各函数在整个生命周期中的调用顺序和调用时机如图 2-3 所示。

表 2-1 Activity 生命周期事件回调函数

方　法	描　述	是否可终止	可后续方法
onCreate()	Activity 第一次被创建时调用，方法内部可以用于完成 Activity 的初始化、创建 View 控件和绑定数据等	否	onStart()
onStart()	Activity 显示在屏幕上，对用户可见时调用此方法	否	onResume() 或者 onStop()
onRestart()	Activity 从停止状态进入活动状态前，调用此方法	否	onStart()
onResume()	Activity 开始和用户交互，用户可输入信息时调用此方法	否	onPause()
onPause()	当系统将重新启动前一个 Activity 或者开始新的 Activity 调用当前 Activity 的此方法，即当前 Activity 进入暂停状态。此方法常用于保存改动的数据或者释放必要的内存	是	onResume() 或者 onStop()

（续）

方　法	描　述	是否可终止	可后续方法
onStop( )	当 Activity 对用户不可见时调用，通常由于新的 Activity 启动并覆盖当前的 Activity 或者当前 Activity 被销毁	是	onRestart( ) 或者 onDestroy( )
onDestroy( )	Activity 被彻底销毁前最后调用的方法，通常由于用户主动使用 finish( )方法或者系统释放必要内存	是	无

Activity 除了七个生命周期的回调函数外，还有两个函数在整个生命周期中同样具有举足轻重的低位，分别是 onRestoreInstanceState( )方法和 onSaveInstanceState( )方法。这两个函数不是生命周期的回调函数，在 Android 系统由于资源紧张需要终止一个 Activity 而此 Activity 会在稍后一段时间仍要显示给用户时的情况下调用。onSaveInstanceState( )方法在系统资源不足终止 Activity 前调用，用以保存 Activity 状态信息，供 onRestoreInstanceState( )或者 onCreate( )方法恢复 Activity 用。onRestoreInstanceState( )方法恢复 onSaveInstanceState( )保存的 Activity 状态信息，在 onStart( )和 onResume( )之间调用。

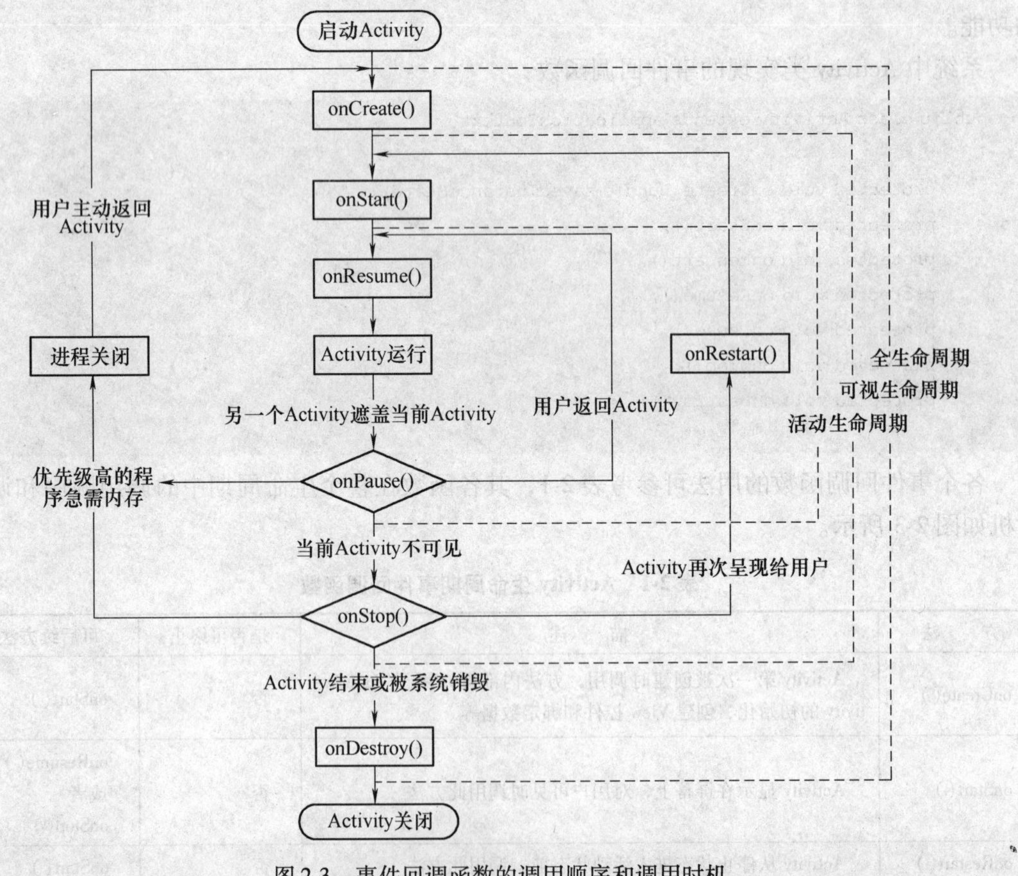

图 2-3　事件回调函数的调用顺序和调用时机

Activity 的完整生命周期还可以分为完整生存期、可视生存期和活动生存期，其中每个生存期中都包含相应的事件回调函数。三个生存期的关系是包含与被包含的关系，在整个 Activity 生命周期中由外到内，由完整到具体。

完整生存期（Entire Lifetime）几乎和 Activity 的完整生命周期等价，属于从 onCreate( ) 方法的第一次调用开始到 onDestroy( )方法的最后一次调用为止的生命段。此生存期存在 Activity 进程被终止而没有调用 onDestroy( )方法的特殊情况。onCreate( )方法完成初始化活动，例如扩展用户界面、分配对类变量的引用、将数据绑定到控件并创建服务和线程。onCreate( )方法传递了一个包含 UI 状态的 Bundle 对象，该对象是在最后一次调用 onSaveInstanceState( )时保存的。使用这个 Bundle 对象将用户界面恢复为上一次的状态，这里既可以通过 onCreate( )方法，也可以通过重写 onRestoreInstanceState( )方法来实现。另外，通过重写 onDestroy( )方法来清除 onCreate( )创建的所有资源，并保证所有的诸如网络或者数据库链接等外部资源被关闭。在 Android 中尽量避免创建短期的对象，对象的快速创建和销毁会导致额外的垃圾收集过程，建议只在 onCreate( )中生成对象一次。

可视生存期（Visible Lifetime）是指从调用 onStart( )开始到调用 onStop( )结束之间的周期阶段。在这段时间里，Activity 对用户是可见的，但是它有可能不是用户所关注的 Activity，或者它可能被部分遮挡了。Activity 在它的完整生存期内可能会经历多个可见生存期，因为 Activity 可能在前台和后台之间进行切换。在个别极端情况下，Android 运行时可能会在一个 Activity 位于可见生存期就将其销毁，而并不调用 onStop( )方法。onStop( )方法用来暂停或者停止动画、线程、计时器、服务或者其他专门用于更新用户界面的进程。当 Activity 界面不可见的时候更新它是没有意义的，因为这样消耗了资源却没有起到实际的作用。当 Activity 界面再次可见的时候，可以使用 onStart( )或者 onRestart( )来恢复或者重启这些进程。onRestart( )方法在 onStart( )方法前被调用，用于 Activity 从不可见变为可见的过程，进行特定处理。Activity 经常在可见和不可见的状态多次转换，所以 onStart( )和 onStop( )多次被调用。另外，onStart( )和 onStop( )同样也用于注册和销毁 BroadcastReceiver。

活动生存期（Foreground Lifetime）指调用 onResume( )及其对应的 onPause( )之间的那段生存期，此时，Activity 在屏幕的最上层，能够与用户直接交互。在活动生存期内可以安全地假设 onSaveInstanceState( )和 onPause( )会被调用，大部分 Activity 至少需重写 onPause( ) 方法来提交未保存的改动，因为在 onPause( )之外 Activity 可能在没有任何警告的情况下被终止。Activity 不在前台的时候也可以根据应用程序的架构，选择挂起线程、进程或者广播接收器。onResume( )方法可以是轻量级的，因为要求加载 Activity 界面的时候可以由 onCreate( )和 onRestoreIntanceState( )方法处理。使用 onResume( )可以重新注册已经使用 onPause( )停止的广播接收器或者其他进程。onPause( )最好也是轻量级的，因为下一个要显示到前台的 Activity 的 onRsume( )要等当前 Activity 的 onPause( )返回后才执行。所以，尽量让 onPause( )方法中的代码执行迅速，让 onResume( )方法中的代码尽可能少，以保证在前台和后台之间进行切换的时候程序能够保持响应。

下面通过编写代码查看 Activity 生命周期中主要事件回调函数的执行顺序。

【例 2-2】 监听按钮启动另一个生命进程的 MainActivity. java 代码文件。

```
......
1     //主窗体
2     public class MainActivity extends Activity {
3       private Button m_btnMainAct = null;
4       private TextView m_txtMainAct = null;
```

```
5     /** Called when the activity is first created. */
6     @Override
7     public void onCreate(Bundle savedInstanceState) {
8         System.out.println("MainActivity-onCreate");
9         super.onCreate(savedInstanceState);
10        setContentView(R.layout.main);
11        m_btnMainAct = (Button)findViewById(R.id.btnMainAct);
12        m_txtMainAct = (TextView)findViewById(R.id.txtMainAct);
13        m_btnMainAct.setOnClickListener(new ButtonListener());
14        m_btnMainAct.setText("启动 SecondActivity");
15        m_txtMainAct.setText("这是 MainActivity");
16    }
17    @Override
18    protected void onDestroy() {
19        //TODO Auto-generated method stub
20        System.out.println("MainActivity-onDestroy");
21        super.onDestroy();
22    }
23    @Override
24    protected void onPause() {
25        //TODO Auto-generated method stub
26        System.out.println("MainActivity-onPause");
27        super.onPause();
28    }
29    @Override
30    protected void onRestart() {
31        //TODO Auto-generated method stub
32        System.out.println("MainActivity-onRestart");
33        super.onRestart();
34    }
35    @Override
36    protected void onResume() {
37        //TODO Auto-generated method stub
38        System.out.println("MainActivity-onResume");
39        super.onResume();
40    }
41    @Override
42    protected void onStart() {
43        //TODO Auto-generated method stub
44        System.out.println("MainActivity-onStart");
45        super.onStart();
46    }
47    @Override
48    protected void onStop() {
49        //TODO Auto-generated method stub
50        System.out.println("MainActivity-onStop");
```

```
51          super.onStop();
52      }
53      class ButtonListener implements OnClickListener{
54          @Override
55          public void onClick(View v) {
56              Intent intent = new Intent();
57              intent.setClass(MainActivity.this, SecondActivity.class);
58              MainActivity.this.startActivity(intent);
59          }
60      }
61  }
```

MainActivity 是主窗体类，此类中重写了 Activity 的七个生命周期回调函数，并均使用 System.out.println()方法，此方法向控制台输出相应信息并换行。代码 53 行通过内部类实现按钮单击监听器。代码 56 ~58 行，通过 Intent 实现不同 Activity 的交互，Intent 会在后面的内容中详细介绍。ButtonListener 类实现的功能是按钮单击后的响应，即启动另一个 Activity。

SecondActivity.java 代码文件：

```
......
1   //第二屏
2   public class SecondActivity extends Activity {
3   private Button m_btnSecondAct = null;
4   private TextView m_txtSecondAct = null;
5   /** Called when the activity is first created. */
6   @Override
7   public void onCreate(Bundle savedInstanceState) {
8           System.out.println("SecondActivity-onCreate");
9           super.onCreate(savedInstanceState);
10          setContentView(R.layout.second);
11          m_btnSecondAct = (Button)findViewById(R.id.btnSecondAct);
12          m_txtSecondAct = (TextView)findViewById(R.id.txtSecondAct);
13          m_btnSecondAct.setOnClickListener(new ButtonListener());
14          m_btnSecondAct.setText("返回 MianActivity");
15          m_txtSecondAct.setText("这是 SecondActivity");
16  }
17  @Override
18  protected void onDestroy() {
19          //TODO Auto-generated method stub
20          System.out.println("SecondActivity-onDestroy");
21          super.onDestroy();
22  }
23  @Override
24  protected void onPause() {
25          //TODO Auto-generated method stub
26          System.out.println("SecondActivity-onPause");
27          super.onPause();
```

```
28      }
29      @Override
30      protected void onRestart() {
31              //TODO Auto-generated method stub
32              System.out.println("SecondActivity-onRestart");
33              super.onRestart();
34      }
35      @Override
36      protected void onResume() {
37              //TODO Auto-generated method stub
38              System.out.println("SecondActivity-onResume");
39              super.onResume();
40      }
41      @Override
42      protected void onStart() {
43              //TODO Auto-generated method stub
44              System.out.println("SecondActivity-onStart");
45              super.onStart();
46      }
47      @Override
48      protected void onStop() {
49              //TODO Auto-generated method stub
50              System.out.println("SecondActivity-onStop");
51              super.onStop();
52      }
53      class ButtonListener implements OnClickListener{
54      @Override
55      public void onClick(View v) {
56              Intent intent = new Intent();
57              intent.setClass(SecondActivity.this, MainActivity.class);
58              SecondActivity.this.startActivity(intent);
59              }
60        }
61      }
```

通过查看日志信息可以跟踪上面代码的运行情况。在 Eclipse 中切换到 DDMS 界面，其中的 LogCat 用于显示系统日志信息，能够捕获的信息包括 Dalvik 虚拟机产生的信息、进程信息、ActivityManager 信息、PackagerManager 信息、Homeloader 信息、WindowsManager 信息、Android 运行时信息和应用程序信息等。通过在 LogCat 中添加日志过滤器可以跟踪查看 System.out 输出信息。

运行程序，启动 MainActivity 完成后，查看日志过滤器中的 System.out 信息如图 2-4a 所示。如果此时再按下模拟器的返回键，则相当于关闭了 MainActivity，日志过滤器中的 System.out 信息变成了如图 2-4b 所示。可以看出，在 Activity 启动过程中，分别调用了 onCreate()、onStart()、onResume()，首先调用 onCreate()方法进行初始化，然后调用 onStart()方法使 Activity 显示在屏幕上，最后调用 onResume()方法获取屏幕焦点，此时 Activity 能够与用户正常交互并接收用户输入信息了，即 Activity 的启动完成。当用户按返回键或者主动调用 finish( )方法试图关闭 Activity 时，系统相继调用 onPause( )、onStop( )和 onDestroy( )，Activity 被销毁并释放进程资源。显然，一个完整的 Activity 生存期，生命周期时间回调函数的调用顺序依次为 onCreate( )、onStart( )、onResume( )、onPause( )、onStop( )和 onDestroy( )。另外，由于 Activity 直接关闭，onRestart( )方法未被调用。

Log (641) System.out

Time		pid	tag	Message
01-18 13:56:07.606	I	224	System.out	MainActivity-onCreate
01-18 13:56:07.756	I	224	System.out	MainActivity-onStart
01-18 13:56:07.765	I	224	System.out	MainActivity-onResume

a)MainActivity 启动日志

Log (644) System.out

Time		pid	tag	Message
01-18 13:56:07.606	I	224	System.out	MainActivity-onCreate
01-18 13:56:07.756	I	224	System.out	MainActivity-onStart
01-18 13:56:07.765	I	224	System.out	MainActivity-onResume
01-18 13:56:43.735	I	224	System.out	MainActivity-onPause
01-18 13:56:44.395	I	224	System.out	MainActivity-onStop
01-18 13:56:44.395	I	224	System.out	MainActivity-onDestroy

b)MainActivity 全生命周期日志

图 2-4　MainActivity 启动日志和全生命周期日志

如果一个 Activity 启动后调用其他 Activity 而导致前一个 Activity 被完全遮盖或者部分遮盖时，事件回调函数会按照相应的各自 Activity 的状态发生变化。重新运行上面的工程，MainActivity 启动完成后，System. out 信息如图 2-4a 所示，此时单击上面的按钮启动 SecondActivity，查看 System. out 信息，如图 2-5 所示。从图中可以看出，前三行信息是 MainActivity 的启动日志，后面五行是 MainActivity 中启动 SecondActivity 的日志。SecondActivity 的启动过程中，首先调用 MainActivity 的 onPause( )方法，再依次调用 SecondActivity 的 onCreate( )、onStart( )和 onResume( )，最后调用 MainActivity 的 onStop( )方法，此时 SecondActivity 完全遮盖 MainActivity。

Log (640) System.out

Time		pid	tag	Message
01-18 14:19:09.527	I	237	System.out	MainActivity-onCreate
01-18 14:19:09.596	I	237	System.out	MainActivity-onStart
01-18 14:19:09.617	I	237	System.out	MainActivity-onResume
01-18 14:21:31.137	I	237	System.out	MainActivity-onPause
01-18 14:21:31.187	I	237	System.out	SecondActivity-onCreate
01-18 14:21:31.207	I	237	System.out	SecondActivity-onStart
01-18 14:21:31.217	I	237	System.out	SecondActivity-onResume
01-18 14:21:31.547	I	237	System.out	MainActivity-onStop

图 2-5　MainActivity 调用 SecondActivity 日志

如果这时按模拟器的返回键，从 SecondActivity 返回到 MainActivity，这个过程的日志信息如图 2-6 所示。此过程中首先调用 SecondActivity 的 onPause( )方法，接着调用 MainActivity 的 onRestart( )、onStart( )和 onResume 方法，由于是返回键返回，此时 SecondActivity 被销毁，依次调用了 onStop( )和 onDestroy( )方法。

Time		pid	tag	Message
01-18 14:19:09.527	I	237	System.out	MainActivity-onCreate
01-18 14:19:09.596	I	237	System.out	MainActivity-onStart
01-18 14:19:09.617	I	237	System.out	MainActivity-onResume
01-18 14:21:31.137	I	237	System.out	MainActivity-onPause
01-18 14:21:31.187	I	237	System.out	SecondActivity-onCreate
01-18 14:21:31.207	I	237	System.out	SecondActivity-onStart
01-18 14:21:31.217	I	237	System.out	SecondActivity-onResume
01-18 14:21:31.547	I	237	System.out	MainActivity-onStop
01-18 14:35:24.927	I	237	System.out	SecondActivity-onPause
01-18 14:35:24.967	I	237	System.out	MainActivity-onRestart
01-18 14:35:24.967	I	237	System.out	MainActivity-onStart
01-18 14:35:24.967	I	237	System.out	MainActivity-onResume
01-18 14:35:25.267	I	237	System.out	SecondActivity-onStop
01-18 14:35:25.287	I	237	System.out	SecondActivity-onDestroy

图 2-6　返回键从 SecondActivity 返回 MainActivity 日志

如果不是返回键返回而是单击按钮返回，SecondActivity 返回到 MainActivity，这个过程的日志信息如图 2-7 所示。此过程中首先调用 SecondActivity 的 onPause( ) 方法，接着调用 MainActivity 的 onCreate( )、onStart( ) 和 onResume 方法，最后调用了 SecondActivity 的 onStop ( ) 方法。比较图 2-6，此时的 MainActivity 被全新重建，与此同时 SecondActivity 也并没有被销毁，其资源仍占据着系统资源。

Log (594)　System.out

Time		pid	tag	Message
01-18 14:46:47.929	I	229	System.out	MainActivity-onCreate
01-18 14:46:48.129	I	229	System.out	MainActivity-onStart
01-18 14:46:48.179	I	229	System.out	MainActivity-onResume
01-18 14:46:52.109	I	229	System.out	MainActivity-onPause
01-18 14:46:52.159	I	229	System.out	SecondActivity-onCreate
01-18 14:46:52.189	I	229	System.out	SecondActivity-onStart
01-18 14:46:52.189	I	229	System.out	SecondActivity-onResume
01-18 14:46:52.539	I	229	System.out	MainActivity-onStop
01-18 14:47:00.229	I	229	System.out	SecondActivity-onPause
01-18 14:47:00.259	I	229	System.out	MainActivity-onCreate
01-18 14:47:00.279	I	229	System.out	MainActivity-onStart
01-18 14:47:00.309	I	229	System.out	MainActivity-onResume
01-18 14:47:00.739	I	229	System.out	SecondActivity-onStop

图 2-7　主动从 SecondActivity 返回 MainActivity 日志

以上是 SecondActivity 完全遮挡 MainActivity 的情形，下面讨论部分遮挡的情况。在 AndroidManifest.xml 文件中为 SecondActivity 添加属性 android:theme 并且设置其值为 android:theme = " @android:style/Theme.Dialog"。此属性定义 SecondActivity 显示方式为窗口而不是整个屏幕。运行程序从 MainActivity 中启动 SecondActivity，此过程 System.out 输出信息如图 2-8所示。比较图 2-5，可以看出，部分遮挡的情况下，MainActivity 不会调用自身的 onStop( ) 方法，这是因为 MainActivity 仍可见，只是不能直接和用户交互。

Log (594)　System.out

Time		pid	tag	Message
01-19 14:20:17.743	I	220	System.out	MainActivity-onCreate
01-19 14:20:17.963	I	220	System.out	MainActivity-onStart
01-19 14:20:17.993	I	220	System.out	MainActivity-onResume
01-19 14:22:50.273	I	220	System.out	MainActivity-onPause
01-19 14:22:50.323	I	220	System.out	SecondActivity-onCreate
01-19 14:22:50.343	I	220	System.out	SecondActivity-onStart
01-19 14:22:50.353	I	220	System.out	SecondActivity-onResume

图 2-8　SecondActivity 部分遮挡 MainActivity 输出日志

上面详细介绍了 Activity 的生命周期，在 Android 应用开发中，Activity 的生命周期中还需注意以下几点：

（1）当 Activity 处于暂停或停止状态下，操作系统内存缺乏可能会销毁 Activity，或者其他意外突发情况，Activity 被操作系统销毁，内存回收时 onSaveInstanceState( )会被调用，但是当用户主动销毁一个 Activity 时（例如按返回键）onSaveInstanceState( )就不会被调用。onSaveInstanceState( )适合保存一些临时性的数据，onPause( )适合保存一些持久化的数据。

（2）onRestoreInstanceState( )是在 onStart( )和 onCreate( )之间执行用户恢复 Activity UI 状态。

（3）如果数据比较重要但是数据仍在运算当中，则应该缓存它们，如果运算结束并且得到了结果，则应该对其进行持久化操作。

（4）只要 Activity 被覆盖一定会调用 onPause( )方法，只要 Activity 重新回到前台一定会调用 onResume( )方法。

2.2.3 Task 与 Activity 栈

1. Task

一个 Task 是用户可以完成一个特定目标的一组 Activity，与 Activity 属于哪个 Application 无关。除非明确地新建一个 Task，否则用户启动的所有 Activity 都默认是当前 Task 的一部分。这些 Activity 可能属于任何一个 Application，属于同一个 Application 或者属于不同的 Application。例如，从联系人列表（第一个 Activity）开始，然后选择一个邮箱地址（第二个 Activity），然后附加一个照片（第三个 Activity），联系人列表、邮箱和图片，这些都存在于不同的 Activity 中，但却属于同一个 Task。

启动 Task 的 Activity 被称作根 Activity。通常，Task 是从应用管理器、主屏或者最近的 Task（长按 HOME 键）开始的。用户可以通过单击根 Activity 的图标回到 Task 里去，就像启动这个 Activity 一样。在这个 Task 中，使用 BACK 键可以回到这个 Task 的前一个 Activity 里，Activity 栈可以由一个或多个 Task 组成。

Task 的一个重要的特性就是，用户可以中断其当前正在进行的任务，去进行另一个 Task，然后可以返回到原来的那个 Task 去完成它，即打断 Task。这个特性的目的，就是用户可以同时运行多个任务，并且可以在这些任务间切换。离开一个 Task 有两种主要的情形：①用户被 Notification 打断。例如来了一个通知，用户开始关注处理这个通知；②用户决定开始另一个任务。例如用户按了 HOME 键，然后开始了另一个 Application。遇到这两种情况时，应该注意能让用户返回到离开的那个任务。

除了上面提到的两种方法，还有一种方法开始一个新任务，即在代码中启动 Activity 的时候，定义它要开始一个新 Task。地图和浏览器这两个应用就是这么做的，例如，在电子邮件中单击一个地址，会在新 Task 调出地图 Activity，在电子邮件中单击一个链接，会在新的 Task 中调出浏览器。在这种情况下，BACK 键会回到上一个 Activity，即另一个 Task 中的电子邮件 Activity，因为它不是从主屏启动的。

2. Activity 栈

当用户在 Application 中，从一个 Activity 跳到另一个 Activity 时，Android 系统会保存一个用户访问 Activity 的线性导航历史，这就是 activity 栈，也被称为返回栈。一般来说，当用户运行一个新的 Activity，这个 Activity 就会被加到 Activity 栈里。因此，当用户按 BACK 键的时候，栈中的上一个 Activity 就会被展示出来，用户可以一直按 BACK 键，直到返回主屏 Activity。把 Activity 加入到当前栈里的操作，与 Activity 是否启动了一个新 Task 无关，但是返回操作可以使用户从当前 Task 回到上一个 Task。用户可以在应用管理器、主屏或者“最近 Task”屏幕，恢复到刚刚的 Task。

只有 Activity 可以加到 Activity 栈里去，View、Window、Menu 或者 Dialog 都能进行此种操作。假设界面 A 跳到界面 B，然后用户可以用 BACK 跳回界面 A。这种情况下，界面 A 和界面 B 都要被实现成 Activity。这个规则有一个例外的情况，那就是其应用控制了 BACK 键并且自行管理界面导航。

下面通过图示介绍多个 Activity 互相调用时 Activity 栈的变化。如图 2-9 所示，假设一个 Application 中包含四个 Activity，为 Activity1 ~ Activity4。应用程序启动之后，运行第一个 Activity1，Activity1 对象被压入到 Stack 当中。在 Activity1 中启动第二个 Activity2，Activity2 对象被压入到 Stack 当中，由于手机显示的总是位于 Stack 顶部的 Activity，所以此时用户看到的屏幕是 Activity2。在 Activity2 中启动第三个 Activity3，Activity3 对象被压入到 Stack 当中，用户看到的屏幕变成 Activity3。在 Activity3 中启动最后一个 Activity4，Activity4 对象被压入到 Stack 当中，手机屏幕变成 Activity4。单击 BACK 按钮，这时 Activity4 对象在栈中被弹出，第三个 Activity 置于栈顶，依次单击 BACK 按钮，最后可返回到主屏幕 Activity1。

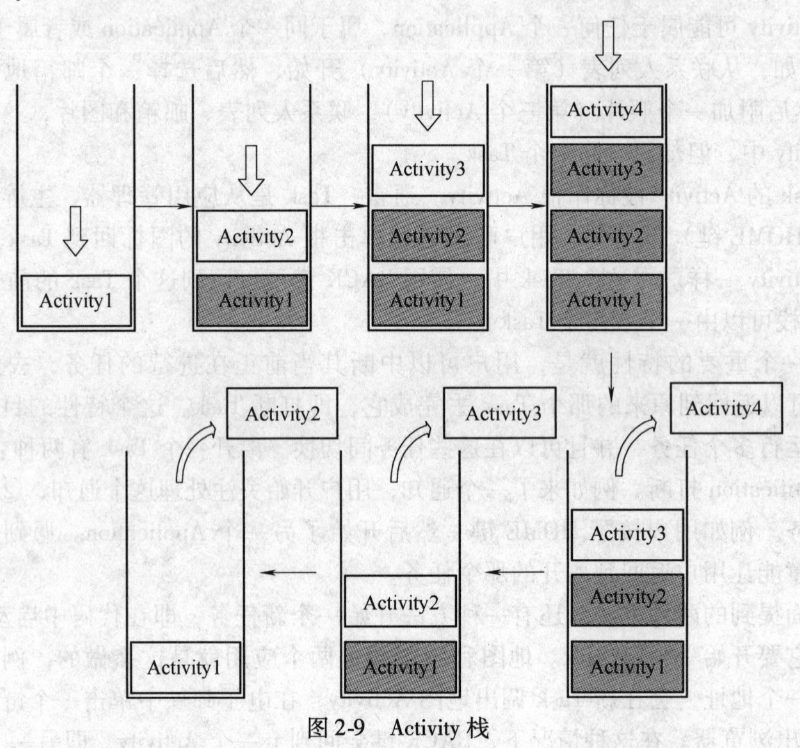

图 2-9　Activity 栈

2.2.4　Activity 基本状态

在 Android 系统中，Activity 拥有四种基本状态：活动态、暂停态、停止态和非活动态。

活动态（Active/Running）：Activity 在用户界面中处于最上层，完全能被用户看到，能够与用户进行交互。

暂停态（Paused）：Activity 界面上被部分遮挡，该 Activity 不再处于用户界面的最上层，且不能够与用户进行交互。

停止态（Stopped）：Activity 在界面上完全不能被用户看到，也就是说这个 Activity 被其他 Activity 全部遮挡。

非活动态（Killed）：又称为死亡态，是指不在上面三种状态中之内的 Activity 状态。

当 Activity 实例被创建、销毁或者启动另外一个 Activity 时，它在这四种状态之间进行转换，这种转换的发生依赖于用户程序的动作，各个状态之间的转换关系如图 2-10 所示，状态转换与事件回调函数的关系如图 2-11 所示。

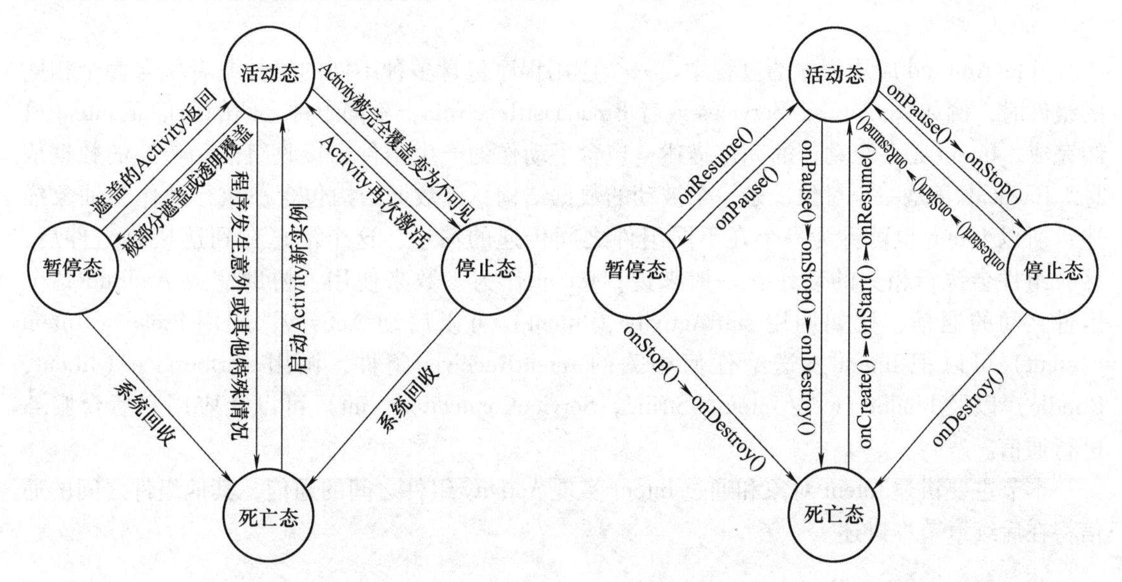

图 2-10　Activity 状态之间的转换关系　　　　图 2-11　Activity 状态转换与事件回调函数的关系

（1）Activity 活动态与暂停态的转换

活动态到暂停态：Activity 被别的窗体遮住了部分界面或者被透明窗体覆盖，失去了用户焦点，但仍可见，onPause( )方法被调用。

暂停态到活动态：上述情况下的 Activity 重新获得焦点，onResume( )方法被调用。

活动态与暂停态的转换过程中，Activity 的实例总是存在的。

（2）Activity 暂停态与死亡态的转换

暂停态到死亡态：系统由于资源紧张，需要回收部分资源用于其他高优先级进程使用，此过程依次调用 onStop( )和 onDestroy( )方法。

死亡态到暂停态：转换不可实现。

（3）Activity 活动态与死亡态的转换

活动态到死亡态：一般情况下，这种转换依次经历了暂停态和停止态，但 Activity 在这两种状态下并不存在保持的阶段，而是直接过渡到死亡态，此过程依次调用 onPause( )、onStop( )和 onDestroy( )方法。

死亡态到活动态：这种转换发生在新实例的启动，依次调用 onCreate( )、onStart( )和 onResume( )方法。

（4）Activity 活动态与停止态的转换

活动态到停止态：这个过程发生在 Activity 的界面完全被别的 Activity 遮住，当然也失去了用户焦点，这个过程中 Activity 的实例仍然存在，依次调用 onPause( )和 onStop( )方法。

停止态到活动态：Activity 再次被激活时，产生这个转换，依次调用 onResart( )、onStart( )和 onResume( )方法。

(5) Activity 停止态与死亡态的转换

停止态到死亡态：系统销毁 Activity，调用 onDestroy( )方法。

2.3 Intent 信使

开发 Android 应用程序的过程中，一个应用程序包含多种不同的组件或者包含多个相同的组件时，例如 Activities、Services 或者 BroadcastReceivers，不同组件之间的通信靠 Intent 机制完成。Intent 是一个动作的完整描述，包含了动作的产生组件、接收组件和传递的数据信息。Intent 本身是一个对象，是一个被动的数据结构，该数据结构包含被执行动作的抽象描述，所以 Intent 也可称为一个在不同组件之间传递的消息，这个消息在到达接收组件后，接收组件会执行相关的动作。一般来说，Intent 作为参数来使用，协助完成 Android 各个组件之间的通信，比如调用 startActivity（Intent）可以启动 Activity，调用 broadcastIntent（Intent）可以把 Intent 发送给任何相关的 IntentReceiver 组件，调用 startService（Intent, Bundle）以及 bindService（Intent, String, ServiceConnection, int）可以让应用和后台服务进行通信。

本节主要讲述 Intent 对象和通过 Intent 实现 Activity 组件之间的通信，其他组件之间的通信将在后续章节中讲述。

2.3.1 Intent 基本构成

Intent 对象抽象地描述了要执行的动作，其描述的基本内容分为六部分：组件名称（Component Name）、动作（Action）、数据（Data）、类别（Category）、附加信息（Extra）和标志（Flag）。

1. 组件名称（Component Name）

组件名称是指 Intent 目标组件的名称，组件名称是一个 ComponentName 对象，这种对象名称是目标组件完全限定类名和目标组件所在应用程序的包名的组合。组件中包名不一定要和 AndroidManifest. xml 文件中的包名完全匹配。

组件名称是一个可选项，如果 Intent 消息中指明了目标组件的名称，这就是一个显式消息，Intent 会传递给指明的组件。如果目标组件名称并没有指定，Android 则通过 Intent 内的其他信息和已注册的 IntentFilter 的比较来选择合适的目标组件。组件名字通过 setComponent( )、setClass( )或 setClassName( )设置，通过 getComponent( )读取。

2. 动作（Action）

动作描述 Intent 所触发动作名字的字符串，对于 BroadcastIntent 来说，Action 指被广播出去的动作。理论上 Action 可以为任何字符串，而与 Android 系统应用有关的 Action 字符串以静态字符串常量的形式定义在 Intent 类中。Android 系统支持的用于 Activity 组件常见的 Action 字符串常量如表 2-2 所示。类似于一个方法名决定了参数和返回值，动作很大程度上决定了接下来 Intent 如何构建，特别是数据和附加字段。一个 Intent 对象的动作通过 setAction( )方法设置，通过 getAction( )方法读取。

表 2-2 Android 系统支持的用于 Activity 组件常见的 Action 字符串常量

动　作	说　明
ACTION_ANSWER	打开接听电话的 Activity,默认为 Android 内置的拨号盘界面
ACTION_CALL	打开拨号盘界面并拨打电话,使用 URI 中的数字部分作为电话号码
ACTION_DELETE	打开一个 Activity,对所提供的数据进行删除操作
ACTION_DIAL	打开内置拨号盘界面,显示 URI 中提供的电话号码
ACTION_EDIT	打开一个 Activity,对所提供的数据进行编辑操作
ACTION_INSERT	打开一个 Activity,在提供数据的当前位置插入新项
ACTION_PICK	启动一个子 Activity,从提供的数据列表中选取一项
ACTION_SEARCH	启动一个 Activity,执行搜索动作
ACTION_SENDTO	启动一个 Activity,向数据提供的联系人发送信息
ACTION_SEND	启动一个可以发送数据的 Activity
ACTION_VIEW	最常用的动作,对以 URI 方式传送的数据,根据 URI 协议部分以最佳方式启动相应的 Activity 进行处理。对于 http:address 将打开浏览器查看;对于 tel:address 将打开拨号呼叫指定的电话号码
ACTION_WEB_SEARCH	打开一个 Activity,对提供的数据进行 Web 搜索

3. 数据（Data）

数据是描述 Intent 要操作的数据 URI 和数据 MIME 类型，不同的动作有不同的数据规格。例如，如果动作字段是 ACTION_EDIT，数据字段将包含显示用于编辑的文档 URI；如果动作是 ACTION_CALL，数据字段将是一个 tel:URI 和将拨打的号码；如果动作是 ACTION_VIEW，数据字段是一个 http:URI，接收 Activity 将被调用去下载和显示 URI 指向的数据。

匹配一个 Intent 到一个能够处理数据的组件，通常需要知道数据的类型（它的 MIME 类型）和它的 URI。例如，一个组件能够显示图像数据，不应该被调用去播放一个音频文件。在许多情况下，数据类型能够从 URI 中推测，特别是 content:URIs，它表示位于设备上的数据且被内容提供者（Content Provider）控制。但是类型也能够显式地设置，通常使用 setData()方法指定数据的 URI，使用 setType()指定 MIME 类型，使用 setDataAndType()指定数据的 URI 和 MIME 类型，使用 getData()读取 URI，getType()读取类型。

4. 类别（Category）

类别指定了将要执行 Action 的其他一些额外的信息，Android 系统支持的常见 Category 字符串常量如表 2-3 所示。通常使用 addCategory()方法添加一个种类到 Intent 对象中，使用 removeCategory()方法删除一个之前添加的种类，使用 getCategories()方法获取 Intent 对象中的所有种类。

表 2-3 Android 系统支持的常见 Category 字符串常量

值	说　明
ALTERNATIVE	Intent 数据默认动作的一个可替换的执行方法
SELECTED_ ALTERNATIVE	和 ALTERNATIVE 类似，但替换的执行方法不是指定的，而是被解析出来的
BROWSABLE	声明 Activity 可以由浏览器启动
DEFAULT	为 Intent 过滤器中定义的数据提供默认动作
HOME	设备启动后显示的第一个 Activity
LAUNCHER	在应用程序启动时首先被显示

5. 附加信息（Extra）

附加信息是其他所有附加信息的集合。使用 Extra 可以为组件提供扩展信息，当使用 Intent 连接不同的组件时，有时需要在 Intent 中附加额外的信息，以便将数据传递给目标 Activity。例如 ACTION_TIMEZONE_CHANGED 需要带有附加信息表示新的时区。

Extra 采用键值对结构保存在 Intent 对象当中，Intent 对象通过调用方法 putExtras( ) 和 getExtras( ) 来存储和获取 Extra。Extra 是以 Bundle 对象的形式来保存的，Bundle 对象提供了一系列 put 和 get 方法来设置、提取相应的键值信息。在 Intent 类中，Android 系统支持的常见 Exrta 字符串常量，如表 2-4 所示。

表 2-4　Android 系统支持的常见 Extra 字符串常量

值	说　明
EXTRA_BCC	装有邮件密送地址的字符串数组
EXTRA_CC	装有邮件抄送地址的字符串数组
EXTRA_EMAIL	装有邮件发送地址的字符串数组
EXTRA_INTENT	使用 ACTION_PICK_ACTIVITY 动作时装有 Intent 选项的键
EXTRA_KEY_EVENT	触发该 Intent 的按键的 KeyEvent 对象
EXTRA_PHONE_NUMBER	使用拨打电话相关的 Action 时，电话号码字符串的键，类型为 String
EXTRA_SHORTCUT_ICON	使用 ACTION_CREATE_SHORTCUT 在 Activity 创建快捷方式时，对快捷方式的描述信息。ICON 和 ICON_RESOURCE 描述的是快捷方式的图标，类型分别为 Bitmap 和 ShortcutIconResource。INTENT 描述的是快捷方式相对应的 Intent 对象。NAME 描述的是快捷方式的名字
EXTRA_SHORTCUT_ICON_RESOURCE	
EXTRA_SHORTCUT_INTENT	
EXTRA_SHORTCUT_NAME	
EXTRA_SUBJECT	描述信息主体的键
EXTRA_TEXT	使用 ACTION_SEND 动作时，用来描述要发送的文本信息，类型是 ChatSequence
EXTRA_TITLE	使用 ACTION_CHOOSER 动作时，描述对话框标题的键，类型是 ChatSequence
EXTRA_UID	使用 ACTION_UID_REMOVED 动作时，描述删除用户 Id 的键，类型为 int

6. 标记（Flag）

Flag 指示 Android 系统如何去启动一个 Activity 和启动之后的处理。例如，活动应该属于那个任务，是否属于最近的活动列表。通常使用 setFlags( ) 方法和 addFlags( ) 方法设置和添加 Flag。

2.3.2　Intent 形式

Android 系统中，明确指出了目标组件名称的 Intent，称之为显式 Intent，没有明确指出目标组件名称的 Intent，则称之为隐式 Intent。

显式 Intent（Explicit Intents）指定了目标组件，一般调用 setComponent( ) 或者 setClass (Context, Class) 方法设定 Intents 的 component 属性，指定具体的组件类。这些 Intent 一般不包括其他任何信息，通常用于应用程序内部消息，如一个 Activity 启动从属的服务或启动另一个 Activity。

隐式 Intent（Implicit Intents）没有明确指明目标组件，经常用于启动其他应用程序中的组件。

在应用程序中，一般存在多个 Activity，Intent 实现不同 Activity 的切换和数据传递，根据 Intent 形式的不同，Activity 的启动方式可以分为显式启动和隐式启动。显式启动必须在 Intent 中指明启动的 Activity 所在的类，隐式启动则由 Android 系统根据 Intent 的动作和数据来决定启动哪一个 Activity，也就是说在隐式启动时，Intent 中只包含需要执行的动作和所包含的数据，而无需指明具体启动哪一个 Activity，选择权有 Android 系统和最终用户来决定。

Intent 显式启动 Activity 时，首先需要创建一个 Intent 对象，并需要显式指明启动的目标 Activity。下面通过实例说明如何使用 Intent 显式启动 Activity。

【例 2-3】 用 Intent 信使启动另一个生命周期。

建立工程 Chp02_IntentStartDirectly，添加两个 Activity，分别为 MainActivity 和 SecondActivity，MainActivity 作为主屏，在其中使用 Intent 显式启动 SecondActivity。

AndroidManifest. xml 文件代码如下：

```
    ......
1   <application android:icon="@drawable/icon"android:label="@string/app_name">
2   <activity android:name=".MainActivity"
3           android:label="@string/app_name">
4   <intent-filter>
5   <action android:name="android.intent.action.MAIN"/>
6   <category android:name="android.intent.category.LAUNCHER"/>
7   </intent-filter>
8   </activity>
9   <activity android:name=".SecondActivity"
10          android:label="@string/app_name">
11  </activity>
12  </application>
    ......
```

前面提到，Android 应用程序中，用户使用的每个组件都必须在 AndroidManifest. xml 文件中的 <application> 节点内注册，上面代码中在 AndroidManifest. xml 文件中使用 <activity> 标签注册 MainActivity 和 SecondActivity，嵌套在 <application> 标签内部。

MainActivity. java 文件代码如下：

```
    ......
1   public class MainActivity extends Activity{
2       private Button m_btnMainAct = null;
3   /** Called when the activity is first created. */
4   @Override
5   public void onCreate(Bundle savedInstanceState){
6   super.onCreate(savedInstanceState);
7   setContentView(R.layout.main);
8   m_btnMainAct = (Button)findViewById(R.id.btnMainAct);
9   m_btnMainAct.setOnClickListener(new OnClickListener(){
10          public void onClick(View view){
11                      //声明一个 Intent 对象
12      Intent inttMainAct = new Intent(MainActivity.this,SecondActivity.class);
```

```
13                      //Intent 显示启动 Activity
14      startActivity(inttMainAct);
15      }
16  });
17  }
18  }
    ......
```

代码 12 行声明了一个 Intent 对象，并进行了初始化。Android 系统中，Intent 类定义了以下 6 种构造函数：

（1）Intent( );

（2）Intent（Intent o）;

（3）Intent（String action）;

（4）Intent（String action，Uri uri）;

（5）Intent（Context packageContext，Class <？ >cls）;

（6）Intent（String action，Uri uri，Context packageContext，Class <？ >cls）;

第一种为空构造函数，调用此构造函数的 Intent 为一个空 Intent 对象。第二种的参数也是一个 Intent 对象，通过复制该对象初始化新的 Intent 对象。第三种构造函数指定了动作 Action 的类型。第四种构造函数有两个参数分别指定了动作 Action 和 URI Data。第五种构造函数指定了主调组件和目标组件。第六种构造函数拥有三个参数，描述了动作 Action、数据 Data 和目标组件。

代码 12 行采用了第五种构造函数形式，指明了主调组件 MainActivity 和目标组件 SecondActivity。startActivity( )方法启动 Intent 指向的 Activity。上面代码使用了 Activity 显示启动的方式，直接指明了需要启动的 Activity。

在 Android 应用程序中，启动哪一个 Activity 并不需要在程序内部直接指明，很多时候也不能明确获知将要启动的 Activity，而需要由 Android 系统根据需要自行决定，这种情况下，就需要隐式启动 Activity。隐式启动 Activity 时，Android 系统在应用程序运行时对 Intent 进行解析，并根据一定的规则对 Intent 和 Activity 进行匹配，使 Intent 上的动作、数据与 Activity 完全吻合。匹配的 Activity 可以是应用程序本身的，也可以是 Android 系统内置的，还可以是第三方应用程序提供的。因此，这种方式更加强调了 Android 应用程序中组件的可复用性。

下面代码为隐式启动 Activity 的 Intent 示例。

```
Intent inttMainAct = new Intent(Intent.ACTION_VIEW,Uri.parse("http://www.baidu.com"));
startActivity(inttMainAct);
```

Intent 的动作是 Intent. ACTION_VIEW，根据 URI 的数据类型来匹配动作。数据部分的 URI 是 Web 地址，使用 Uri. parse（urlString）方法，可以简单地把一个字符串解释成 URI 对象。显然 Intent 构造函数采用的是第四种格式，此代码的意义就是隐式调用系统内部的 Web 浏览器，打开网址 www. baidu. com。

2.3.3 Intent 过滤器

Intent 过滤器，即 Intent Filter，是 Android 系统提供的一种机制，根据 Intent 中的动作

（Action）、类别（Category）和数据（Data）等内容，对适合接收该 Intent 的组件进行匹配和筛选，用于隐式启动组件过程中。Intent 过滤器同样可以匹配数据类型、路径和协议，还包括可以用来确定多个匹配项顺序的优先级。Android 应用程序中 Activity 组件、Service 组件和 BroadcastReceiver 都可以注册 Intent 过滤器，可以注册一个也可以注册多个。组件如果没有注册任何 Intent 过滤器，则只能接收显式的 Intent，而注册了 Intent 过滤器的组件既可以显式使用 Intent 也可以隐式使用 Intent。只有当一个 Intent 对象的动作（Action）、数据（Data）和类别（Category）同时符合 Intent 过滤器时，才被考虑，附加信息（Extra）和标志（Flag）在此过程中不起作用。

组件注册 Intent 过滤器常见的方法是在 AndroidManifest. xml 文件中用节点 <Intent-Filter> 描述。当然也可以在代码中动态地为组件设置 Intent 过滤器。在节点 <Intent-Filter> 中声明 <action>、<category> 和 <data> 标签分别定义 Intent 过滤器的动作（Action）、类别（Category）和数据（Data），其支持的属性如表 2-5 所示。

表 2-5　<Intent-Filter> 节点支持的属性

标　签	属　性	说　明
<action>	android:name	指定组件所能响应的动作，用字符串表示，通常使用 Java 类名和包的完全限定名构成
<category>	android:category	指定以何种方式去服务 Intent 请求的动作
<data>	android:host	指定一个有效的主机名
	android:mimetype	指定组件能处理的数据类型
	android:path	有效的 URI 路径名
	android:port	主机的有效端口号
	android:scheme	所需要的特定的协议

前面提到，一个 Intent 过滤器有对应于 Intent 对象的动作（Action）、数据（Data）和种类（Category）的字段。Intent 过滤器要检测隐式 Intent 的这三个字段，其中任何一个失败，Android 系统都不会传递 Intent 给组件。然而，因为一个组件可以有多个 Intent 过滤器，一个 Intent 通不过组件的过滤器检测，其他的过滤器可能通过检测。

在执行动作检测时，虽然一个 Intent 对象仅是单个动作，但是一个 Intent 过滤器可以列出不止一个 <action>。另外，<Intent-Filter> 中 <action> 列表不能够为空，一个 Intent 过滤器必须至少包含一个 <action> 子元素，否则它将阻塞所有的 Intent。要通过动作检测，Intent 对象中指定的动作必须匹配过滤器的 <action> 列表中的一个。如果过滤器没有指定动作，那么没有一个 Intent 将与之匹配，所有的 Intent 将检测失败，即没有 Intent 能够通过过滤器，如果 Intent 对象没有指定动作，将自动通过检查。

在执行类别检测时，一个 Intent 要通过类别检测，Intent 对象中的每个类别必须匹配过滤器中的一个，即过滤器可以列出额外的类别，但是 Intent 对象中的类别都必须能够在过滤器中找到。<category> 标签可以定义多个，可使用自定义的类别。原则上如果一个 Intent 对象中没有类别应该总是通过类别检测，而不管过滤器中有什么类别。但是有个例外，Android 对待所有传递给 startActivity()的隐式 Intent，它们至少包含 android. intent. category. DEFAULT（对应 CATEGORY_DEFAULT 常量）。因此，Activity 想要接收隐式 Intent 必须要在 Intent 过滤器中包含 android. intent. category. DEFAULT。

在执行数据检测时，每个 <data>元素指定一个 URI 和数据类型（MIME 类型），它有四个属性 scheme、host、port 和 Path 对应于 URI 的每个部分，格式如下：

scheme://host:port/path

例如，content://com. proj:80/folder/subfolder/etc

scheme 是 content，host 是/com. proj，port 是 80，path 是 folder/subfolder/etc。host 和 port 一起构成 URI 的凭据（Authority），如果 host 没有指定，port 也被忽略。这四个属性都是可选的，但它们之间并不都是完全独立的，要让凭据有意义，scheme 必须也要指定，要让 Path 有意义，scheme 和 authority 也都必须要指定。当比较 Intent 对象和过滤器的 URI 时，仅仅比较过滤器中出现的 URI 属性。例如，如果一个过滤器仅指定了 scheme，所有拥有此 scheme 的 URI 都匹配过滤器；如果一个过滤器指定了 scheme 和 authority，但没有指定 Path，所有匹配 scheme 和 authority 的 URI 都通过检测，而不管它们的 Path；如果四个属性都指定了，要都匹配才能算是匹配。然而，过滤器中的 Path 可以包含通配符来要求匹配 Path 中的一部分。<data>元素的 type 属性指定数据的 MIME 类型，Intent 对象和过滤器都可以用"*"通配符匹配子类型字段，例如"text/*"，"audio/*"表示任何子类型。数据检测既要检测 URI，也要检测数据类型。规则如下：

如果一个 Intent 对象既不包含 URI，也不包含数据类型，仅当过滤器不指定任何 URI 和数据类型时，才不能通过检测，否则都能通过。

如果一个 Intent 对象包含 URI，但不包含数据类型，仅当过滤器不指定数据类型，同时它们的 URI 匹配时，才能通过检测。

如果一个 Intent 对象包含数据类型，但不包含 URI，仅当过滤只包含数据类型且与 Intent 相同，才通过检测。

如果一个 Intent 对象既包含 URI，也包含数据类型（或数据类型能够从 URI 推断出），则数据类型部分，只有与过滤器中之一匹配才算通过；URI 部分，它的 URI 要出现在过滤器中，或者它有 content：或 file：URI，又或者过滤器没有指定 URI。换句话说，如果它的过滤器仅列出了数据类型，组件假定支持 content:和 file:。

如果一个 Intent 能够通过不止一个 Activity 或服务器的过滤，用户可能会被问哪个组件被激活。如果没有找到目标，则会抛出一个异常。

结合动作检测、类别检测和数据检测，Android 系统中 Intent 过滤器的匹配规则可以概括成以下四点：

（1）Android 系统把所有应用程序包中的 Intent 过滤器集合在一起，形成了一个完整的 Intent 过滤器列表。

（2）在 Intent 与 Intent 过滤器进行匹配时，Android 系统会将列表中所有 Intent 过滤器的动作和类别与 Intent 进行匹配，任何不匹配的 Intent 过滤器都将被过滤掉。没有指定动作的 Intent 过滤器可以匹配任何的 Intent，但是没有指定类别的 Intent 过滤器只能匹配没有类别的 Intent。

（3）把 Intent 数据 URI 的每个子部与 Intent 过滤器的 <data>标签中的属性进行匹配，如果 <data>标签指定了协议、主机名、路径名或 MIME 类型，那么这些属性都要与 Intent 的 URI 数据部分进行匹配，任何不匹配的 Intent 过滤器均被过滤掉。

（4）如果 Intent 过滤器的匹配结果多于一个，则可以根据在 <intent-filter> 标签中定义的优先级标签来对 Intent 过滤器进行排序，优先级最高的 Intent 过滤器将被选择。

下面代码简单地示例出了 Android 系统中 Intent 过滤器的匹配规则。

【例 2-4】 用 Intent 信使直接启动生命周期的 AndroidManifest. xml 代码文件。

```
    ......
1      <application android:icon="@drawable/icon"android:label="@string/app_name">
2      <activity android:name=".MainActivity"
3                 android:label="@string/app_name">
4          <intent-filter>
5          <action android:name="android.intent.action.MAIN"/>
6      <category android:name="android.intent.category.LAUNCHER"/>
7          </intent-filter>
8          </activity>
9          <activity android:name=".SecondActivity"
10                    android:label="@string/app_name">
11         <intent-filter>
12         <action android:name="android.intent.action.VIEW"/>
13         <category android:name="android.intent.category.DEFAULT"/>
14         <data android:scheme="schemevalue"android:host="com.Project"/>
15         </intent-filter>
16     </activity>
17     </application>
    ......
```

第 4 ~ 7 行是 MainActivity 的 Intent 过滤器，动作是 android. intent. action. MAIN，类别是 android. intent. category. LAUNCHER。由 Intent 过滤器的动作和类别可知，MainActivity 是应用程序启动后显示的默认用户界面。第 11 ~ 15 行是 SecondActivity 的 Intent 过滤器，过滤器的动作是 android. intent. action. VIEW，表示根据 URI 协议，以最佳的方式启动相应的 Activity；类别是 android. intent. category. DEFAULT，表示数据的默认动作；数据的协议部分是 android：scheme = "schemevalue"，数据的主机名称部分是 android：host = "com. Project"。

使用 Intent 隐式启动的代码如下：

```
Intent inttMainAct=new Intent(Intent.ACTION_VIEW,Uri.parse("schemevalue://com.Project/path"));
startActivity(inttMainAct);
```

代码中定义的 Intent，动作为 Intent. ACTION_VIEW，与 Intent 过滤器的动作 android. int-ent. action. VIEW 匹配；URI 数据是 "schemevalue://com. Project/path"，其中的协议部分为 "schemevalue"，主机名部分为 "com. Project"，也与 Intent 过滤器定义的数据要求完全匹配。所以，在 Android 系统与 Intent 过滤器列表进行匹配时，会与 AndroidManifest. xml 文件中 SecondActivity 定义的 Intent 过滤器完全匹配。

2.3.4 Activity 信息传递

在开发 Android 应用程序过程中，Activity 之间进行信息传递不可避免。Activity 之间信息传递最常见的情形有两种：①获取子 Activity 返回值；②传递消息给子 Activity。

下面通过一个例子具体实现上述两种 Activity 之间进行信息传递的情形。

【例 2-5】用 Intent 信使间接启动生命周期。

建立工程 Chp02_IntentExchangeMessage，实现三个 Activity，分别为 MainActivity、SecondActivity 和 ThirdActivity，其中 MainActivity 为父 Activity，获取 SecondActivity 的返回值，传递信息给 ThirdActivity。

MainActivity. java 文件代码如下：

```
    ......
1   public class MainActivity extends Activity{
2   private Button m_btnGetActReturnMainAct = null;
3   private Button m_btnBroadCastMainAct = null;
4   private EditText m_etxtGetActReturnMainAct = null;
5   private EditText m_etxtBroadCastMainAct = null;
6   private static final int INFORMATIONACT = 1;
7   /* * Called when the activity is first created. * /
8   @ Override
9   public void onCreate(Bundle savedInstanceState){
10      super. onCreate(savedInstanceState);
11      setContentView(R. layout. main);
12      m_btnGetActReturnMainAct = (Button)findViewById(R. id. btnGetActReturnMainAct);
13      m_btnBroadCastMainAct = (Button)findViewById(R. id. btnBroadCastMainAct);
14      m_etxtGetActReturnMainAct = (EditText)findViewById(R. id. etxtGetActReturnMainAct);
15      m_etxtBroadCastMainAct = (EditText)findViewById(R. id. etxtBroadCastMainAct);
16      m_btnGetActReturnMainAct. setOnClickListener(new OnClickListener(){
17  public void onClick(View view){
18          //获取 Activity 返回值启动 SecondActivity
19      Intent inttMainAct = new Intent(MainActivity. this,SecondActivity. class);
20      startActivityForResult(inttMainAct,INFORMATIONACT);
21      }
22      });
23      m_btnBroadCastMainAct. setOnClickListener(new OnClickListener(){
24          public void onClick(View view){
25          //传递信息值
26      Intent inttMainAct = new Intent(MainActivity. this,ThirdActivity. class);
27      String strMessage = m_etxtBroadCastMainAct. getText(). toString();
28      inttMainAct. putExtra("strMessage",strMessage);
29      startActivity(inttMainAct);
30  }
31      });
32  }
33  //
34  @ Override
35  protected void onActivityResult(int requestCode,int resultCode,Intent data){
36      //TODO Auto-generated method stub
37      super. onActivityResult(requestCode,resultCode,data);
38      if(requestCode = = INFORMATIONACT && resultCode = = RESULT_OK)
```

```
39          {
40              Uri uriInfo = data.getData();
41              m_etxtGetActReturnMainAct.setText(uriInfo.toString());
42          }
43          else
44          {
45              m_etxtGetActReturnMainAct.setText("未获取到任何信息");
46          }
47      }
48      }
    ……
```

MainActivity 中定义了两个 Button 控件，用于启动 SecondActivity 和 ThirdActivity，另外还定义了两个 EditText 控件用于显示 SecondActivity 返回值和设置 ThirdActivity 传递值，界面如图 2-12 所示。

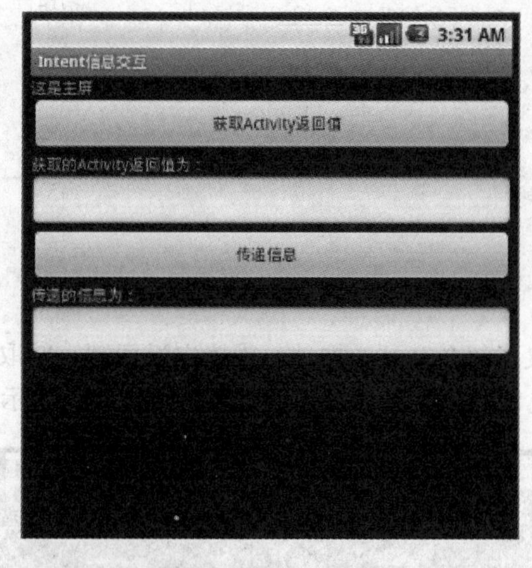

图 2-12　MainActivity

SecondActivity.java 文件代码如下：

```
    ……
1   public class SecondActivity extends Activity{
2       private Button m_btnSetActOKReturnSecondAct = null;
3       private Button m_btnSetActCancelReturnSecondAct = null;
4       private EditText m_etxtSetActReturnSecondAct = null;
5       /** Called when the activity is first created. */
6   @Override
7   public void onCreate(Bundle savedInstanceState){
8       super.onCreate(savedInstanceState);
9       setContentView(R.layout.secondact);
10      m_btnSetActOKReturnSecondAct = (Button)findViewById
11              (R.id.btnSetActOKReturnSecondAct);
```

```
12         m_btnSetActCancelReturnSecondAct = (Button)findViewById
13                  (R.id.btnSetActCancelReturnSecondAct);
14         m_etxtSetActReturnSecondAct = (EditText)findViewById
15                  (R.id.etxtSetActReturnSecondAct);
16         //确定按钮执行事件
17         m_btnSetActOKReturnSecondAct.setOnClickListener(new OnClickListener(){
18     public void onClick(View view){
19         String strSetInfo = m_etxtSetActReturnSecondAct.getText().toString();
20         Uri uriSetInfo = Uri.parse(strSetInfo);
21         Intent inttSetInfo = new Intent(null,uriSetInfo);
22         setResult(RESULT_OK,inttSetInfo);
23         finish();
24     }
25         });
26         //取消按钮执行事件
27         m_btnSetActCancelReturnSecondAct.setOnClickListener(new OnClickListener(){
28     public void onClick(View view){
29         setResult(RESULT_CANCELED,null);
30         finish();
31     }
32           });
33         }
34     }
       ......
```

SecondActivity 中定义了两个 Button 控件：①确定返回键；②取消返回键。还声明了一个 EditText 控件用于设置 SecondActivity 返回值，界面如图 2-13 所示。

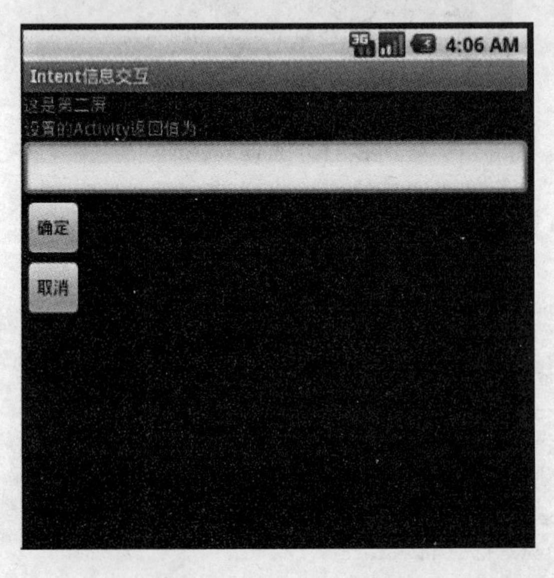

图 2-13　SecondActivity

ThirdActivity.java 文件代码如下：

```
......
1     public class ThirdActivity extends Activity{
2     private EditText m_etxtBroadCastSecondAct = null;
3     /* * Called when the activity is first created. * /
4     @ Override
5     public void onCreate(Bundle savedInstanceState){
6         super.onCreate(savedInstanceState);
7         setContentView(R.layout.thirdact);
8         m_etxtBroadCastSecondAct = (EditText)findViewById(R.id.etxtBroadCastThirdAct);
9         //获取传递过来的信息值
10        Intent inttSecondAct = getIntent();
11        String strMessage = inttSecondAct.getStringExtra("strMessage");
12        m_etxtBroadCastSecondAct.setText(strMessage);
13        }
14    }
......
```

ThirdActivity 中仅声明了一个 EditText 控件用于显示从 MainActivity 传递过来的信息。

Activity 之间通信的第一种情形中，获取子 Activity 返回值通常首先需要启动子 Activity，在这里需要调用 startActivityForResult（Intent，requestCode）方法来启动。参数 Intent 用于决定启动哪个 Activity，既可以是显式 Intent 也可以是隐式 Intent，参数 requestCode 是唯一的标识子 Activity 的请求码，使用 requestCode 确定是哪个子 Activity 的返回值，在这里 requestCode 值需要是非负整型。

上例中启动子 Activity 代码：

```
Intent inttMainAct = new Intent(MainActivity.this,SecondActivity.class);
startActivityForResult(inttMainAct,INFORMATIONACT);
```

其次，在设置子 Activity 返回值时，需要在子 Activity 调用 finish( ) 函数关闭前，调用 setResult( ) 函数将所需的数据返回给父 Activity。setResult( ) 函数有两个参数：①结果码，定义子 Activity 的返回状态，通常为 Activity. RESULT_OK 或者 Activity. RESULT_CANCELED，或自定义的结果码，结果码均为整数类型；②参数是返回值，通常封装在 Intent 中，子 Activity 通过 Intent 将需要返回的数据传递给父 Activity，数据主要是 URI 形式，可以附加一些额外信息，这些额外信息用 Extra 的集合表示。

上例中设置子 Activity 返回值代码：

```
String strSetInfo = m_etxtSetActReturnSecondAct.getText().toString();
Uri uriSetInfo = Uri.parse(strSetInfo);
Intent inttSetInfo = new Intent(null,uriSetInfo);
setResult(RESULT_OK,inttSetInfo);
finish();
```

最后在父 Activity 中获取返回值，当子 Activity 关闭时，父 Activity 的 onActivityResult( ) 函数将被调用，如果需要在父 Activity 中处理子 Activity 的返回值，则需重载此函数。此函数拥有三个参数，第一个参数 requestCode，用来表示是哪一个子 Activity 的返回值，第二个参数 resultCode 用于表示子 Activity 的返回状态，第三个参数 data 是子 Activity 的返回数据，返回数据类型是 Intent，根据返回数据的用途不同，URI 数据的协议对应不同，也可以使用 Extra 方法返回一些原始类型的数据。

上例中父 Activity 中获取返回值代码：

```
1       protected void onActivityResult(int requestCode,int resultCode,Intent data){
2               //TODO Auto-generated method stub
3               super.onActivityResult(requestCode,resultCode,data);
4               if(requestCode = = INFORMATIONACT && resultCode = = RESULT_OK)
5               {
6                   Uri uriInfo = data.getData();
7                   m_etxtGetActReturnMainAct.setText(uriInfo.toString());
8               }
9               else
10              {
11                  m_etxtGetActReturnMainAct.setText("未获取到任何信息");
12              }
13      }
```

Activity 之间通信的第二种情形中，传递消息给子 Activity 过程中，首先在父 Activity 中通过调用 Intent 对象的 putExtra( )方法采用键值对以 Intent 中 Extra 内容的形式将信息传递给子 Activity。putExtra( )是为 Intent 添加扩展数据的方法，有多种重载，一般含有两个参数，第一个参数是附加数据名称，第二个参数是相应的值，值的类型支持很多种，根据重载方法的不同可选择相应类型。

上例中传递消息给子 Activity 代码：

```
1       Intent inttMainAct = new Intent(MainActivity.this,ThirdActivity.class);
2       String strMessage = m_etxtBroadCastMainAct.getText().toString();
3       inttMainAct.putExtra("strMessage",strMessage);
4       startActivity(inttMainAct);
```

子 Activity 获取传递过来的信息首先需要调用 getIntent( )方法获得传递的 Intent 对象，Intent 中获取字符串附加信息的方法使用 getStringExtra( )方法，其参数就是前面提到的 putExtra( )方法的第一个参数。

上例中获取传递过来的信息值：

```
1       Intent inttSecondAct = getIntent();
2       String strMessage = inttSecondAct.getStringExtra("strMessage");
3       m_etxtBroadCastSecondAct.setText(strMessage);
```

运行上面的工程，MainActivity 作为第一屏的界面如图 2-12 所示，单击第一个按钮，启动 SecondActivity，其画面如图 2-13 所示。在其编辑文本框中可以输入 SecondActivity 返回给 MainActivity 的值，例如“SecondActivity 的返回值”，单击下面的“确定”按钮，将会再次回到上面 MainActivity 画面并在第一个文本框中显示“SecondActivity 的返回值”，如图 2-14 所示。在 MainActivity 的第二个文本编辑框中可以设置传递给 ThirdActivity 的值，例如“传递给 ThirdActivity”单击 MainActivity 第二个按钮，启动 ThirdActivity 如图 2-15 所示，显示出传递过来的信息。

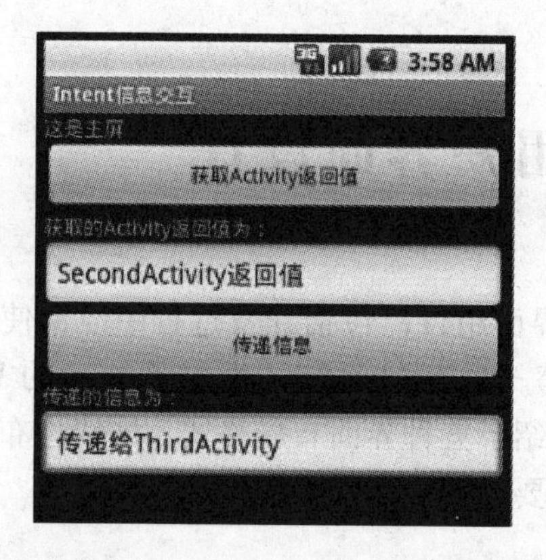

图 2-14 获取子 Activity 返回值的 MainActivity

图 2-15 ThirdActivity

本章小结

本章主要讲述了 Android 生命周期和组件之间的通信，生命周期主要讲述了 Android 四大组件之一的 Activity 生命周期，包括生命周期函数、栈结构和基本状态三方面。组件的通信靠 Intent 实现，Intent 由六个基本部分构成，分别为组件名称（Component Name）、动作（Action）、数据（Data）、类别（Category）、附加信息（Extra）和标志（Flag）。Intent 存在显式和隐式两种形式，隐式启动时主要靠 Intent 过滤器机制完成。

习　题

2-1 Android 应用程序中主要包括哪几种类型的组件？

2-2 在 Android 的 Activity 组件中有几个生命周期，它们的回调函数分别是什么？

2-3 Intent 的作用是什么，它主要包括哪些内容？

2-4 Activity 之间信息传递最常见的情形有哪两种？试通过创建一个应用程序来实现其中一种方式，并在模拟器中显示该结果。

第3章 Android用户界面设计

Android系统为我们提供了丰富的可视化用户界面组件，包括开发过程中经常使用到的控件，如菜单、常用基础控件、对话框与消息框等。Android系统还借用了Java的UI设计思想，包括布局管理及事件响应机制，本章详细介绍了五种界面布局格式，即线性布局、相对布局、表格布局、绝对布局和框架布局，以及必要外部操作的响应处理机制。

3.1 菜单

菜单（Menu）是软件界面设计中最常见的UI元素，提供了具有亲和力的人机交互接口，也为Android应用程序开发提供了简便的接口。Android系统支持三种菜单，分别为选项菜单（Option Menu）、上下文菜单（Context Menu）和子菜单（Sub Menu）。

本节只要讲述三种菜单的创建和使用，以及如何使用布局XML文件生成菜单。

3.1.1 选项菜单（Option Menu）

在应用程序中，选项菜单（Option Menu）是最常用的菜单，当用户按下手机的MENU键时，屏幕底端弹出的就是选项菜单。不过，这个功能需要编程实现，如果未开发实现这个功能，则MENU键不会响应。选项菜单分为图标菜单（Icon Menu）和扩展菜单（Expanded Menu）。如图3-1所示，图标菜单能够同时显示图标和文字，图标菜单在显示时最多只能显示六个，如图3-2所示，当图标菜单多于六个时，将只显示前五个和一个扩展菜单，单击扩展菜单将弹出其余的图标菜单。扩展菜单采用垂直排列方式放置包括第六个图标菜单在内的所有其余菜单，扩展菜单不能显示图标，但是可以显示单选框和复选框，图标菜单不支持单选按钮和复选框。

OptionMenuActivity. java文件代码：

```
    ……
1   public class OptionMenuActivity extends Activity{
2       private static final int GROUP1 =1;
3       private static final int GROUP2 =2;
4       private static final int ITEM1 =1;
5       private static final int ITEM2 =2;
6       private static final int ITEM3 =3;
7       private static final int ITEM4 =4;
8       private static final int ITEM5 =5;
9       private static final int ITEM6 =6;
10      private static final int ITEM7 =7;
11      private static final int ITEM8 =8;
12      /* * Called when the activity is first created. * /
13      @Override
```

```
14        public void onCreate(Bundle savedInstanceState){
15            super.onCreate(savedInstanceState);
16            setContentView(R.layout.main);
17        }
18        @Override
19        public boolean onCreateOptionsMenu(Menu menu){
20            super.onCreateOptionsMenu(menu);
21            menu.add(GROUP1,ITEM1,Menu.NONE,"春")
22              .setShortcut('0','a').setIcon(R.drawable.icon);
23            menu.add(GROUP1,ITEM2,Menu.NONE,"夏")
24              .setShortcut('1','b').setIcon(R.drawable.icon);
25            menu.add(GROUP1,ITEM3,Menu.NONE,"秋")
26              .setShortcut('2','c').setIcon(R.drawable.icon);
27            menu.add(GROUP1,ITEM4,Menu.NONE,"冬")
28              .setShortcut('3','d').setIcon(R.drawable.icon);
29            menu.add(GROUP2,ITEM5,Menu.NONE,"东")
30              .setShortcut('4','e').setIcon(R.drawable.icon);
31            menu.add(GROUP2,ITEM6,Menu.NONE,"南")
32              .setShortcut('5','f');
33            menu.add(GROUP2,ITEM7,Menu.NONE,"西")
34              .setShortcut('6','g');
35            menu.add(GROUP2,ITEM8,Menu.NONE,"北")
36              .setShortcut('7','h');
37            return true;
38        }
39        //事件方法:各个选项的响应事件
40        @Override
41        public boolean onOptionsItemSelected(MenuItem item){
42            String strTitle=item.getTitle().toString();
43            showAlertDialog(strTitle);
44            return super.onOptionsItemSelected(item);
45        }
46        //功能函数:消息提示
47        private void showAlertDialog(String strMessage){
48            new AlertDialog.Builder(this)
49            .setTitle("您的选择:")
50            .setMessage(strMessage)
51            .setPositiveButton("关闭",null)
52            .show();
53        }
54    }
```

程序运行的结果如图 3-1 所示，单击扩展菜单 More 弹出界面如图 3-2 所示。

从上面代码中可以看出，在 Android 系统中创建菜单和处理相应的事件都存在相应的回调方法，例如代码 19 行 onCreateOptionsMenu(Menu menu)方法和 41 行 onOptionsItemSelected(MenuItem item)方法，除了这两个方法外，常用的方法还有 onOptionsMenuClosed(Menu menu)

方法和 onPrepareOptionsMenu(Menu menu)方法。

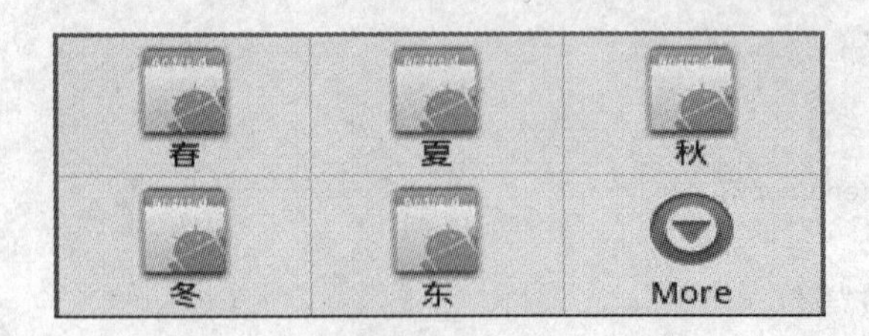

图 3-1 图标菜单(Icon Menu)内容

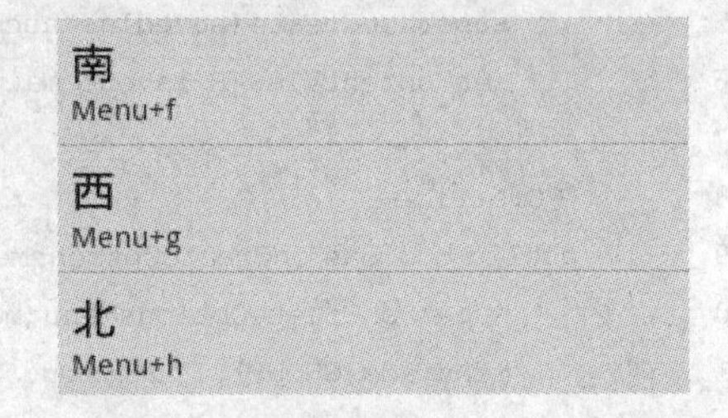

图 3-2 扩展菜单(Expanded Menu)内容

boolean onCreateOptionsMenu(Menu menu)：初始化选项菜单，该方法只在第一次使用菜单时调用，一般用来初始化菜单子项的相关内容，设置菜单子项自身的子项的 ID 和组 ID，设置菜单子项显示的文字和图片等。如果需要每次显示菜单时更新菜单项，则需要重写 onPrepareOptionsMenu(Menu)函数。

boolean onOptionsItemSelected(MenuItem item)：当选项菜单中某个选项被选中时调用该方法，默认的是一个返回 false 的空方法。

void onOptionsMenuClosed(Menu menu)：当选项菜单关闭时，例如，用户按下了返回键或者选择了某个菜单选项，调用该方法。

boolean onPrepareOptionsMenu(Menu menu)：为程序准备选项菜单，每次选项菜单显示前会调用该方法，可以通过该方法设置某些菜单项可用或者不可用再或者修改菜单项的内容。重写该方法时需要返回 true，否则选项菜单将不会显示。

代码 21 ~35 行，主要用到了 Menu 类的 add()函数，用于添加选项菜单子项，add()方法提供了四种重载方式：

```
(1) MenuItem add(int titleRes);
(2) MenuItem add(int groupId, int itemId, int order, CharSequence title);
(3) MenuItem add(int groupId, int itemId, int order, int titleRes);
(4) MenuItem add(CharSequence title);
```

各参数的意义说明如下：

titleRes——String 对象的资源标识符；

groupId——菜单子项所在的组 ID，通过分组的目的是便于对菜单子项进行批量操作，如果菜单子项不需要属于任何组，传入 Menu. NONE；

itemId——唯一标识菜单子项的 ID，可传入 Menu. NONE；

order——菜单子项在选项菜单中的排列顺序；

title——菜单子项所显示的标题。

Menu 类使用 add()函数添加菜单子项的同时，还提供了 addSubMenu()函数添加子菜单，addSubMenu()方法同样提供了对应的四种重载方式：

```
(1) SubMenu addSubMenu(int titleRes);
(2) SubMenu addSubMenu(int groupId, int itemId, int order, int titleRes);
(3) SubMenu addSubMenu(CharSequence title);
(4) SubMenu addSubMenu(int groupId, int itemId, int order, CharSequence title);
```

各参数的意义和上面 add( )方法的各个参数的意义完全相同。

Menu 类还提供了多种其他常用的方法：

void clear( )：清除选项菜单中的所有菜单子项。

void close( )：菜单处于显示状态时关闭选项菜单。

MenuItem findItem(int id)：返回指定 ID 的 MenuItem，参数 ID 为 MenuItem 的标识符。

void removeGroup(int groupId)：移除指定组的所有菜单子项，参数 groupId 为组 ID。

void removeItem(int id)：移除指定 ID 的菜单子项，参数 ID 为 MenuItem 的标识符。

int size( )：获取菜单子项的数目。

int addIntentOptions(int groupId, int itemId, int order, ComponentName caller, Intent[ ] specifics, Intent intent, int flags, MenuItem[ ] outSpecificItems)：用来动态产生选项菜单，参数 groupId、itemId 和 order 的意义如前所述，参数 caller 为发起 Activity 的 Activity 组件，参数 specifics 指以 action + uri 的具体方式来添加激活相应 Activity 的菜单项，参数 intent 指以 categroy + uri 这种一般形式来添加激活相应 Activity 的菜单项，参数 intent 和 specifics 的区别是，一个用 categroy + uri 来匹配 Activity，一个用 action + uri 来匹配 Activity，参数 outSpecificItems 是返回的 MenuItem 值，对应以 specifics 方式匹配的菜单项。

代码 22 行的 setShortcut( )方法和 setIcon( )方法为菜单子项 MenuItem 类提供的方法，MenuItem 类提供的常用方法和用法如表 3-1 所示。

表 3-1 MenuItem 类提供的常用方法和用法

方 法	参 数	功 能
MenuItem setAlphabeticShortcut（char alphaChar）	alphaChar 字母快捷键	设置菜单子项的字母快捷键
MenuItemset NumericShortcut（char numericChar）	numericChar 数字快捷键	设置菜单子项的数字快捷键
MenuItem setIcon（int iconRes）	iconRes 图标 ID	设置菜单子项的图标
MenuItem setIcon（Drawable icon）	icon 图标 Drawable 对象	
MenuItem setIntent（Intent intent）	intent 与菜单子项绑定的 Intent 对象	为菜单子项绑定 Intent 对象，当被选中时将会调用 startActivity( )方法处理相应的 Intent
MenuItem setOnMenuItemClickListener（MenuItem. On MenuItemClickListener menuItemClickListener）	menuItemClickListener 监听器	为菜单子项设置自定义的监听器，一般使用回调方法 onOptionsItemSelected 取代
MenuItem setShortcut（char numericChar, char alphaChar）	numericChar 数字快捷键 alphaChar 字母快捷键	设置菜单子项的字母快捷键和数字快捷键
MenuItem setTitle（int title）	title 标题资源 ID	设置菜单子项的标题
MenuItem setTitle（CharSequence title）	title 标题名称	
MenuItemset TitleCondensed（CharSequence title）	title 缩略标题	设置菜单子项的缩略标题，当菜单子项不能显示完全标题时，显示缩略标题

3.1.2 上下文菜单（Context Menu）

上下文菜单（Context Menu）是个浮动式列表菜单，类似于其他应用程序中的右键菜单。如果说选项菜单服务于 Activity 组件，那么上下文菜单不同于选项菜单，上下文菜单绑定在 View 控件对象上，当用户长按此控件对象约 2 秒后，将启动绑定在此对象上的上下文菜单。另外，上下文菜单并不支持绑定快捷键，其菜单子项中也不能附带图标，但是上下文菜单的标题可以指定图标。

ContextMenu 类继承自 Menu 类，所以 Menu 类的很多方法均可以在 ContextMenu 对象中使用，例如使用 add( )方法添加菜单子项等。

ContextMenuActivity. java 文件代码：

```
……
1    public class ContextMenuActivity extends Activity {
2    final static int ITEM1 =1;
3    final static int ITEM2 =2;
4    final static int ITEM3 =3;
5    private TextView m_txtContextAct = null;
6    private Button m_btnContextAct = null;
7    /* * Called when the activity is first created. * /
8    @Override
9    public void onCreate(Bundle savedInstanceState){
10       super.onCreate(savedInstanceState);
11       setContentView(R.layout.main);
12       m_txtContextAct = (TextView)findViewById(R.id.txtContextAct);
13       m_btnContextAct = (Button)findViewById(R.id.btnContextAct);
14       registerForContextMenu(m_btnContextAct);
15   }
16   @Override
17   public boolean onContextItemSelected(MenuItem item){
18       //TODO Auto-generated method stub
19       switch(item.getItemId()){
20       case ITEM1:
21           m_txtContextAct.setText("选中:苹果");
22           return true;
23       case ITEM2:
24           m_txtContextAct.setText("选中:梨子");
25           return true;
26       case ITEM3:
27           m_txtContextAct.setText("选中:香蕉");
28           return true;
29       }
30       return false;
31   }
32   @Override
33   public void onCreateContextMenu(ContextMenu menu, View v,ContextMenuInfo menuInfo) {
34       //TODO Auto-generated method stub
35       super.onCreateContextMenu(menu, v, menuInfo);
36       menu.setHeaderTitle("上下文菜单");
```

```
37            menu.add(0, ITEM1, 0,"苹果");
38            menu.add(0, ITEM2, 1,"梨子");
39            menu.add(0, ITEM3, 2,"香蕉");
40        }
41    }
```

在 Activity 中使用上下文菜单时，首先要将上下文菜单注册绑定到某个控件上，例如上面的 Button 控件，为指定的 View 控件注册上下文菜单使用 registerForContextMenu（View view）方法，其参数 view 就是要显示上下文菜单的 View 控件。类似于选项菜单在第一次启动时调用 onCreateOptionsMenu（Menu menu）方法，上下文菜单会调用 onCreateContext-Menu( )方法，不同的是上下文菜单在每次启动时均调用该方法，而不仅仅在第一次启动时调用，在该方法内部可以设置上下文菜单标题，添加菜单子项等操作，该方法的语法格式为：

```
void onCreateContextMenu(ContextMenu menu, View v,ContextMenuInfo menuInfo)
```

各参数的意义说明如下：

menu——创建的上下文菜单；

v ——上下文菜单依附的控件；

menuInfo——上下文菜单需要额外显示的信息。

上下文菜单内菜单子项的选择事件通过重写方法 onContextItemSelected（MenuItem item）响应，类似于选项菜单中的 onOptionsItemSelected（MenuItem item）方法，参数 item 即为响应的被选中的菜单子项，例如上面代码中完成响应后将选择的菜单子项信息显示出来。另外，在上下文菜单的开发过程中，还有一个方法经常用到，那就是 onContextMenuClosed（Menu menu），此方法在上下文菜单被关闭时调用。

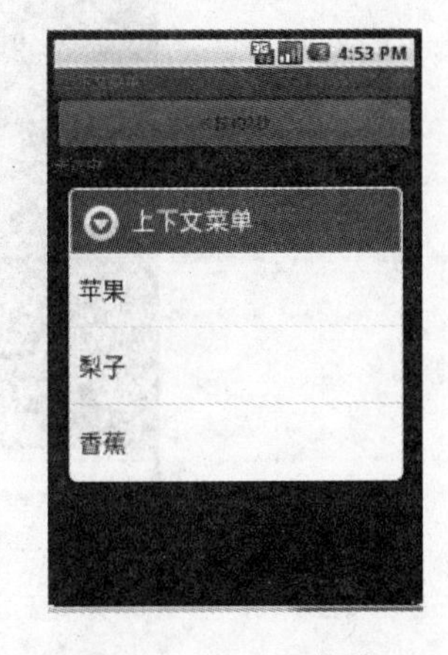

图 3-3　上下文菜单

运行上面的程序，运行结果如图 3-3 所示。

3.1.3　子菜单（Sub Menu）

子菜单（Sub Menu）也是个浮动式列表菜单，提供了一种自然的组织菜单项的方式，使用非常灵活，可以在选项菜单或快捷菜单中使用子菜单，有利于将相同或相似的菜单子项组织在一起，便于显示和分类。子菜单不支持嵌套，不能在子菜单中再嵌套子菜单。子菜单也不支持图标，但同样支持标题图标。

子菜单（Sub Menu）类同样继承自 Menu 类，在选项菜单或快捷菜单中添加子菜单通过 addSubMenu( )函数实现，如下列代码所示，其中添加了两个子菜单，分别为子菜单 1 和子菜单 2，如图 3-4a 所示。子菜单 1 添加了四个菜单子项，子菜单 2 添加了两个菜单子项，如图 3-4b 所示。菜单子项的选择事件处理函数，仍然使用 onOptionsItemSelected ( )函数。

```
1    @Override
2    public boolean onCreateOptionsMenu(Menu menu) {
3        //TODO Auto-generated method stub
4        SubMenu subMenu1 =menu.addSubMenu(GROUP,
```

```
5                       SUBITEM1, Menu.NONE, "子菜单 1");
6           SubMenu subMenu2 = menu.addSubMenu(GROUP,
7                       SUBITEM2, Menu.NONE, "子菜单 2");
8           subMenu1.add(GROUP, ITEM1, Menu.NONE, "苹果");
9           subMenu1.add(GROUP, ITEM2, Menu.NONE, "梨子");
10          subMenu1.add(GROUP, ITEM3, Menu.NONE, "香蕉");
11          subMenu1.add(GROUP, ITEM4, Menu.NONE, "菠萝");
12          subMenu2.add(GROUP, ITEM5, Menu.NONE, "老虎");
13          subMenu2.add(GROUP, ITEM6, Menu.NONE, "大象");
14          return true;
15      }
```

a)

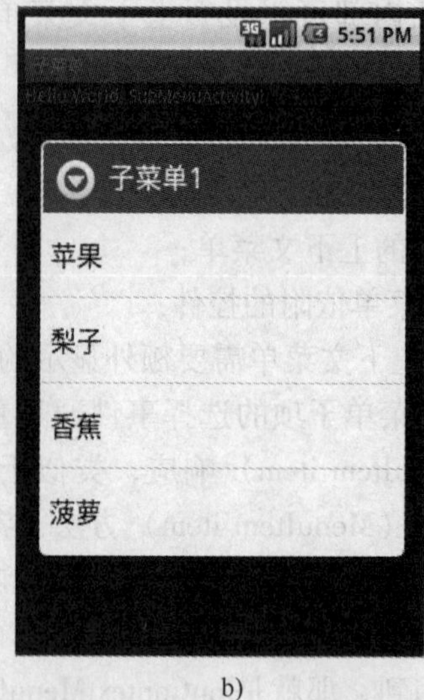

b)

图 3-4　子菜单

上述三种菜单的创建均是采用在 Activity 类中动态创建 Menu 的方法，Android 系统提供了另一种创建菜单的方法，即为将菜单定义为 XML 资源。这种方法有利于减少用户界面和功能代码之间的耦合，使应用程序的开发更具独立性。使用这种方法创建菜单，一个菜单对应一个 XML 文件，并要求一个 XML 文件只能有一个根节点 <menu>，下面的代码采用了 XML 生成上下文菜单。

menu.xml 文件代码如下：

```
1   <?xml version="1.0" encoding="utf-8"?>
2   <menu xmlns:android="http://schemas.android.com/apk/res/android">
3       <group android:id="@+id/GROUP1"
4               android:checkableBehavior="single">
5       <item android:id="@+id/GROUP1ITEM1"
6               android:title="大象 1"
7               android:checked="false"
8               />
9       <item android:id="@+id/GROUP1ITEM2"
```

```
10                android:title = "大象 2"
11                android:checked = "true"
12                />
13          </group>
14          <group android:id = "@ +id/GROUP2" >
15            <item android:id = "@ +id/GROUP2ITEM1"
16                android:title = "老虎 1"
17                android:checkable = "true"
18                android:checked = "false"
19                />
20            <item android:id = "@ +id/GROUP2ITEM2"
21                android:title = "老虎 2"
22                android:checkable = "true"
23                android:checked = "true"
24                />
25          </group>
26          <item android:id = "@ +id/Submenu"
27                android:title = "狮子 (Submenu)" >
28                <menu >
29                  <item android:id = "@ +id/SubmenuITEM1"
30                        android:title = "狮子 1"
31                        />
32                  <item android:id = "@ +id/SubmenuITEM2"
33                        android:title = "狮子 2"
34                        />
35                </menu>
36          </item>
37      </menu>
```

XML 中的标签属性与在代码中动态生成菜单时相对应，其意义也相同。使用 XML 定义菜单后，要在 Menu 生成时调用 getMenuInflater( )方法获取 MenuInflater 对象。MenuInflater 是用来实例化 Menu 目录下的 Menu 布局文件的，MenuInflater 获取方法只有 Activity. getMenuInflater( )一种。将 XML 资源文件传递给菜单对象使用 MenuInflater. inflater（int menuRes，Menu menu）方法，另外，MenuInflater 只有一个 void inflater（int menuRes，Menu menu）非构造方法。

```
1    @Override
2    public void onCreateContextMenu(ContextMenu menu, View v,
3            ContextMenuInfo menuInfo) {
4        // TODO Auto-generated method stub
5        super.onCreateContextMenu(menu, v, menuInfo);
6        getMenuInflater().inflate(R.layout.menu, menu);
7    }
```

上面采用 XML 创建的 XML 生成上下文菜单运行结果如图 3-5 所示。

图 3-5　XML 生成上下文菜单

3.2　常用基础控件

Android 系统非常重视应用程序中用户界面的友好性，在设计应用程序界面时系统提供了一整套比较完善的基础控件。这些控件使用起来非常方便，直接引入系统库中支持的控件类就得以完成创建，这让 Android 开发人员减轻了工作量，同样也有利于应用程序界面风格的一致。

本节主要讲解十二种常用的基础界面控件，包括列表视图（ListView）控件。

3.2.1　列表视图

列表视图（ListView）是一种视图控件，采用垂直显示方式显示视图中包含的条目信息，条目多于屏幕最大显示数量时，会自动添加垂直滚动条。列表中的条目支持选中单击事件响应，可以很方便地过渡到条目中所指示的内容上进行数据内容的显示和进一步处理。

使用列表视图只需要向布局文件中添加 <ListView> 标签即可，其常用的 XML 属性如表 3-2所示。

表 3-2　ListView 中常用的 XML 属性

属　性	说　明
android:choiceMode	规定此 ListView 所使用的选择模式；默认状态下，list 没有选择模式，属性值必须设置为下列常量之一：none，值为 0，表示无选择模式；singleChoice，值为 1，表示最多可以有一项被选中；multipleChoice，值为 2，表示可以多项被选中
android:divider	规定 List 项目之间用某个图形或颜色来分隔，可以用" @ [+] [package:] type: name" 或者"? [package:] [type:] name" （主题属性）的形式来指向某个已有资源；也可以用" #rgb"," #argb"," #rrggbb" 或者" #aarrggbb" 的格式来表示某个颜色

（续）

属　性	说　明
android:dividerHeight	分隔符的高度，若没有指明高度，则用此分隔符固有的高度，必须为带单位的浮点数，如" 14.5sp"。可用的单位如 px（pixel 像素），dp（density-independent pixels 与密集度无关的像素），sp（scaled pixels based on preferred font size 基于字体大小的固定比例的像素），in（inches 英寸），mm（millimeters 毫米）。 可以用" @ [package:] type: name" 或者"? [package:] [type:] name"（主题属性）的格式来指向某个包含此类型值的资源
android:entries	引用一个将使用在此 ListView 里的数组。若数组是固定的，使用此属性将比在程序中写入更为简单。必须以" @ [+] [package:] type: name" 或者"? [package:] [type:] name" 的形式来指向某个资源
android:VfooterDividersEnabled	设成 flase 时，此 ListView 将不会在页脚视图前画分隔符，此属性默认值为 true。属性值必须设置为 true 或 false。 可以用" @ [package:] type: name" 或者"? [package:] [type:] name"（主题属性）的格式来指向某个包含此类型值的资源
android:headerDividersEnabled	设成 flase 时，此 ListView 将不会在页眉视图后画分隔符。此属性默认值为 true。属性值必须设置为 true 或 false。 可以用" @ [package:] type: name" 或者"? [package:] [type:] name"（主题属性）的格式来指向某个包含此类型值的资源

下面代码是一个使用列表视图的简单例子，其运行结果如图 3-6 所示。

```
1   public class MainListViewAct extends Activity {
2       private Object[]Acts = {
3           "文本框",          TextViewAct.class,
4           "按钮",            ButtonAct.class,
5           "时钟",            ClockAct.class,
6           "日期时间",        DateAndTimeAct.class,
7           "计时",            ChronometerAct.class,
8           "进度条",          ProgressBarAct.class,
9           "拖动条",          SeekbarAndRatingbarAct.class,
10          "下拉表",          SpinnerAct.class,
11      };
12      /** Called when the activity is first created. */
13      @Override
14      public void onCreate(Bundle savedInstanceState) {
15          super.onCreate(savedInstanceState);
16          setContentView(R.layout.mainlistviewact);
17          //菜单名称的数组
18      CharSequence[]strArrItemsNames = new CharSequence[Acts.length/2];
19      for(int i = 0;i < strArrItemsNames.length;i ++){
20          strArrItemsNames[i] = (String)Acts[i* 2];
21      }
22      //设置菜单名称
23      ArrayAdapter < CharSequence > adpItemsNames = new ArrayAdapter < CharSequence >
24                      (this, android.R.layout.simple_list_item_1, strArrItemsNames);
25      ListView listMainListViewAct = (ListView)findViewById(R.id.listMainListViewAct);
```

```
26        listMainListViewAct.setAdapter(adpItemsNames);
27        //按下菜单名称指向相关的界面
28        listMainListViewAct.setOnItemClickListener(new OnItemClickListener(){
29            public void onItemClick(AdapterView<? >parent, View view,
30                            int position, long id){
31                Intent intent =new Intent(MainListViewAct.this,
32                            (Class<? >)Acts[position* 2 +1]);
33                startActivity(intent);
34            }
35        });
36        }
37    }
```

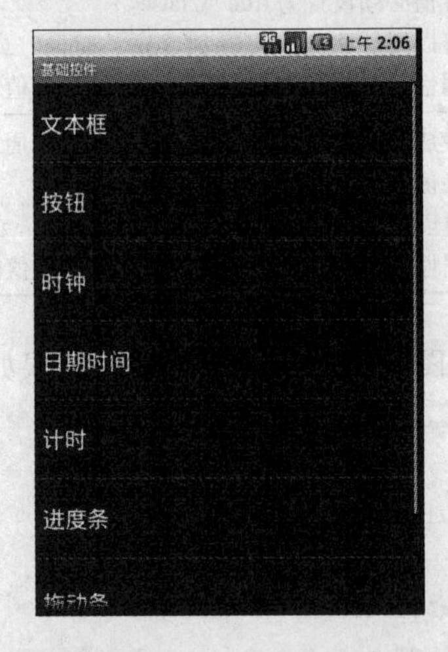

图 3-6　列表视图

3.2.2　文本框类

提供文本信息显示或者编辑功能的控件总称之为文本框，这是广义上的文本框，狭义的文本框控件指仅有文本显示功能的文本框，例如 Android 系统中 TextView 控件和 .NET 中的 Lable 控件。在 Android 系统中，常用文本框类有四种，分别为：

（1）文本框（TextView）

（2）编辑框（EditText）

（3）自动完成文本框（AutoCompleteTextView）

（4）多项自动完成文本框（MultiAutoCompleteTextView）

四种文本框中，TextView 是基本的文本显示框，也是只读文本显示控件，其他三种文本框均可提供用户编辑功能。

1. 文本框（TextView）

文本框（TextView）是最常用的显示字符串的控件，类似于很多其他语言平台中的 Lable 控件，只支持显示信息功能而不接受用户输入。TextView 类是文本编辑的基类，很多其他文本框都继承自该类，例如编辑框 EditText。向界面中添加文本框非常简单，在布局文

件中添加 <TextView> 标签即可，例如下面 XML 中添加文本框代码：

```
1    <TextView
2        android:layout_width = "fill_parent"
3        android:layout_height = "wrap_content"
4        android:id = "@ + id/txtTextViewAct"
5    />
```

文本框具有很多有用的属性，例如上面代码中的 android：layout_width、android：layout_height 和 android：id。属性的设置既可以在 XML 布局文件中设置也可以调用相应的方法在 Java 代码中动态设置，例如，动态设置文本框背景颜色和显示文本，Java 代码为：

```
1    TextView txtTextViewAct = (TextView)findViewById(R.id.txtTextViewAct);
2    txtTextViewAct.setBackgroundColor(Color.WHITE);
3    txtTextViewAct.setText("这是文本框");
```

文本框常用属性和对应方法如表 3-3 所示。

表 3-3　文本框常用属性和对应方法

属　性	方　法	说　明
android:autoLink	setAutoLinkMask（int）	设置是否当文本为 URL 链接/email/电话号码/map 时，文本显示为可单击的链接，可选值 none/web/email/phone/map/all
android:ellipsize	setEllipsize（TextUtils.TruncateAt）	设置当文字过长时，该控件该如何显示。有如下值设置："start"——省略号显示在开头；"end"——省略号显示在结尾；"middle"——省略号显示在中间；"marquee"——以跑马灯的方式显示（动画横向移动）
android:gravity	setGravity（int）	设置文本位置，如设置成"center"，文本将居中显示
android:hint	setHint（int）	Text 为空时显示的文字提示信息
android:linksClickable	setLinksClickable（boolean）	设置链接是否单击链接，即使设置了 autoLink。
android:marqueeRepeatLimit	setMarqueeRepeatLimit（int）	在 ellipsize 指定 marquee 的情况下，设置重复滚动的次数，当设置为 marquee_forever 时表示无限次
android:text	setText（CharSequence）	设置显示文本
android:textColor	setTextColor（ColorStateList）	设置文本颜色
android:textColorHighlight	setHighlightColor（int）	被选中文字的底色，默认为蓝色
android:textColorHint	setHintTextColor（int）	设置提示信息文字的颜色，默认为灰色
android:textColorLink	setLinkTextColor（int）	文字链接的颜色
android:textScaleX	setTextScaleX（float）	设置文字缩放，默认为 1.0f
android:textSize	setTextSize（float）	设置文字大小
android:textStyle	setTypeface（Typeface）	设置字形
android:typeface	setTypeface（Typeface）	设置文本字体
android:height	setHeight（int）	设置 TextView 的高度
android:maxHeight	setMaxHeight（int）	设置 TextView 的最大高度

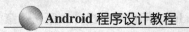

（续）

属　性	方　法	说　明
android:minHeight	setMinHeight（int）	设置 TextView 的最小高度
android:width	setWidth（int）	设置 TextView 的宽度
android:maxWidth	setMaxWidth（int）	设置 TextView 的最大宽度
android:minWidth	setMinWidth（int）	设置 TextView 的最小宽度

2. 编辑框（EditText）

编辑框（EditText）类继承自 TextView 类，但不同于文本框，编辑框支持用户输入和编辑。编辑框在界面设计中也是经常使用的文本控件之一，例如，应用程序中实现登录界面，需要用户输入用户名和密码等信息，用到的正是编辑框。开发应用程序时，在界面上添加编辑框，只要在界面布局文件添加编辑框 <EditText> 标签即可。

```
1    <EditText
2        android:id="@ +id/etxtEditTextAct"
3        android:layout_width="wrap_content"
4        android:layout_height="wrap_content"
5    />
```

使用编辑框的属性和方法类似于文本框，由于继承自 TextView 类，EditText 的属性和方法大部分也来自 TextView 类，其中经常用到的属性和方法可参考表 3-4。

表 3-4　编辑框常用属性和对应方法

属　性	方　法	说　明
android:cursorVisible	setCursorVisible（boolean）	设定光标为显示/隐藏，默认显示
android:lines	setLines（int）	设置文本的行数，例如，设置两行就显示两行，即使第二行没有数据
android:maxLines	setMaxLines（int）	设置文本的最大显示行数，与 width 或者 layout_width 结合使用，超出部分自动换行，超出行数将不显示
android:minLines	setMinLines（int）	设置文本的最小行数，与 lines 类似
android:password	setTransformationMethod（TransformationMethod）	设置文本框的内容是否显示为密码
android:phoneNumber	setKeyListener（KeyListener）	设置文本框的内容只能是电话号码
android:scrollHorizontally	setHorizontallyScrolling（boolean）	设置文本超出宽度的情况下，是否出现横拉条
android:selectAllOnFocus	setSelectAllOnFocus（boolean）	如果文本是可选中的，获取焦点时自动选中全部文本内容
android:shadowColor	setShadowLayer（float，float，float，int）	指定文本阴影的颜色，需要与 shadowRadius 一起使用
android:shadowDx	setShadowLayer（float，float，float，int）	设置阴影横向坐标开始位置
android:shadowDy	setShadowLayer（float，float，float，int）	设置阴影纵向坐标开始位置
android:shadowRadius	setShadowLayer（float，float，float，int）	设置阴影的半径

（续）

属　性	方　法	说　明
android:singleLine	setTransformationMethod（TransformationMethod）	设置单行显示。如果和 layout_width 一起使用，当文本不能全部显示时，后面用"…"来表示，如果不设置 singleLine 或者设置为 false，文本将自动换行
android:maxLength	setFilters（InputFilter）	设置显示文本长度，超出部分不显示

3. 自动完成文本框（AutoCompleteTextView）

自动完成文本框（AutoCompleteTextView）类继承自 EditText 类，当用户输入文字信息时，会自动列出下拉列表提示与用户输入文字相关的信息条目。类似于在百度或者 Google 中搜索信息所用的输入框，可以在输入少量文字的时候列出下拉菜单显示相关的搜索关键字，可以选择想要搜索的关键字而快速获取需要的信息。

在应用程序开发时，通过在布局文件中添加 <AutoCompleteTextView>标签在界面上生成自动完成文本框：

```
1  <AutoCompleteTextView
2      android:id="@+id/atxtAutoCompleteText"
3      android:layout_width="wrap_content"
4      android:layout_height="wrap_content"
5  />
```

使用自动完成文本框时常用的属性和对应的设置属性方法如表 3-5 所示。

表 3-5　自动完成文本框常用属性和对应方法

属　性	方　法	说　明
android:completionThreshold	setThreshold（int）	设置需要用户输入的字符数
android:dropDownHeight	setDropDownHeight（int）	设置下拉菜单高度
android:dropDownWidth	setDropDownWidth（int）	设置下拉菜单宽度
android:popupBackground	setDropDownBackgroundResource（int）	设置下拉菜单背景

Java 代码中使用 AutoCompleteTextView 的代码如下：

```
1  AutoCompleteTextView atxtAutoCompleteText =
2      (AutoCompleteTextView)findViewById(R.id.atxtAutoCompleteText);
3  ArrayAdapter<CharSequence>arrAutoCompleteText =
4      ArrayAdapter.createFromResource(this, R.array.languages,
5      android.R.layout.simple_dropdown_item_1line);
6  atxtAutoCompleteText.setAdapter(arrAutoCompleteText);//设置适配器
```

代码第 3 行创建了一个适配器，ArrayAdapter. createFromResource( )方法有三个参数，第一个参数是指上下文对象，第二个参数引用了在 string. xml 文件当中定义的 string 数组，第三个参数是用来指定 Spinner 的样式，是一个布局文件 ID，该布局文件由 Android 系统提供，也可替换为自己定义的布局文件。最后通过 setAdapter( )方法为 AutoCompleteTextView 添加适配器，此外 AutoCompleteTextView 类常用的方法还有如下 3 个：

（1） clearListSelection( )清除选中的列表项

（2） ismissDropDown( )关闭下拉菜单

（3） getAdapter( )获取适配器

4. 多项自动完成文本框（MultiAutoCompleteTextView）

多项自动完成文本框（MultiAutoCompleteTextView）类继承自 AutoCompleteTextView 类，延长 AutoCompleteTextView 的长度，在多次输入的情况下可支持选择多个值，分别用分隔符分开，每个值在输入时会自动去匹配。

布局文件中添加 <MultiAutoCompleteTextView>标签自动项用户界面添加多项自动完成文本框。

```
1    <MultiAutoCompleteTextView
2      android:id="@ +id/matxtMultiAutoCompleteText"
3      android:layout_width="wrap_content"
4      android:layout_height="wrap_content"
5    />
```

使用多项自动完成文本框必须提供一个 MultiAutoCompleteTextView. Tokenizer 以用来区分不同的子串，例如下面使用 MultiAutoCompleteTextView 的 Java 代码。

```
1    MultiAutoCompleteTextView matxtMultiAutoCompleteText =
2        (MultiAutoCompleteTextView)findViewById(R.id.matxtMultiAutoCompleteText);
3    matxtMultiAutoCompleteText.setAdapter(arrAutoCompleteText);
4    matxtMultiAutoCompleteText.setTokenizer(new
5                  MultiAutoCompleteTextView.CommaTokenizer());
```

代码第 4 行使用 setTokenizer 方法设置 MultiAutoCompleteTextView. Tokenizer 用户正在输入时，tokenizer 设置用于确定文本相关范围。

MultiAutoCompleteTextView 常用的方法还有 enoughToFilter( )和 performValidation( )。enoughToFilter( )方法的作用是当文本长度超过阈值时过滤，此方法并不是检验什么时候文本的总长度超过了预定的值，而是在仅当从函数 findTokenStart( )到 getSelectionEnd( )函数得到的文本长度为零或者超过了预定值的时候才起作用。performValidation( )方法的意义是代替验证整个文本，并不是用来确定整个文本的有效性，而是用来确定文本中的单个符号的有效性，并且空标记将被移除。

下面具体使用上述的四种文本框。

在布局文件中分别添加文本框、编辑框、自动完成文本框和多项自动完成文本框，布局文件代码如下：

```
1    <? xml version="1.0" encoding="utf-8"? >
2    <TableLayout
3        xmlns:android="http://schemas.android.com/apk/res/android"
4        android:layout_width="wrap_content"
5        android:layout_height="wrap_content"
6        android:layout_gravity="center">
```

```
7    <TextView
8        android:layout_width="wrap_content"
9        android:layout_height="wrap_content"
10       android:text="普通文本控件:"/>
11   <TextView
12       android:id="@+id/txtTextViewAct"
13       android:layout_width="wrap_content"
14       android:layout_height="wrap_content"
15   />
16   <TextView
17       android:layout_width="wrap_content"
18       android:layout_height="wrap_content"
19       android:text="可编辑文本控件:"/>
20   <EditText
21       android:id="@+id/etxtEditTextAct"
22       android:layout_width="wrap_content"
23       android:layout_height="wrap_content"
24   />
25   <TextView
26       android:layout_width="wrap_content"
27       android:layout_height="wrap_content"
28       android:text="自动完成文本框(单项):"
29   />
30   <AutoCompleteTextView
31       android:id="@+id/atxtAutoCompleteText"
32       android:layout_width="wrap_content"
33       android:layout_height="wrap_content"
34   />
35   <TextView
36       android:layout_width="wrap_content"
37       android:layout_height="wrap_content"
38       android:text="自动完成文本框(多项):"
39   />
40   <MultiAutoCompleteTextView
41       android:id="@+id/matxtMultiAutoCompleteText"
42       android:layout_width="wrap_content"
43       android:layout_height="wrap_content"
44   />
45   </TableLayout>
```

代码 11～15 行使用 <TextView> 标签定义所要观察的文本框，另外还添加了四个文本框用于显示信息。代码 20～24 行使用 <EditText> 标签定义编辑框，30～34 使用 <AutoCompleteTextView> 标签定义自动完成文本框，40～44 行使用 <MultiAutoCompleteTextView> 定义多项自动完成文本框。

包含处理四种文本框的 Activity 的 Java 代码如下：

```
......
1     @ Override
2     protected void onCreate(Bundle savedInstanceState) {
3         //TODO Auto-generated method stub
4         super.onCreate(savedInstanceState);
5         setContentView(R.layout.textviewact);
6         //1 普通文本框 TextView
7         TextView txtTextViewAct = (TextView) findViewById(R.id.txtTextViewAct);
8         txtTextViewAct.setBackgroundColor(Color.WHITE);
9         txtTextViewAct.setText("这是文本框");
10        //2 编辑框 EditText
11        EditText etxtEditTextAct = (EditText) findViewById(R.id.etxtEditTextAct);
12        etxtEditTextAct.setText("编辑框");
13        //3 自动完成文本框(单项)-AutoCompleteTextView
14        AutoCompleteTextView atxtAutoCompleteText =
15            (AutoCompleteTextView) findViewById(R.id.atxtAutoCompleteText);
16        ArrayAdapter<CharSequence> arrAutoCompleteText =
17           ArrayAdapter.createFromResource(this, R.array.languages,
18           android.R.layout.simple_dropdown_item_1line);
19        atxtAutoCompleteText.setAdapter(arrAutoCompleteText); //设置适配器
20        //4 自动完成文本框(多项)-MultiAutoCompleteTextView
21        MultiAutoCompleteTextView matxtMultiAutoCompleteText =
22        (MultiAutoCompleteTextView) findViewById(R.id.matxtMultiAutoCompleteText);
23        matxtMultiAutoCompleteText.setAdapter(arrAutoCompleteText);
24        matxtMultiAutoCompleteText.setTokenizer(new
25                  MultiAutoCompleteTextView.CommaTokenizer());
26        }
27    }
......
```

代码 6 ~ 9 行处理文本框，设置背景色为白色并设置了显示信息；代码 10 ~ 12 行处理编辑框，设置编辑框默认显示信息；代码 13 ~ 19 行处理自动完成文本框，创建适配器和绑定适配器，自动匹配的内容项列表数据为 Value 文件下的 array.xml 数据表，表中数据格式如下代码所示：

```
<?xml version="1.0" encoding="utf-8"?>
<resources>
    <string-array name="languages">
        <item>Austria</item>
        <item>Brazil</item>
        <item>China</item>
        <item>Colombia</item>
        <item>Cuba</item>
        <item>Denmark</item>
        <item>Egypt</item>
        <item>Estonia</item>
```

```
            <item>Finland</item>
            <item>France</item>
            <item>Germany</item>
            <item>Greece</item>
            <item>Hong Kong</item>
            <item>Ireland</item>
            <item>Italy</item>
            <item>Japan</item>
            <item>Malaysia</item>
            <item>Poland</item>
            <item>Romania</item>
            <item>Russia</item>
            <item>Spain</item>
            <item>Switzerland</item>
            <item>Taiwan</item>
        </string-array>
    </resources>
```

代码 21～25 行处理多项自动完成文本框，绑定适配器并设置 Tokenizer。

运行上面代码，其结果如图 3-7 所示，可以在编辑框中输入文本，也可以在自动完成文本框中输入文本，在输入时会自动寻找近似匹配列表。在多项自动完成文本框中可自动匹配输入多项，每项之间自动添加逗号区分文本项，其效果如图 3-8 所示。

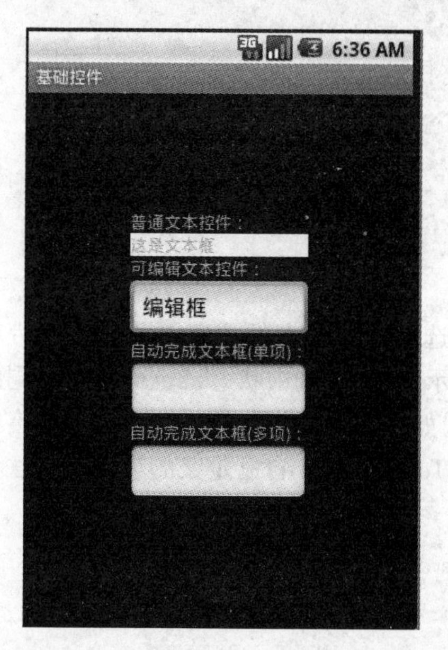

图 3-7　四种文本框

图 3-8　多项自动完成文本框效果

3.2.3　按钮类

按钮是最常见的控件，几乎在所有成熟的应用程序中都含有按钮控件，应用程序中的单击动作一度也被用户认为是专属于按钮的操作。在 Android 系统中，常用的按钮控件有五种。

（1）普通按钮（Button）

（2）开关按钮（ToggleButton）

（3）图片按钮（ImageButton）

（4）复选按钮（CheckBox）

（5）单选按钮（RadioButton）

五种按钮的效果如图 3-9 所示，下面分别介绍这五种按钮的用法和操作。

图 3-9　按钮类效果

1. 普通按钮（Button）

普通按钮（Button）类即为 Button 类，其类结构图如图 3-10 所示，可以看出 Button 类直接继承自 TextView 类，很多属性和方法亦来自 TextView。界面中添加 Button 控件只需要在布局文件中添加 <Button> 标签即可生成普通按钮控件。使用普通按钮的主要目的是为其设置单击事件后的处理，按钮单击后的响应主要通过设置 Button. OnClickListener 监听器实现，设置监听器后，还需重写其中的 onClick( ) 方法，在此方法内即可根据需要处理单击，如下面代码所示。

```
1    final Button button = (Button) findViewById(R. id. button_id);
2    button. setOnClickListener(new View. OnClickListener() {
3        public void onClick(View v) {
4        //处理单击
5        }
6    });
```

2. 开关按钮（ToggleButton）

开关按钮（ToggleButton）是一个具有选中和未选中两种状态的按钮，通过一个带有亮度指示同时默认文本为“ON”或“OFF”的按钮显示选中/未选中状态，也可以为不同的状态设置不同的显示文本。ToggleButton 类的继承机构如图 3-11 所示，ToggleButton 直接继承自 CompoundButton 类，XML 属性除了继承而来还有 ToggleButton 自己定义的属性和方法，如表 3-6 所示。

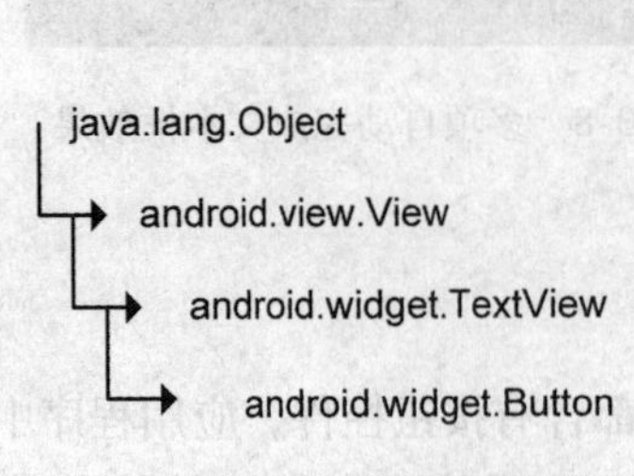

图 3-10　Button 类的继承结构

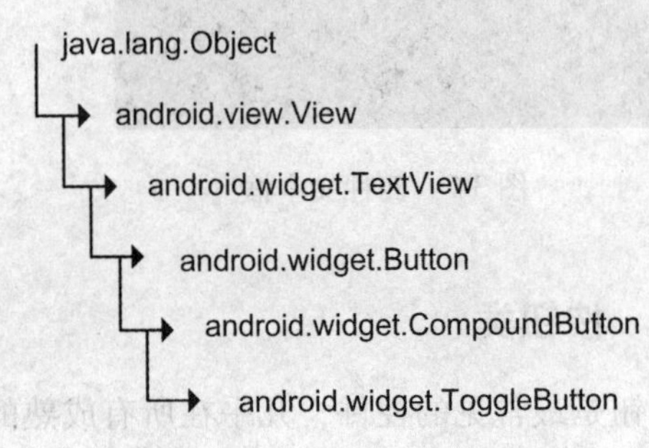

图 3-11　ToggleButton 类继承结构

表 3-6 ToggleButton 常用 XML 属性

属性/方法	说明
android：disabledAlpha	设置按钮在禁用时透明度
android：textOff	设置未选中时按钮的文本
android：textOn	设置选中时按钮的文本
CharSequence getTextOff( )	返回按钮未选中时的文本
CharSequence getTextOn( )	返回按钮选中时的文本
void setBackgroundDrawable（Drawable d）	设置背景图片
void setChecked（boolean checked）	改变按钮的选中状态
void setTextOff（CharSequence textOff）	设置按钮未选中时显示的文本
void setTextOn（CharSequence textOn）	设置按钮选中时显示的文本

3. 图片按钮（ImageButton）

图片按钮（ImageButton）类继承结构如图 3-12 所示，ImageButton 类继承自 ImageView。默认情况下，图片按钮看起来像一个普通的按钮，但是图片按钮没有 text 属性，按钮中将显示图片来代替文本，可以定义自己的背景图片或设置背景为透明。按钮的图片可用通过 XML <ImageButton>元素的 android:src 属性或 setImageResource(int)方法指定。为了表示不同的按钮状态，可以为各种状态定义不同的图片，这种情况下可以通过编写 XML 文件来实现。

图 3-12 ImageButton 类继承结构

```
1  <?xml version="1.0" encoding="utf-8"?>
2  <selector xmlns:android="http://schemas.android.com/apk/res/android">
3      <item android:state_pressed="true"
4          android:drawable="@drawable/button_pressed" />
5      <item android:state_focused="true"
6          android:drawable="@drawable/button_focused"/>
7      <item android:drawable="@drawable/button_normal"/>
8  </selector>
```

将该文件名作为一个参数设置到 ImageButton 的 android：src 属性，例如将上面文件命名为 imagebuttonselector.xml 并保存在 layout 文件夹下，那么这里设置为"@layout/imagebuttonselector"，设置 android：background 也是可以的，但效果不太一样。Android 根据按钮的状态改变会自动地去 XML 中查找相应的图片以显示。<item>元素的顺序很重要，因为是根据这个顺序判断是否适用于当前按钮状态，这也是为什么正常（默认）状态指定的图片放在最后，是因为它只会在 pressed 和 focused 都判断失败之后才会被采用。另外，如果按钮被按下时是同时获得焦点的，但是获得焦点并不一定按了按钮，所以这里会按顺序查找，找到匹配的就不再继续往下查找。例如按钮被单击了，那么第一个将被选中，且不再在后面查找其他状态。如图 3-13 所示，为图片按钮的默认状态和单击状态的效果图。

图 3-13 ImageButton 效果图

4. 复选按钮（CheckBox）

复选按钮（CheckBox）是一种支持多选的按钮控件，其类继承结构如图 3-14 所示。使用复选按钮经常需要根据其选择状态的变化而引发不同的处理，通过设置复选按钮状态变化监听器可以跟踪这种变化并根据相应的变化而做出相应的操作处理，复选按钮设置状态变化监听器的方法如下面代码所示。

```
1    final CheckBox checkbox = (CheckBox) findViewById(R.id.CheckBox1);
2    checkbox.setOnCheckedChangeListener(new CheckBox.OnCheckedChangeListener(){
3        @Override
4        public void onCheckedChanged(CompoundButton buttonView,boolean isChecked){
5                //TODO Auto-generated method stub
6        }
7    });
```

5. 单选按钮（RadioButton）

单选按钮（RadioButton）在同组内仅支持单选，其类继承结构如图 3-15 所示。单选按钮一般包含在 RadioGroup 中，RadioGroup 是 RadioButton 的承载体，程序运行时不可见，一个应用程序中可以包含一个或多个 RadioGroup。一个 RadioGroup 包含多个 RadioButton，在每个 RadioGroup 中，用户仅能够选择其中一个 RadioButton。类似于复选框，获取状态的改变并进行相应处理同样依靠设置态变化监听器完成，其实现方法通过 setOnCheckedChange-Listener( )方法实现。RadioButton 类还有一些常用的其他方法，如表 3-7 所示。

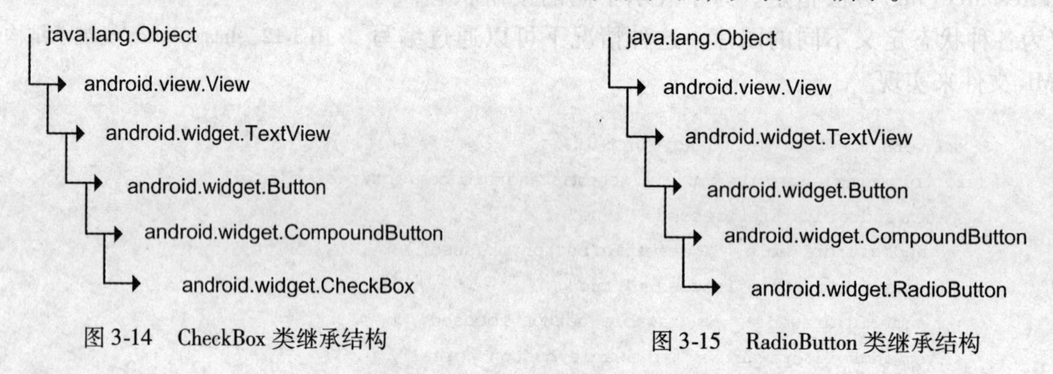

图 3-14　CheckBox 类继承结构

图 3-15　RadioButton 类继承结构

表 3-7　RadioButton 常用方法

方　法	说　明
Boolean isChecked( )	判断是否被选中，如果选中返回 true，否则返回 false
performClick( )	调用 onClickListener 监听器，模拟一次单击
setChecked（boolean checked）	设置控件状态（选中与否）
toggle( )	置反控件当前选中状态

下面具体实现图 3-9 所示的界面效果，并实现必要的操作处理。

布局文件中添加五种按钮控件的代码如下：

```
1    <Button android:id = "@ + id/btnNormalButton"
2            android:layout_width = "wrap_content"
```

```
3          android:layout_height = "wrap_content"
4          android:text = "普通按钮"/>
5    <ToggleButton android:id = "@ +id/tgbtnToggleButton"
6          android:layout_width = "wrap_content"
7          android:layout_height = "wrap_content"
8          android:textOff = "打开"
9          android:textOn = "关闭"/>
10   <ImageButton android:id = "@ +id/imgbtnImageButton"
11         android:layout_width = "wrap_content"
12         android:layout_height = "wrap_content"
13         android:src = "@ layout/imagebuttonselector"/>
14   <CheckBox android:id = "@ +id/ckbCheckBox1"
15         android:layout_width = "wrap_content"
16         android:layout_height = "wrap_content"
17         android:text = "甲"/>
18   <CheckBox android:id = "@ +id/ckbCheckBox2"
19         android:layout_width = "wrap_content"
20         android:layout_height = "wrap_content"
21         android:text = "乙"/>
22   <RadioGroup android:id = "@ +id/radgActButton01"
23         android:layout_width = "wrap_content"
24         android:layout_height = "wrap_content"
25         android:orientation = "horizontal">
26          <RadioButton android:id = "@ +id/radbtnRadioButton1"
27               android:layout_width = "wrap_content"
28               android:layout_height = "wrap_content"
29               android:text = "A"/>
30           <RadioButton android:id = "@ +id/radbtnRadioButton2"
31               android:layout_width = "wrap_content"
32               android:layout_height = "wrap_content"
33               android:text = "B"/>
34   </RadioGroup>
```

代码1~4行为添加普通按钮，4~9行添加开关按钮并设置状态显示文本为“打开”“关闭”。代码10~14行添加了一个图片按钮并设置android：src为layout下的imagebuttonselector.xml文件，其内容如下：

```
1   <?xml version = "1.0" encoding = "utf-8"? >
2   <selector xmlns:android = "http://schemas.android.com/apk/res/android">
3     <item android:state_pressed = "true"
4          android:drawable = "@ drawable/pressed" />
5     <item android:state_focused = "true"
6          android:drawable = "@ drawable/focused" />
7     <item android:drawable = "@ drawable/normal" />
8   </selector>
```

代码14~21行添加了两个复选按钮，提供两个选项“甲”和“乙”，在代码22~24行定义了一个 RadioGroup，其中包含两个单选按钮，提供“A”和“B”选择项。

除了上面五种按钮，还添加了一个文本框用于显示被单击或选择按钮的信息。

```
1    <TextView android:id="@+id/txtButtonAct"
2        android:layout_width="wrap_content"
3        android:layout_height="wrap_content"
4        android:text="此处显示选中按钮的结果"
5    />
```

接下来处理相应的按钮操作，普通按钮、开关按钮和图片按钮设置单击处理事件，在单击相应按钮时 TextView 控件显示所选择的是哪一个按钮，处理代码如下。

```
1    final Button btnNormalButton = (Button)findViewById(R.id.btnNormalButton);
2    btnNormalButton.setOnClickListener(new Button.OnClickListener(){
3        public void onClick(View v) {
4            m_txtButtonAct.setText("普通按钮单击");
5        }
6    });
7    //开关按钮
8    final ToggleButton tgbtnToggleButton =
9                        (ToggleButton)findViewById(R.id.tgbtnToggleButton);
10   tgbtnToggleButton.setOnClickListener(new Button.OnClickListener() {
11       public void onClick(View v) {
12           m_txtButtonAct.setText("开关按钮:" + tgbtnToggleButton.getText());
13       }
14   });
15   //图示按钮
16   final ImageButton imgbtnImageButton =
17                       (ImageButton)findViewById(R.id.imgbtnImageButton);
18   imgbtnImageButton.setOnClickListener(new Button.OnClickListener() {
19       public void onClick(View v) {
20           m_txtButtonAct.setText("图示按钮");
21       }
22   });
```

复选按钮实现选中状态发生改变引发的事件处理，其处理结果就是显示选中项，单选按钮状态改变事件处理机制和复选按钮相同，其中复选按钮相应的处理代码如下：

```
1    final CheckBox ckbCheckBox1 = (CheckBox)findViewById(R.id.ckbCheckBox1);
2    final CheckBox ckbCheckBox2 = (CheckBox)findViewById(R.id.ckbCheckBox2);
3    ckbCheckBox1.setOnCheckedChangeListener(new CheckBox.OnCheckedChangeListener(){
4        @Override
5        public void onCheckedChanged(CompoundButton buttonView,
6                boolean isChecked) {
7            // TODO Auto-generated method stub
8            String strMsg = "复选项:";
9            if (ckbCheckBox1.isChecked()) {
10                   strMsg = strMsg + ckbCheckBox1.getText() + ",";
```

```
11              }
12              if (ckbCheckBox2.isChecked()) {
13                  strMsg = strMsg + ckbCheckBox2.getText() + ",";
14              }
15              m_txtButtonAct.setText(strMsg);
16              }
17      });
18      ckbCheckBox2.setOnCheckedChangeListener(new CheckBox.OnCheckedChangeListener(){
19      @Override
20      public void onCheckedChanged(CompoundButton buttonView,
21                                  boolean isChecked) {
22              // TODO Auto-generated method stub
23              String strMsg = "复选项：";
24              if (ckbCheckBox1.isChecked()) {
25                                  strMsg = strMsg + ckbCheckBox1.getText() + ",";
26              }
27              if (ckbCheckBox2.isChecked()) {
28                                  strMsg = strMsg + ckbCheckBox2.getText() + ",";
29              }
30              m_txtButtonAct.setText(strMsg);
31      }
32  });
```

3.2.4 时钟控件类

Android 系统中存在两种时钟显示方式，即模拟时钟和数字时钟，如图 3-16 所示，模拟时钟是一个带有时针和分针的转动圆盘，不能显示秒针，而数字时钟的时、分和秒均采用数字显示。Android 系统类库中模拟时钟来自 AnalogClock 类，数字时钟来自 DigitalClock 类，其各自类继承结构如图 3-17 所示。

图 3-16 时钟控件

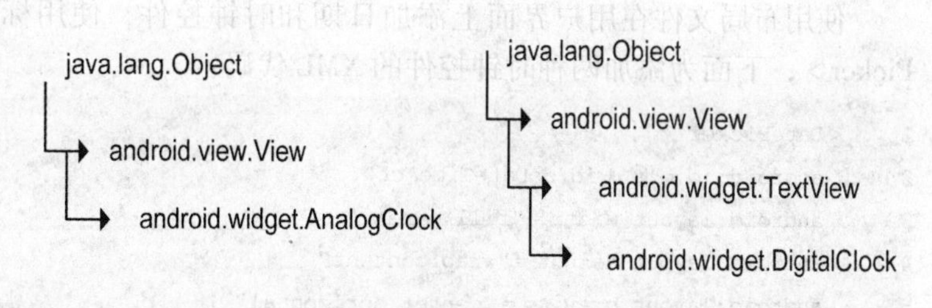

图 3-17 AnalogClock 和 DigitalClock 类继承结构

使用布局文件在用户界面上添加模拟时钟和数字时钟控件，使用标签 <AnalogClock> 和 <DigitalClock>，下面为添加两种时钟控件的 XML 代码：

```
1   <AnalogClock android:id="@+id/aclockClockAct"
2       android:layout_width="wrap_content"
3       android:layout_height="wrap_content"
4   />
```

```
1    <DigitalClock android:id="@ +id/dclockClockAct"
2        android:layout_gravity="center_horizontal"
3        android:layout_height="wrap_content"
4        android:layout_width="wrap_content"
5        android:textSize="20dp"
6    />
```

3.2.5 日期与时间类

日期控件类为 DatePicker，提供年月日日期数据的显示和其他操作。时间控件类为 TimePicker，用于选择一天中时间的视图，支持 24 小时及上午/下午模式。小时、分钟及上午/下午都可以用垂直滚动条来控制，也可以用键盘来输入小时，两个数的小时数可以通过输入两个数字来实现，例如，在一定时间框输入‘1’和‘2’即选择了 12 点，分钟也能显示输入的数字。另外，在 AM/PM 模式下，用户可以输入'a'，'A'或'p'，'P'来选取上下午。两种控件的显示效果如图 3-18 所示，DatePicker 类和 TimePicker 类的继承结构如图 3-19 所示。

图 3-18 日期与事件控件

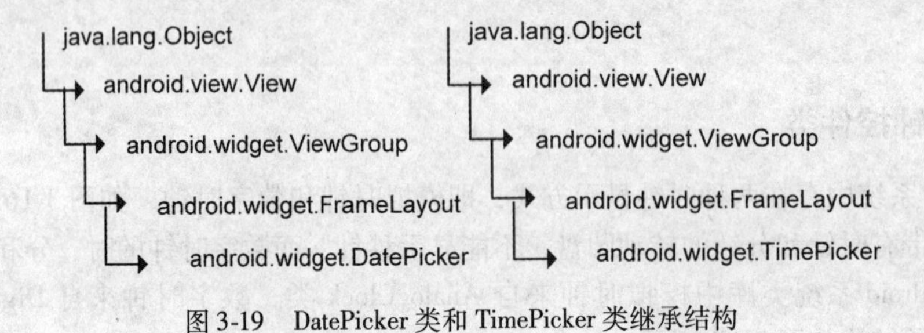

图 3-19 DatePicker 类和 TimePicker 类继承结构

使用布局文件在用户界面上添加日期和时钟控件，使用标签 <DatePicker> 和 <TimePicker>，下面为添加两种时钟控件的 XML 代码：

```
1    <DatePicker
2      android:id="@ +id/dpDateTimeAct"
3      android:layout_width="fill_parent"
4      android:layout_height="wrap_content"
5      android:layout_gravity="center_horizontal"
6    />
7    <TimePicker
8      android:id="@ +id/tpDateTimeAct"
9      android:layout_width="fill_parent"
10     android:layout_height="wrap_content"
11     android:layout_gravity="center_horizontal"
12   />
```

在使用日期和时间控件时，通常情况下会需要捕捉用户修改日期和时间的事件，Android 系统提供的用于捕捉日期和时间控件修改数据的事件响应的监听器是 OnDateChangedListener 和 OnTimeChangedListener。设置监听器的方法及 DatePicker 类和 TimePicker 类提供的其他常用方法可参考表 3-8。

表 3-8 DatePicker 类和 TimePicker 类常用方法

方 法	说 明
int getDayOfMonth( )	获取 DatePicker 选择的天数
int getMonth( )	获取 DatePicker 选择的月份
int getYear( )	获取 DatePicker 选择的年份
void init (int year, int monthOfYear, int dayOfMonth, DatePicker. OnDateChangedListener onDateChangedListener)	初始化 DatePicker 状态，其中 onDateChangedListener 为日期改变时通知用户的事件监听期
void setEnabled (boolean enabled)	设置 DatePicker 是否可用
void updateDate (int year, int monthOfYear, int dayOfMonth)	更新 DatePicker 控件数据
int getBaseline( )	返回 TimePicker 文本基准线到其顶边界的偏移量
integer getCurrentHour( )	获取 TimePicker 小时部分
integer getCurrentMinute( )	获取 TimePicker 分钟部分
boolean is24HourView( )	获取 TimePicker 是否是 24 小时制
void setCurrentHour (Integer currentHour)	设置 TimePicker 小时部分
void setCurrentMinute (Integer currentMinute)	设置 TimePicker 分钟部分
void setEnabled (boolean enabled)	设置 TimePicker 是否可用
void setIs24HourView (Boolean is24HourView)	设置 TimePicker 是 24 小时制还是上午/下午制
void setOnTimeChangedListener (TimePicker. OnTimeChangedListener onTimeChangedListener)	设置 TimePicker 时间调整事件监听器 OnTimeChangedListener

3.2.6 计时控件

计时控件类（Chronometer）实际上是个简单的定时器，默认情况下，定时器的值的显示形式为“分：秒”或“时：分：秒”，或者可以使用的 Set( )方法设置相应的自定义格式。Chronometer 类继承结构如图 3-20 所示，从类继承结构上可以看出，Chronometer 类直接继承自 TextView 类，很多属性和方法也同样来自于 TextView 类，另外 Chronometer 还有自己定义的 XML 属性和方法，如表 3-9 所示。在使用计时控件时，只需要在布局文件中添加 <Chronometer>标签即可向用户界面上添加一个计时控件类，如图 3-21 所示。

```
1    <Chronometer android:id="@+id/chroChronometerAct"
2        android:layout_width="wrap_content"
3        android:layout_height="wrap_content"
4        android:layout_gravity="center"
5        android:textSize="20dp"
6    />
```

java.lang.Object
→ android.view.View
→ android.widget.ProgressBar

图 3-20　Chronometer 类继承结构

图 3-21　计时控件效果

表 3-9　Chronometer 类常用 XML 属性和方法

方　　法	说　　明
android:format	格式化字符串。计时器将根据这个字符串来显示，替换字符串中第一个“%s”为当前“MM：SS”或“H：MM：SS”格式的时间显示，如果不指定，计时器将简单地显示“MM：SS”或者“H：MM：SS”格式的时间
long getBase( )	获取由 setBase（long）设置的基准时间
String getFormat( )	获取由 setFormat（String）设置的格式化字符串
Chronometer. OnChronometerTickListener getOnChronometerTickListener( )	获取用于监听计时器变化的事件的监听器
void setBase（long base）	设置基准时间
void setFormat（String format）	设置用于显示的格式化字符串，如果指定了格式化字符串，计时器将根据这个字符串来显示，替换字符串中第一个“%s”为当前“MM：SS”或“H：MM：SS”格式的时间显示；如果这个格式化字符串为空，或者从未调用过 setFormat( )方法，计时器将简单地显示“MM：SS”或者“H：MM：SS”格式的时间
void setOnChronometerTickListener（OnChronometerTickListener listener）	设置计时器变化时监听器
void start( )	开始计时
void stop( )	停止计时

3.2.7　进度条控件

进度条（ProgressBar）控件用于显示相应应用的进度，例如，当一个应用在后台执行时，前台界面就不会有什么信息，这种情况下用户根本不知道应用程序是否在执行，也不知道执行进度如何和是否遇到异常错误而终止等。此时使用进度条控件来提示用户后台程序执行进度显得非常有必要，也对界面的友好性非常重要。Android 系统库中提供了两种进度条样式，长形进度条和圆形进度条。

进度条 ProgressBar 类直接继承自 View 类，其类继承结构如图 3-22 所示。使用进度条时常使用的 XML 属性和 ProgressBar 类方法如表 3-10 所示。

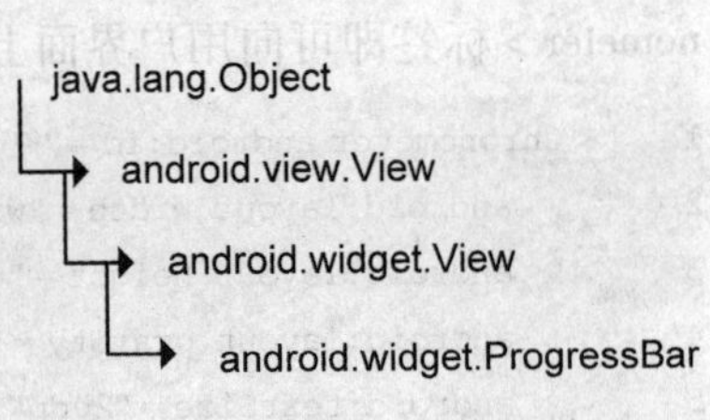

图 3-22　ProgressBar 类继承结构

在布局文件中添加进度条使用 < ProgressBar >标签，如下列代码所示，在 XML 布局文件中另外添加了一个 TextView 控件，用于模拟下载进度条。

表 3-10 ProgressBar 常用的 XML 属性和方法

方 法	说 明
android：progressBarStyle	默认进度条样式
android：progressBarStyleHorizontal	水平进度条样式
android：progressBarStyleLarge	圆形进度条样式，圆圈较大
android：progressBarStyleSmall	圆形进度条样式，圆圈较小
int getMax( )	返回进度条的范围上限
int getProgress( )	返回进度条的当前进度
int getSecondaryProgress( )	返回次要进度条的当前进度
void incrementProgressBy(int diff)	增加进度条进度
boolean isIndeterminate( )	指示进度条是否在不确定模式下
void setIndeterminate(boolean indeterminate)	设置不确定模式
void setVisibility(int v)	设置该进度条是否可视

```
1    <? xml version = "1.0" encoding = "utf-8"? >
2    <TableLayout
3       xmlns:android = "http://schemas.android.com/apk/res/android"
4       android:layout_width = "fill_parent"
5       android:layout_height = "wrap_content"
6       android:layout_gravity = "center" >
7    <TextView android:id = "@ + id/txtProgressBarAct"
8       android:layout_width = "wrap_content"
9       android:layout_height = "wrap_content"
10     android:textSize = "30dp"
11     android:text = "正在下载..."/>
12   <ProgressBar android:id = "@ + id/pbarProgressBarAct"
13       android:layout_height = "wrap_content"
14       android:layout_width = "fill_parent"
15       style = "? android:attr/progressBarStyleHorizontal"
16       android:max = "100"/>
17   </TableLayout>
```

上面代码中，进度条使用水平进度条，并设置最大范围为 100，使用进度条的 Activity 的 Java 代码如下：

```
1    public class ProgressBarAct extends Activity{
2    @Override
3    protected void onCreate(Bundle savedInstanceState) {
4    // TODO Auto-generated method stub
5    super.onCreate(savedInstanceState);
6    requestWindowFeature(Window.FEATURE_PROGRESS);
7    setContentView(R.layout.progressbaract);
8    final ProgressBar pbarProgressBarAct =
9                                  (ProgressBar)findViewById(R.id.pbarProgressBarAct);
10   setProgressBarVisibility(true);
```

```
11    final Handler handler = new Handler();//设定 Handle 类别
12    final Runnable callback1 = new Runnable() {
13      public void run() {
14        pbarProgressBarAct.incrementSecondaryProgressBy(1);//第二个进度条进度值增加 1
15                        setSecondaryProgress(100* pbarProgressBarAct.getSecondaryProgress());
16            }
17           };
18    final Runnable callback2 = new Runnable() {
19         public void run() {
20          pbarProgressBarAct.incrementProgressBy(25);
21          setProgress(100* pbarProgressBarAct.getProgress());
22          pbarProgressBarAct.incrementSecondaryProgressBy(-100);
23          setSecondaryProgress(100* pbarProgressBarAct.getSecondaryProgress());
24          if(pbarProgressBarAct.getProgress() == pbarProgressBarAct.getMax())
25          { TextView txtProgressBarAct = (TextView)findViewById(R.id.txtProgressBarAct);
26              txtProgressBarAct.setText("下载完毕!");
27              }}
28            };
29            Thread thread = new Thread() {      //建立一个 Thread 来 Run
30              @Override
31              public void run() {
32                 try {
33                   for (int i = 0; i < pbarProgressBarAct.getMax() / 10; i++) {
34                       for (int j = 0; j < pbarProgressBarAct.getMax(); j++) {
35                           Thread.sleep(10);
36                           handler.post(callback1);
37                   }
38                   handler.post(callback2);
39              }
40          } catch (InterruptedException e) {
41              e.printStackTrace();
42          }
43     }
44    };
45    thread.start();
46. }
47 }
```

代码第 6 行使用 requestWindowFeature( )方法启用窗体的扩展特性，很多情况下，例如应用程序装载资源和网络连接时，可以使用进度条提示用户稍等，这类进度条只能代表应用程序中某一部分程序的执行进度，而整个应用程序执行情况则需要通过应用程序标题栏来显示进度，这就需要对窗口的显示风格进行设置，这正是 requestWindowFeature( )方法的基本功能。requestWindowFeature( )的参数是 Window 类中定义的常量，具体如下：

DEFAULT_FEATURES：系统默认状态，一般不需要指定；

FEATURE_CONTEXT_MENU：启用 ContextMenu，默认该项已启用，一般无需指定；

FEATURE_CUSTOM_TITLE：自定义标题，当需要自定义标题时必须指定，如标题是一个按钮时；

FEATURE_INDETERMINATE_PROGRESS：不确定的进度；

FEATURE_LEFT_ICON：标题栏左侧的图标；

FEATURE_NO_TITLE：无标题；

FEATURE_OPTIONS_PANEL：启用“选项面板”功能，默认已启用；

FEATURE_PROGRESS：进度指示器功能；

FEATURE_RIGHT_ICON：标题栏右侧的图标。

代码第 10 行设置进度条可见，代码第 11 行定义了一个 Handler 对象，Android UI 操作并不是线程安全的并且这些操作必须在 UI 线程中执行。Android 是利用 Handler 来实现 UI 线程的更新的，Handler 是 Android 中的消息发送器，其在哪个 Activity 中创建就属于且只属于该 Activity，还可以说其在哪个线程中实例化的，就是那个线程的 Handler。Handler 主要接收子线程发送的数据，并用此数据配合主线程更新 UI。当应用程序启动时，Android 首先会开启一个主线程，也就是 UI 线程，主线程为管理界面中的 UI 控件，进行事件分发，比如说，单击一个 Button，Android 会分发事件到 Button 上，来响应相应的操作。如果此时需要一个耗时的操作，不能把这些操作放在主线程中，如果放在主线程中的话，界面会出现假死现象，如果 5 秒还没有完成，会收到 Android 系统的一个错误提示“强制关闭”。这个时候需要把这些耗时的操作，放在一个子线程中，因为子线程涉及 UI 更新，Android 主线程是不安全的，也就是说，更新 UI 只能在主线程中更新，在子线程中操作是危险的，因此，Handler 用于解决这个复杂的问题。由于 Handler 运行在主线程中（UI 线程中），它与子线程可以通过 Message 对象来传递数据，此时 Handler 就承担着接收子线程传过来的子线程用 sendMessage( )方法传递 Message 对象，把这些消息放入主线程队列中，配合主线程进行更新 UI。

代码 12 ~ 17 行和代码 18 ~ 28 行，分别声明了 Runnable 对象，用于进度条次要进度显示处理和主要进度显示处理。代码 29 ~ 44 行，通过线程来改变进度条进度。

运行上面的代码，其效果如图 3-23 所示。

图 3-23 进度条效果

3.2.8 拖动条控件

Android 提供了两种常用的拖动条，一类是连续拖动条 SeekBar，另一类是星级评分条 RatingBar，其类的继承结构如图 3-24 所示。

图 3-24 SeekBar 和 RatingBar 类继承结构

SeekBar 是 ProgressBar 的扩展，在其基础上增加了一个可滑动的滑片。用户可以触摸滑片并向左或向右拖动，或者也可以使用方向键设置当前的进度等级。当然，在使用过程中尽量不要把可以获取焦点的其他控件放在 SeekBar 的左边或右边。

RatingBar 是基于 SeekBar 和 ProgressBar 的扩展，用星星来显示等级评定。使用 RatingBar 的默认大小时，用户可以触摸/拖动或使用键来设置评分，它有两种样式，小风格用 ratingBarStyleSmall，大风格用 ratingBarStyleIndicator，一般情况下大风格适合指示，不适用于用户交互。当使用可以支持用户交互的 RatingBar 时，同样不适合将其他控件放在它的左边或者右边。在使用 RatingBar 过程中，只有当布局的宽被设置为“wrap_content”时，通过函数 setNumStars（int）或者在 XML 的布局文件中设置的星星数量才将完全显示出来，否则显示结果并不确定。

SeekBar 和 RatingBar 常用的 XML 属性和方法如表 3-11 所示。

表 3-11　SeekBar 和 RatingBar 常用的 XML 属性和方法

属性或方法	说　明
android：thumb	Seekbar 上绘制的 thumb（可拖动的那个图标）
android：isIndicator	RatingBar 是否是一个指示器
android：numStars	RatingBar 显示的星星数量，必须是一个整形值
android：rating	RatingBar 默认的评分，必须是浮点类型
android：stepSize	RatingBar 评分的步长，必须是浮点类型
void setOnSeekBarChangeListener(SeekBar. OnSeekBarChangeListener l)	Seekbar 设置一个监听器以接收进度改变时的通知
int getNumStars()	RatingBar 返回显示的星星数量
RatingBar. OnRatingBarChangeListener getOnRatingBarChangeListener()	RatingBar 返回监听器（可能为空）监听评分改变事件
float getRating()	获取当前的评分
float getStepSize()	获取评分条的步长
boolean isIndicator()	判断当前的评分条是否仅仅是一个指示器
void setIsIndicator（boolean isIndicator）	设置当前的评分条是否仅仅是一个指示器
synchronized void setMax（int max）	设置评分等级的范围，从 0 到 max
void setNumStars（int numStars）	设置显示的星星的数量
void setOnRatingBarChangeListener（RatingBar. OnRatingBarChangeListener listener）	设置当评分等级发生改变时回调的监听器
void setRating（float rating）	设置星星的数量
void setStepSize（float stepSize）	设置当前评分条的步长

下面通过代码具体使用两种拖动条，在布局文件（seekbarandratingbaract. xml）中分别添加 RatingBar 控件和 SeekBar 控件，设置 RatingBar 共有五个星星且单步前进半个星级，设置 SeekBar 最大进度量为 100，默认显示进度 30，代码如下：

```
1    <RatingBar
2            android:id="@+id/rbarRatingBarAct"
3            android:layout_width="wrap_content"
```

```
4           android:layout_height = "wrap_content"
5           android:numStars = "5"
6           android:stepSize = "0.5"/>
7   <SeekBar
8           android:id = "@ + id/sbarSeekBar"
9           android:layout_width = "wrap_content"
10          android:layout_height = "wrap_content"
11          android:max = "100"
12          android:progress = "30"/>
```

Activity 处理代码如下：

```
1   public class SeekbarAndRatingbarAct extends Activity{
2       @Override
3       protected void onCreate(Bundle savedInstanceState) {
4           // TODO Auto-generated method stub
5           super.onCreate(savedInstanceState);
6           setContentView(R.layout.seekbarandratingbaract);
7           final TextView txtRatingBar = (TextView)findViewById(R.id.txtRatingBar);
8           final TextView txtSeekBar = (TextView)findViewById(R.id.txtSeekBar);
9           RatingBar rbarRatingBarAct = (RatingBar)findViewById(R.id.rbarRatingBarAct);
10          SeekBar sbarSeekBar = (SeekBar)findViewById(R.id.sbarSeekBar);
11
12          txtRatingBar.setText("Rating Bar: = " + rbarRatingBarAct.getProgress());
13          txtSeekBar.setText("Seek Bar: = " + sbarSeekBar.getProgress());
14          //当 RatingBar 进度条的进度发生变化时调用该方法
15          rbarRatingBarAct.setOnRatingBarChangeListener(new
16                                          OnRatingBarChangeListener(){
17              @Override
18              public void onRatingChanged(RatingBar ratingBar, float rating,
19                          boolean fromUser) {
20                  txtRatingBar.setText("Rating Bar: = " + ratingBar.getProgress());
21              }
22          });
23          //当 SeekBar 进度条的进度发生变化时调用该方法
24          sbarSeekBar.setOnSeekBarChangeListener(new OnSeekBarChangeListener() {
25          public void onProgressChanged(SeekBar sekbActMiscActivity01, int progress,
26              boolean fromTouch) {
27              txtSeekBar.setText("Seek Bar: = " + sekbActMiscActivity01.getProgress());
28              }
29              public void onStartTrackingTouch(SeekBar arg0) {}
30              public void onStopTrackingTouch(SeekBar seekBar) {}
31          });
32      }
33  }
```

代码 15 行设置 RatingBar 进度条变化监听器，检测 RatingBar 进度条值的变化，监听器必须重写 onRatingChanged()方法，监听器检测到改变后使用文本控件显示动态变化值。代码 24 行设置 SeekBar 进度条变化监听器，检测 SeekBar 进度条的变化，监听器必须同时重写 onProgressChanged()、onStartTrackingTouch()和 onStopTrackingTouch()方法，同样监听器检测到改变后使用文本控件显示动态变化值。

运行结果如图 3-25 所示。

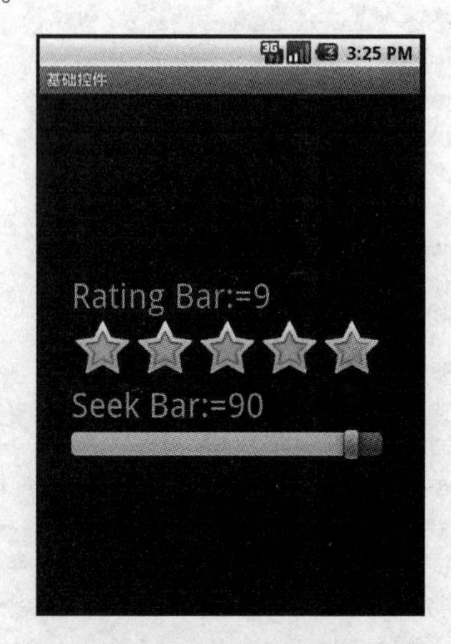

图 3-25　两种拖动条

3.2.9　下拉列表控件

下拉列表控件通过 Spinner 类实现，Spinner 类似于一组单选框，是一个每次只能选择所有项中一项的部件，选项来自于与之相关联的适配器中，选项采用浮动菜单呈现给用户，用户通过弹出的浮动选项界面选择相应的选项值。

Spinner 类提供的常用 XML 属性和常用方法如表 3-12 所示。

表 3-12　Spinner 类提供的常用 XML 属性和常用方法

属性或方法	说　明
android：prompt	在下拉列表对话框显示时提示信息
int getBaseline()	返回控件文本基线的偏移量，如果这个控件不支持基线对齐，那么方法返回 -1
CharSequence getPrompt()	返回下拉列表对话框显示时提示信息
Boolean performClick()	模拟一次单击事件
void setOnItemClickListener(AdapterView.OnItemClickListener I)	设置选项单击监听器

布局文件使用下拉表的 <Spinner> 标签，例如下面 XML 代码，添加下拉表的同时还向界面添加了一个文本框。文本框用于提示下拉框的功能，下拉框的选项信息为“东”、“南”、“西”、“北”、“东南”、“东北”、“西南”、“西北”等方向选项，提供方向信息的选择。

```
1    <TextView android:id="@+id/txtSpinnerAct"
2    android:layout_height="wrap_content"
3    android:layout_width="fill_parent"
4    android:textSize="30dp"
5    android:text="请选择方向:"/>
6    <Spinner android:id="@+id/spiSpinnerAct"
7    android:layout_width="fill_parent"
8    android:layout_height="wrap_content"/>
```

界面 Activity 的 Java 代码如下：

```
1   public class SpinnerAct extends Activity{
2       private Spinner m_spiSpinnerAct=null;
3       protected void onCreate(Bundle savedInstanceState) {
4           super.onCreate(savedInstanceState);
5           setContentView(R.layout.spinneractivity);
6           m_spiSpinnerAct = (Spinner)findViewById(R.id.spiSpinnerAct);
7           //设定字符序列数组
8           ArrayAdapter<CharSequence>  adapter=ArrayAdapter.createFromResource(this,
9                                   R.array.orientation, android.R.layout.simple_spinner_item);
10          //设置下拉菜单
11          adapter.setDropDownViewResource(android.R.layout.simple_spinner_dropdown_item);
12          m_spiSpinnerAct.setAdapter(adapter);
13          m_spiSpinnerAct.setOnItemSelectedListener(new MyOnItemSelectedListener());
14      }
15      //下拉选单的选项处理
16      public class MyOnItemSelectedListener implements OnItemSelectedListener {
17          public void onItemSelected(AdapterView<?> parent, View view, int position, long id) {
18                  Toast.makeText(parent.getContext(),"所选的方向是-" +
19                    parent.getItemAtPosition(position).toString(), Toast.LENGTH_LONG).show();
20          }
21          public void onNothingSelected(AdapterView<?> parent) {}
22      }
23  }
```

代码第 8 行定义了下拉表的适配器，适配器中用于下拉表选项信息的数据定义在单独的 XML 文件中（values/arrays.xml），定义代码如下：

```
1   <string-array name="orientation">
2       <item>A-东</item>
3       <item>B-南</item>
4       <item>C-西</item>
5       <item>D-北</item>
6       <item>E-东南</item>
7       <item>F-东北</item>
8       <item>G-西南</item>
9       <item>H-西北</item>
10  </string-array>
```

代码 16～22 行处理下拉菜单单击事件响应，并采用信息提示框显示用户选择的方向信息，如图 3-26 和图 3-27 所示。

图 3-26　Spinner 菜单

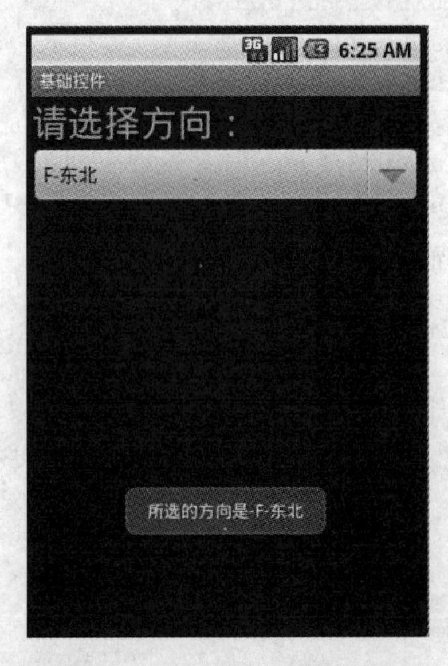

图 3-27　Spinner 菜单选择方向信息

3.3　对话框和消息框

对话框和消息框在应用程序中非常有用，经常用于程序中遇到异常信息或者其他需要通知用户知晓的事件信息等展现给用户。

3.3.1　对话框

对话框是程序运行中弹出的窗口，Android 系统中有四种默认的对话框，分别是警告对话框（AlertDialog）、进度对话框（ProgressDialog）、日期选择对话框（DatePickerDialog）和时间选择对话框（TimePickerDialog）。除了四种默认的对话框，通过继承对话框基类 Dialog 还可实现自定义的对话框。

Dialog 类是一切对话框的基类，Dialog 类虽然可以在界面上显示，但是并非继承自 View 类，而是直接从 java. lang. Object 开始构造出来的，类似于 Activity，Dialog 也是有生命周期的，它的生命周期由 Activity 来维护，在生命周期的每个阶段都有一些回调函数供系统调用。在 Activity 当中用户可以常用的主动调用的函数有两个 showDialog（int id）和 dismissDialog（int id）。showDialog( )负责显示指定 ID 的 Dialog，这个函数如果调用后，系统将反向调用 Dialog 的回调函数 onCreateDialog(int id)。dismissDialog( )使标识为 ID 的 Dialog 在界面当中消失。

Dialog 有两个比较常见的回调函数，onCreateDialog(int id) 和 onPrepareDialog(int id, Dialog dialog) 函数。当在 Activity 当中调用 onCreateDialog(int id) 后，如果这个 Dialog 是第一次生成，系统将反向调用 Dialog 的回调函数 onCreateDialog(int id)，然后再调用 onPrepare-

Dialog(int id, Dialog dialog)。如果这个 Dialog 已经生成，只不过没有显示出，那么将不会回调 onCreateDialog(int id)，而是直接回调 onPrepareDialog(int id, Dialog dialog) 方法。

onPrepareDialog(int id, Dialog dialog) 方法提供了这样一套机制，即当 Dialog 生成但是没有显示出来的时候，使得有机会在显示前对 Dialog 做一些修改，例如修改 Dialog 标题等。

警告对话框 AlertDialog 是 Dialog 的一个直接子类，AlertDialog 也是 Android 系统当中最常用的对话框之一。一个 AlertDialog 可以有两个 Button 或三个 Button，可以对一个 AlertDialog 设置 title 和 message。不能直接通过 AlertDialog 的构造函数来生成一个 AlertDialog，一般生成 AlertDialog 都是通过它的一个内部静态类 AlertDialog. builder 来构造的，AlertDialog 可以不通过 onCreateDialog(int)，而直接调用 AlertDialog. builder 类的 show()显示，如图 3-28 所示。AlertDialog. builder 类提供的常用方法如表 3-13 所示。

图 3-28 警告对话框 AlertDialog

表 3-13 AlertDialog. builder 类提供的常用方法

方　法	说　明
setTitle()	设置 title
setIcon()	设置图标
setMessage()	设置对话框的提示信息
setItems()	设置对话框要显示的一个 List，一般用于要显示几个命令时
setSingleChoiceItems()	设置对话框显示一个单选的 List
setMultiChoiceItems()	设置对话框显示一系列的复选框
setPositiveButton()	给对话框添加"Yes"按钮
setNegativeButton()	给对话框添加"No"按钮

下面代码实现了图 3-28 所示的警告对话框，代码第 8 行和第 9 行分别设置标题和设置显示信息，10～15 行为对话框添加"Yes"按钮并做相应单击处理，16～21 行为对话框添加"No"按钮并做相应单击处理。代码 22 行直接使用 show()方法显示对话框。

```
1      Button btnAlertDialog = (Button) findViewById(R.id.btnAlertDialog);
2      final TextView txtMsg = (TextView) findViewById(R.id.txtMsg);
3      btnAlertDialog.setOnClickListener(new OnClickListener() {
4          @Override
5          public void onClick(View v) {
6            // TODO Auto-generated method stub
7            AlertDialog.Builder  bdAlertDialog = new Builder(MainAct.this);
8            bdAlertDialog.setTitle("我是 AlertDialog");
9            bdAlertDialog.setMessage("请选择接下来的操作!");
10           bdAlertDialog.setPositiveButton("Yes", new DialogInterface.OnClickListener() {
11               public void onClick(DialogInterface dialog, int which) {
12                   // TODO Auto-generated method stub
13                   txtMsg.setText("Yes");
```

```
14            }
15        });
16        bdAlertDialog.setNegativeButton("No", new DialogInterface.OnClickListener(){
17            public void onClick(DialogInterface dialog, int which) {
18                // TODO Auto-generated method stub
19                txtMsg.setText("No");
20            }
21        });
22        bdAlertDialog.show();
23        }
24    });
```

进度对话框 ProgressDialog 是 AlertDialog 的扩展，它可以显示一个进度的动画，进度环或者进度条，这个对话框也可以提供按钮。打开一个进度对话框很简单，只需要调用 ProgressDialog.show( )即可。ProgressDialog 对话框同样可以不通过 onCreateDialog（int），而直接显示，例如：

```
ProgressDialog dialog = ProgressDialog.show(this, "忙碌中……","下载中,请等待……", true);
```

第一个参数是应用程序上下文，第二个为对话框的标题（可以为空），第三个为对话框内容，最后一个为该进度是否为不可确定的，进度对话框的默认样式为一个旋转的环（如图 3-29 所示），也可以设置动画等样式来显示进度。

图 3-29　进度对话框 ProgressDialog

```
1   Button btnProgressDialog = (Button)findViewById(R.id.btnProgressDialog);
2   btnProgressDialog.setOnClickListener(new OnClickListener(){
3       @Override
4       public void onClick(View v) {
5           // TODO Auto - generated method stub
6           final ProgressDialog progressDialog = ProgressDialog.show(MainAct.this,
7                   "忙碌中……","下载中,请等待……", true);
8           final Handler handler = new Handler();
9           //建立处理程序 callback
10          final Runnable callback = new Runnable() {
11              public void run() {
12                  progressDialog.dismiss();
13              }
14          };
```

```
15          //建立一个 Thread 来 Run
16          Thread thread = new Thread() {
17            @Override
18            public void run() {
19              try {
20                  Thread.sleep(5000);
21              } catch (InterruptedException e) {
22                      e.printStackTrace();
23              }
24                handler.post(callback);
25            }
26          };
27          thread.start();
28      }
29    });
```

日期选择对话框 DatePickerDialog 和时间选择对话框 TimePickerDialog 允许用户选择日期和时间的对话框，如图 3-30 和图 3-31 所示。

图 3-30 日期选择对话框 DatePickerDialog

图 3-31 时间选择对话框 TimePickerDialog

```
1     public class MainAct extends Activity {
2         private int m_nYear,m_nMonth,m_nDay,m_nHour,m_nMinute;
3             /** Called when the activity is first created. */
4             @Override
5             public void onCreate(Bundle savedInstanceState) {
6                 super.onCreate(savedInstanceState);
7                 setContentView(R.layout.main);
8                 Calendar cCalendar = Calendar.getInstance();
9                 m_nYear = cCalendar.get(Calendar.YEAR);
10                m_nMonth = cCalendar.get(Calendar.MONTH);
11                m_nDay = cCalendar.get(Calendar.DAY_OF_MONTH);
12                m_nHour = cCalendar.get(Calendar.HOUR);
13                m_nMinute = cCalendar.get(Calendar.MINUTE);
14                Button btnDatePickerDialog = (Button)findViewById(R.id.btnDatePickerDialog);
```

```
15              Button btnTimePickerDialog = (Button)findViewById(R.id.btnTimePickerDialog);
16              btnDatePickerDialog.setOnClickListener(new OnClickListener(){
17              @Override
18              public void onClick(View v) {
19                  DatePickerDialog dtDialog = new DatePickerDialog(MainAct.this,
20                              lsOnDateSetListener,m_nYear, m_nMonth, m_nDay);
21                  dtDialog.show();
22              }
23            });
24            btnTimePickerDialog.setOnClickListener(new OnClickListener(){
25              @Override
26              public void onClick(View v) {
27                  TimePickerDialog tpDialog  = new TimePickerDialog(MainAct.this,
28                              lsOnTimeSetListener, m_nHour, m_nMinute, false);
29                  tpDialog.show();
30              }
31            });
32          }
33          private OnDateSetListener lsOnDateSetListener = new OnDateSetListener() {
34          public void onDateSet(DatePicker view, int year, int monthOfYear, int dayOfMonth) {
35              m_nYear = year;
36              m_nMonth = monthOfYear;
37              m_nDay = dayOfMonth;
38              }
39          };
40          private OnTimeSetListener lsOnTimeSetListener = new OnTimeSetListener() {
41          public void onTimeSet(TimePicker view, int hourOfDay, int minute) {
42              m_nHour = hourOfDay;
43              m_nMinute = minute;
44        }
45      };
46  }
```

3.3.2 消息框

Android 系统消息提示主要通过两种方式，一种是 Toast，另一种是 Notification。

Toast 是 Android 中用来显示信息的一种机制，和 Dialog 不一样的是 Toast 是没有焦点的，而且 Toast 显示的时间有限，过一定的时间就会自动消失。Toast 对象的创建很特殊，通过内部静态方法 makeText( )实现，其语法格式为：

```
public static Toast makeText (Context context, int resId, int duration);
public static Toast makeText (Context context, CharSequence text, int duration);
```

参数 context 是使用的上下文，通常是 Application 或 Activity 对象。参数 resId 或者 text 是要使用的字符串资源 ID，可以是已格式化文本。参数 duration 为该信息的存续期间，值为

LENGTH_SHORT 或 LENGTH_LONG。

Notification 是 Android 通知用户有新邮件、新短信息、未接来电等状态的一种机制，这些通知均在 Android 状态栏中显示。创建一个 Notification 通常可以分为以下四步：

首先，通过 getSystemService( ) 方法得到 NotificationManager 对象。

```
NotificationManager notificationManager = (NotificationManager)
    this.getSystemService(android.content.Context.NOTIFICATION_SERVICE);
```

其次，实例化 Notification 对象。

```
Notification notification =new Notification();
```

再次，对 Notification 对象的一些属性进行设置，比如内容、图标、标题、相应 notification 的动作进行处理等。

Notification 类中定义了很多常量和属性字段，常用的属性字段如表 3-14 所示。

表 3-14 Notification 常用的属性字段

字　段	说　明
contentIntent	设置 PendingIntent 对象，单击时发送该 Intent
defaults	添加默认效果； DEFAULT_ALL——使用所有默认值，比如声音，振动，闪屏等等 DEFAULT_LIGHTS——使用默认闪光提示 DEFAULT_SOUNDS—— 使用默认提示声音 DEFAULT_VIBRATE——使用默认手机振动
flags	设置 flag 位； FLAG_AUTO_CANCEL——该通知能被状态栏的清除按钮给清除掉 FLAG_NO_CLEAR——该通知能被状态栏的清除按钮给清除掉 FLAG_ONGOING_EVENT——通知放置在正在运行 FLAG_INSISTENT—— 是否一直进行，比如音乐一直播放，直到用户响应
icon	设置图标
sound	设置声音
tickerText	显示在状态栏中的文字
when	发送此通知的时间戳

设置事件信息使用 setLatestEventInfo( ) 方法，此方法功能是显示在拉伸状态栏中的 Notification 属性，单击后将发送 PendingIntent 对象，该方法语法格式为：

```
    void setLatestEventInfo(Context context , CharSequencecontentTitle, CharSequence contentText,
PendingIntent contentIntent)
```

参数 context 是上下文环境，参数 contentTitle 为状态栏中的大标题，参数 contentText 为状态栏中的小标题，参数 contentIntent 为单击后将发送的 PendingIntent 对象。

最后，通过 NotificationManager 对象的 notify( ) 方法来发出；取消通知使用 NotificationManager 对象的 cancel( ) 方法。这两个方法的语法格式为：

```
public void cancelAll()移除所有通知 (只是针对当前 Context 下的 Notification)
public void cancel(int id)移除标记为 id 的通知 (只是针对当前 Context 下的所有 Notification)
```

public void notify(String tag ,int id, Notification notification) 将通知加入状态栏，标签为 tag,标记为 id

public void notify(int id, Notification notification)将通知加入状态栏,,标记为 id

下面代码实现了一个简单的未接来电通知，其运行结果如图 3-32 所示。

图 3-32 未接来电通知

```
1      public class MainAct extends Activity {
2          /* *  Called when the activity is first created. * /
3          @ Override
4          public void onCreate(Bundle savedInstanceState) {
5              super. onCreate(savedInstanceState);
6              setContentView(R. layout. main);
7              clearNotification();
8          }
9          @ Override
10         protected void onStop() {
11           showNotification();
12           super. onStop();
13         }
14         @ Override
15         protected void onStart() {
16           clearNotification();
17           super. onStart();
18         }
19         //在状态栏显示通知
20         private void showNotification(){
21         //创建一个 NotificationManager 的引用
22         NotificationManager notificationManager = (NotificationManager)
23             this. getSystemService(android. content. Context. NOTIFICATION_SERVICE);
24         //实例化 Notification 对象
25         Notification notification  = new Notification();
26         //定义 Notification 的各种属性
27         notification. icon = R. drawable. icon;
28         notification. tickerText = "未接来电";
29         notification. when = System. currentTimeMillis();
```

```
30          notification.flags |= Notification.FLAG_ONGOING_EVENT;
31          notification.flags |= Notification.FLAG_NO_CLEAR;
32          notification.flags |= Notification.FLAG_SHOW_LIGHTS;
33          notification.defaults = Notification.DEFAULT_LIGHTS;
34          notification.ledARGB = Color.BLUE;
35          notification.ledOnMS =5000;
36
37          // 设置通知的事件消息
38          CharSequence contentTitle = "您有未接来电";
39          CharSequence contentText = "未接电话是 02488888888";
40          Intent notificationIntent = new Intent(MainAct.this, MainAct.class);
41          PendingIntent contentItent = PendingIntent.getActivity(this, 0, notificationIntent, 0);
42          notification.setLatestEventInfo(this, contentTitle, contentText, contentItent);
43          // 把 Notification 传递给 NotificationManager
44          notificationManager.notify(0, notification);
45      }
46      //删除通知
47      private void clearNotification(){
48              NotificationManager notificationManager = (NotificationManager) this
49                      .getSystemService(NOTIFICATION_SERVICE);
50              notificationManager.cancel(0);
51      }
52  }
```

3.4 界面布局

Android 系统中提供了布局管理器用来控制子控件在屏幕中的位置，布局管理器即界面布局（Layout），定义了界面包含的子控件、控件结构和控件间位置关系等信息。Android 系统中提供了五种窗体布局格式，分别为线性布局（Linear Layout）、相对布局（Relative Layout）、表格布局（Table Layout）、绝对布局（Absolute Layout）和框架布局（Frame Layout）。声明应用程序界面布局的方法有两种，一种是使用 XML 布局文件定义界面布局，另一种是在程序中动态添加布局定义或者修改布局格式。各种布局是可以相互嵌套的，可以使用多个布局格式的组合设计适合需求的整体布局。

3.4.1 线性布局

线性布局（Linear Layout）是 Android 界面布局中最简单的布局，也是最常用和最实用的布局。线性布局最大的特点就是布局元素看上去像一条“线”，其形式有两种，一种是横向线性，一种是纵向线性。横向线性布局，每一列只有一个界面元素，由左到右依序排列。纵向线性布局，每一行只有一个界面元素，由上而下依序排列，如图 3-33 所示。

XML 使用线性布局只需要添加 <LinearLayout> 标签即可完成，LinearLayout 类中常用 XML 属性和对应设置方法如表 3-15 所示。

图 3-33 所示的界面线性布局文件代码如下所示。

表 3-15　LinearLayout 类中常用 XML 属性和对应设置方法

XML 属性	对应方法	说　明
android：orientation	setOrientation（int）	设置是横向线性布局还是纵向线性布局，horizontal 设置横向线性布局，vertical 设置纵向线性布局
android：gravity	setGravity（int）	设置线性布局的内部元素的布局方式，可以使用"|"设置多个值。属性值常可以取以下值： top（0x30）—不改变大小，对齐容器顶部； bottom(0x50）—不改变大小，对齐容器底部； left（0x03）—不改变大小，对齐容器左部； right（0x50）—不改变大小，对齐容器右部； center_vertical（0x10）—不改变大小，对齐容器纵向中央线； fill_vertical（0x70）—仅纵向拉伸填满容器； center_horizontal（0x01）—不改变大小，对齐容器横向中央线； fill_horizontal(0x07）—仅横向拉伸填满容器； center(0x11）—不改变大小，对齐容器中央位置； fill(0x77）—纵向并横向拉伸填满容器；

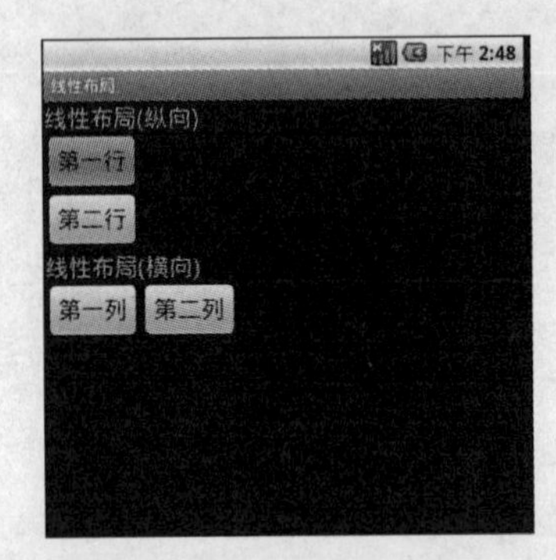

图 3-33　线性布局效果

```
……
1    <TextView
2        android:layout_width = "fill_parent"
3        android:layout_height = "wrap_content"
4        android:textSize = "20dp"
5        android:text = "线性布局(纵向)"/>
6    <LinearLayout
7        android:orientation = "vertical"
8        android:layout_width = "fill_parent"
9        android:layout_height = "wrap_content">
10   <Button
11       android:layout_width = "wrap_content"
12       android:layout_height = "wrap_content"
13       android:textSize = "20dp"
14       android:text = "第一行"/>
15   <Button
16       android:layout_width = "wrap_content"
```

```
17        android:layout_height = "wrap_content"
18        android:textSize = "20dp"
19        android:text = "第二行"/>
20     </LinearLayout>
21     <TextView
22        android:layout_width = "fill_parent"
23        android:layout_height = "wrap_content"
24        android:textSize = "20dp"
25        android:text = "线性布局(横向)"/>
26     <LinearLayout
27        android:orientation = "horizontal"
28        android:layout_width = "fill_parent"
29        android:layout_height = "wrap_content">
30     <Button
31        android:layout_width = "wrap_content"
32        android:layout_height = "wrap_content"
33        android:textSize = "20dp"
34        android:text = "第一列"/>
35     <Button
36        android:layout_width = "wrap_content"
37        android:layout_height = "wrap_content"
38        android:textSize = "20dp"
39        android:text = "第二列"/>
40     </LinearLayout>
   ......
```

3.4.2 相对布局

相对布局（Relative Layout）是一种利用界面元素之间相对关系而设置界面结构的布局定义方式。通常情况下，首先将其中一个界面元素作为参考元素，其他界面元素根据相对于参考元素的位置关系，例如上、下、左、右，设置对应的界面位置，所以相对布局是一种非常灵活的布局设置方式。

RelativeLayout 类常用 XML 属性如表 3-16 所示。

表 3-16 RelativeLayout 类常用 XML 属性

属　性	说　明	值
android:layout_centerHorizontal	当前控件位于父控件的横向中间位置	true/false
android:layout_centerVertical	当前控件位于父控件的纵向中间位置	
android:layout_centerInParent	当前控件位于父控件的中央位置	
android:layout_alignParentBottom	当前控件底端与父控件底端对齐	
android:layout_alignParentLeft	当前控件左侧与父控件左侧对齐	
android:layout_alignParentRight	当前控件右侧与父控件右侧对齐	
android:layout_alignParentTop	当前控件顶端与父控件顶端对齐	
android:layout_alignWithParentIfMissing	参照控件不存在或不可见时参照父控件	

（续）

属　性	说　明	值
android:layout_toRightOf	使当前控件位于给出 ID 控件的右侧	控件的 ID
android:layout_toLeftOf	使当前控件位于给出 ID 控件的左侧	
android:layout_above	使当前控件位于给出 ID 控件的上方	
android:layout_below	使当前控件位于给出 ID 控件的下方	
android:layout_alignTop	使当前控件的上边界与给出 ID 控件的上边界对齐	
android:layout_alignBottom	使当前控件的下边界与给出 ID 控件的下边界对齐	
android:layout_alignLeft	使当前控件的左边界与给出 ID 控件的左边界对齐	
android:layout_alignRight	使当前控件的右边界与给出 ID 控件的右边界对齐	
android:layout_marginLeft	当前控件左侧的空白	像素值
android:layout_marginRight	当前控件右侧的空白	
android:layout_marginTop	当前控件上方的空白	
android:layout_marginBottom	当前控件下方的空白	

图 3-34 为相对布局的一个效果，编辑框相对于“输入信息”文本框位于下面，确定键同样相对于编辑框位于下面，取消键相对于确定键位于左面并与上边界对齐。

图 3-34　相对布局效果

```
1     <RelativeLayout xmlns:android="http://schemas.android.com/apk/res/android"
2        android:layout_width="fill_parent"
3        android:layout_height="fill_parent">
4         <TextView
5            android:id="@+id/txtMsg"
6            android:layout_width="fill_parent"
7            android:layout_height="wrap_content"
8            android:textSize="20dp"
9            android:text="输入信息:"/>
10        <EditText
11           android:id="@+id/etxtMsg"
12           android:layout_width="fill_parent"
```

```
13          android:layout_height="wrap_content"
14          android:layout_below="@id/txtMsg"
15          android:textSize="20dp"/>
16      <Button
17          android:id="@+id/btnOk"
18          android:layout_width="wrap_content"
19          android:layout_height="wrap_content"
20          android:layout_below="@id/etxtMsg"
21          android:layout_alignParentRight="true"
22          android:layout_marginLeft="10dip"
23          android:textSize="20dp"
24          android:text="确定"/>
25      <Button
26          android:id="@+id/btnCancel"
27          android:layout_width="wrap_content"
28          android:layout_height="wrap_content"
29          android:layout_toLeftOf="@id/btnOk"
30          android:layout_alignTop="@id/btnOk"
31          android:textSize="20dp"
32          android:text="取消"/>
33  </RelativeLayout>
```

3.4.3 表格布局

表格布局（Table Layout）是按照行列来组织界面元素的布局。表格布局包含一系列的TableRow对象，用于定义行，表格布局并不为它的行、列和单元格显示表格线，每个行可以包含零个以上（包括零）的单元格；每个单元格可以设置一个View对象。与行包含很多单元格一样，表格包含很多列，表格的单元格可以为空，单元格可以像HTML那样跨列。列的宽度由该列所有行中最宽的一个单元格决定，不过表格布局可以通过setColumnShrinkable()方法或者setColumnStretchable()方法来标记某些列可以收缩或可以拉伸。如果标记为可以收缩，列宽可以收缩以使表格适合容器的大小。如果标记为可以拉伸，列宽可以拉伸以占用多余的空间。表格的总宽度由其父容器决定。列可以同时具有可拉伸和可收缩标记，列可以调整其宽度以占用可用空间，但不能超过限度。另外，通过调用setColumnCollapsed()方法可以隐藏列。

表格布局的子元素不能指定layout_width属性，宽度只能是MATCH_PARENT。不过子元素可以定义layout_height属性，其默认值是WRAP_CONTENT。如果子元素是TableRow，其高度永远是WRAP_CONTENT。

无论是在代码还是在XML布局文件中，单元格必须按照索引顺序加入表格行，列号是从零开始的。如果不为子单元格指定列号，其将自动增值，使用下一个可用列号。如果跳过某个列号，则在表格行中作为空来对待。虽然表格布局典型的子对象是表格行，实际上可以使用任何视图类的子类，作为表格视图的直接子对象。表格布局类TableLayout类的常用XML属性和设置方法如表3-17所示。

表 3-17　TableLayout 类的常用 XML 属性和设置方法

属　性	对应方法	说　明
android：collapseColumns	setColumnCollapsed(int，boolean)	设置指定列号的列为 Collapsed，列号从 0 计算
android：shrinkColumns	setShrinkAllColumns(boolean)	设置指定列号的列为 Shrinkable，列号从 0 计算
android：stretchColumns	setStretchAllColumns(boolean)	设置指定列号的列为 Stretchable，列号从 0 计算

图 3-35 为表格布局的一种常见效果，通过添加 TableRow 来生成各行，并为每种操作设置了快捷键，分别有打开（Ctrl-O）、保存（Ctrl-S）、另存为（Ctrl-Shift-S）、引入（Ctrl-I）、引出（Ctrl-E）和退出（Ctrl-Q）。

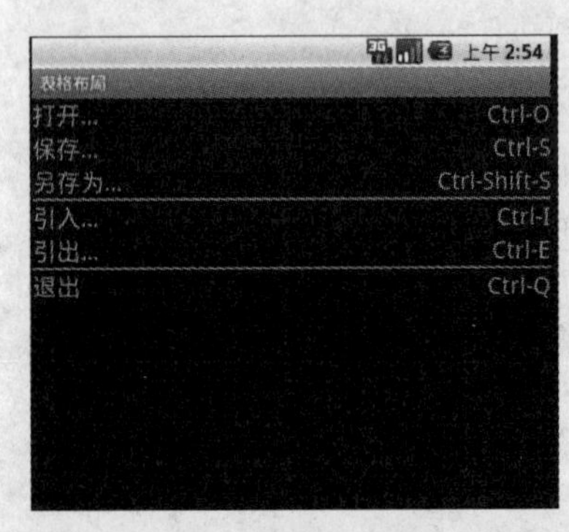

图 3-35　表格布局效果

```
1    <?xml version="1.0" encoding="utf-8"?>
2    <TableLayout xmlns:android="http://schemas.android.com/apk/res/android"
3        android:layout_width="fill_parent"
4        android:layout_height="fill_parent"
5        android:stretchColumns="1">
6    <TableRow>
7        <TextView android:layout_column="1"
8            android:text="打开..."
9            android:textSize="20dp" />
10       <TextView android:text="Ctrl-O"
11           android:gravity="right"
12           android:textSize="20dp"/>
13   </TableRow>
14   <TableRow>
15       <TextView android:layout_column="1"
16           android:textSize="20dp"
17           android:text="保存..."/>
18       <TextView android:text="Ctrl-S"
19           android:textSize="20dp"
20           android:gravity="right" />
21   </TableRow>
22   <TableRow>
```

```
23         <TextView android:layout_column="1"
24             android:text="另存为..."
25             android:textSize="20dp"/>
26         <TextView android:text="Ctrl-Shift-S"
27             android:gravity="right"
28             android:textSize="20dp"/>
29     </TableRow>
30     <View
31             android:layout_height="2dip"
32             android:background="#FF909090"
33             />
34     <TableRow>
35         <TextView android:layout_column="1"
36             android:text="引入..."
37             android:textSize="20dp"/>
38         <TextView android:text="Ctrl-I"
39             android:gravity="right"
40             android:textSize="20dp" />
41     </TableRow>
42     <TableRow>
43         <TextView android:layout_column="1"
44             android:text="引出..."
45             android:textSize="20dp"/>
46         <TextView android:text="Ctrl-E"
47             android:gravity="right"
48             android:textSize="20dp"/>
49     </TableRow>
50     <View
51             android:layout_height="2dip"
52             android:background="#FF909090" />
53     <TableRow>
54         <TextView android:layout_column="1"
55             android:text="退出"
56             android:textSize="20dp"/>
57         <TextView android:text="Ctrl-Q"
58             android:gravity="right"
59             android:textSize="20dp" />
60         </TableRow>
61     </TableLayout>
```

3.4.4 绝对布局

绝对布局（Absolute Layout）将手机屏幕看作是一个二维有限界面，绝对布局的方式是设置所有单元的 x/y 位置，采用绝对位置维护窗体上的单元位置会相当困难，因为没有单元彼此间的关联，绝对布局不是一种比较理想的布局方式。

图 3-36 为绝对布局设置简易登录界面，界面中的所有控件均使用绝对位置（x，y）坐

标生成。

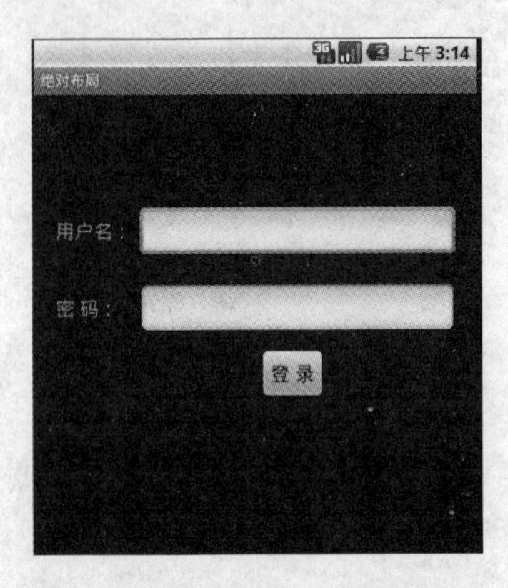

图 3-36　绝对布局设置简易登录界面

```
1     <?xml version="1.0" encoding="utf-8"?>
2     <AbsoluteLayout xmlns:android="http://schemas.android.com/apk/res/android"
3        android:orientation="vertical"
4        android:layout_width="fill_parent"
5        android:layout_height="fill_parent">
6     <TextView  android:layout_x="20dip"
7        android:layout_y="80dip"
8        android:layout_width="wrap_content"
9        android:layout_height="wrap_content"
10       android:textSize="18dp"
11       android:text="用户名:"  />
12    <EditText android:layout_x="100dip"
13       android:layout_y="70dip"
14       android:layout_width="wrap_content"
15       android:width="200px"
16       android:layout_height="wrap_content" />
17       <TextView android:layout_x="20dip"
18       android:layout_y="150dip"
19       android:layout_width="wrap_content"
20       android:layout_height="wrap_content"
21       android:textSize="18dp"
22       android:text="密码:" />
23    <EditText android:layout_x="100dip"
24       android:layout_y="140dip"
25       android:layout_width="wrap_content"
26       android:width="200px"
27       android:layout_height="wrap_content"
28       android:password="true" />
```

```
29    <Button android:layout_x = "160dip"
30       android:layout_y = "200dip"
31       android:layout_width = "wrap_content"
32       android:layout_height = "wrap_content"
33       android:textSize = "18dp"
34       android:text = "登 录"/>
35    </AbsoluteLayout>
```

3.4.5 框架布局

框架布局（Frame Layout）是最简单的布局方式，所有添加到这个布局中的界面元素都以层叠的方式显示。第一个添加的界面元素放到最底层，最后添加到框架中的界面元素显示在最上面，下层控件将会被覆盖。也就是，框架布局在屏幕上开辟出了一块区域，在这块区域中可以添加多个子元素，但是所有的子元素都被对齐到屏幕的左上角，框架布局的大小由子元素中尺寸最大的来决定，如果所有子元素一样大，同一时刻只能看到最上面的子元素。

FrameLayout 类中经常在 XML 中使用 android：foreground 属性设置绘制在所有子控件之上的内容，其对应的设置方法为 setForeground（Drawable）。另外，还会用到 XML 属性 android：foregroundGravity 设置绘制在所有子控件之上内容的 gravity 属性，其对应设置方法为 setForegroundGravity（int）。

图 3-37　采用框架布局的一个效果

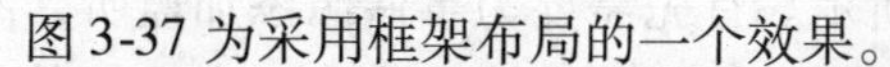

图 3-37 为采用框架布局的一个效果。

```
1     <?xml version = "1.0" encoding = "utf-8"?>
2     <FrameLayout xmlns:android = "http://schemas.android.com/apk/res/android"
3        android:orientation = "vertical"
4        android:layout_width = "fill_parent"
5        android:layout_height = "fill_parent" >
6     <TextView android:layout_width = "250dp"
7        android:layout_height = "250dp"
8        android:background = "#FFFFFF"
9        android:layout_gravity = "center"    />
10    <TextView android:layout_width = "200dp"
11       android:layout_height = "200dp"
12       android:background = "#C0C0C0"
13       android:layout_gravity = "center"    />
14    <TextView android:layout_width = "150dp"
15       android:layout_height = "150dp"
16       android:background = "#696969"
17       android:layout_gravity = "center"    />
18    </FrameLayout>
```

3.5 事件处理机制

不管是桌面应用还是手机应用程序，面对最多的就是用户，经常需要处理的就是用户动作，也就是需要为用户的动作提供响应，这种为用户动作提供响应的机制就是事件处理机制。

3.5.1 事件处理模型

Android 系统中的事件处理机制有两种：第一种是基于回调机制的事件处理，另一种是基于监听接口机制的事件处理。

基于回调机制的事件处理方式是通过重写 View 提供的处理事件回调函数来实现相应事件的响应。这种事件处理模型比较简单，灵活性较差，大多数的界面事件的响应方式采取这种处理模型。常见的回调事件处理方法有 onKeyDown( )、onKeyUp( )、onTouchEvent( )、onTrackBallEvent( )和 onFocusChanged( )。

基于监听接口机制的事件处理方式是一种面向对象的事件处理方式，其中主要涉及三个对象，分别是事件源、事件和事件监听器。

事件源（EventSource）是事件发生的场所，通常就是各个组件，例如窗口、按钮、菜单等。

事件（Event）封装了界面组件上发生的特定事情，通常是一次用户操作，如果程序需要获得界面组件上所发生事件的相关信息，一般通过 Event 对象来取得。

事件监听器（EventListener）负责监听事件源所发生的事件，并对各种事件做出相应的响应。

事件源、事件和事件监听器的关系如图 3-38 所示。首先需要为事件源添加监听，添加成功后，事件触发时系统才能判断事件的目的地，派发处理相应事件的对象。其次，外部动作触发事件源上的事件，事件源判断、生成并封装事件对象后传递给事件监听器。最后，在事件监听器接收到事件对象之后，系统会调用监听器中相应的事件处理方法来处理事件并给出响应。

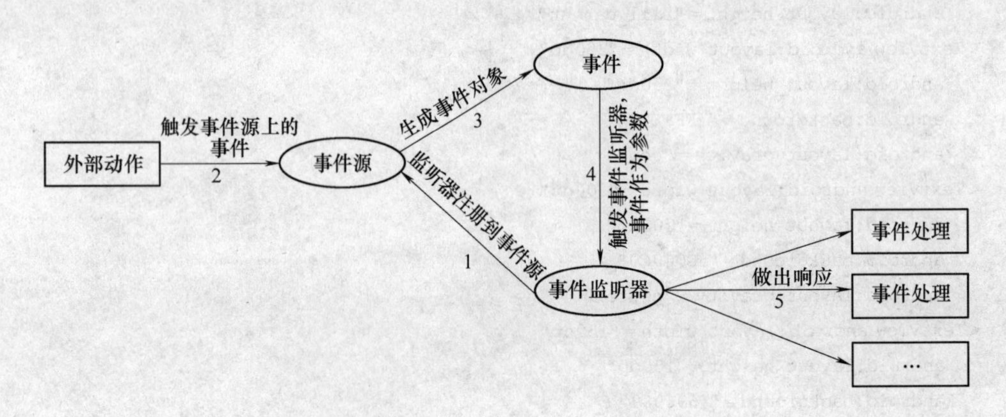

图 3-38　事件源、事件和事件监听器的关系图

基于监听接口机制的常见的监听器有 OnClickListener、OnLongClickListener、OnFocus-ChangeListener、OnKeyListener、OnTouchListener 和 OnCreateContextMenuListener 等。

3.5.2 事件处理函数

1. onKeyDown

onKeyDown 方法是接口 KeyEvent. Callback 中的抽象方法，所有的 View 控件全部实现了该接口并重写了该方法，该方法用来捕捉手机键盘被按下的事件。该方法的语法格式为：

```
public boolean onKeyDown (int keyCode, KeyEvent event)
```

参数 keyCode 为被按下的键值，即键盘码，手机键盘中每个按钮都会有其单独的键盘码，在应用程序中，都是通过键盘码来判断用户按下的是哪个键。

参数 event 为按键事件对象，其中包含了触发事件的详细信息，例如事件的状态、事件的类型、事件发生的时间等。当用户按下按键时，系统会自动将事件封装成 KeyEvent 对象供应用程序使用。

该方法的返回值为一个 boolean 类型的变量，当返回 true 时，表示已经完整地处理了这个事件，并不希望其他的回调方法再次进行处理；当返回 false 时，表示并没有完全处理完该事件，希望其他回调方法继续对其进行处理，例如 Activity 中的回调方法。

2. onKeyUp

onKeyUp 方法同样是接口 KeyEvent. Callback 中的一个抽象方法，并且所有的 View 控件同样全部实现了该接口并重写了该方法，onKeyUp 方法用来捕捉手机键盘按键抬起的事件。该方法的语法格式为：

```
public boolean onKeyUp (int keyCode, KeyEvent event)
```

方法中各参数和返回值的意义同 onKeyDown。

3. onTouchEvent

onTouchEvent 方法在 View 控件类中定义，并且所有的 View 控件子类全部重写了该方法，应用程序可以通过该方法处理手机屏幕的触摸事件。该方法的语法格式为：

```
public boolean onTouchEvent (MotionEvent event)
```

参数 event 为手机屏幕触摸事件封装的对象，其中封装了该事件的所有信息，例如触摸的位置、触摸的类型以及触摸的时间等。该对象会在用户触摸手机屏幕时被创建。

该方法的返回值与键盘响应事件的相同，同样是当已经完整地处理了该事件且不希望其他回调方法再次处理时返回 true，否则返回 false。

该方法并不只处理一种事件，一般情况下，以下三种情况的事件全部由 onTouchEvent 方法处理，只是三种情况中的动作值不同。

（1）当屏幕被按下时，会自动调用该方法来处理事件，此时 MotionEvent. getAction( )的值为 MotionEvent. ACTION_DOWN，如果在应用程序中需要处理屏幕被按下的事件，只需重写该回调方法，然后在方法中进行动作的判断即可。

（2）当触控笔离开屏幕时触发的事件同样需要 onTouchEvent 方法来捕捉，然后在方法中进行动作判断。当 MotionEvent. getAction( )的值为 MotionEvent. ACTION_UP 时，表示屏幕被抬起的事件。

（3）当在屏幕中拖动时，该方法还负责处理触控笔在屏幕上滑动的事件，同样是调用 Motionivent. getAction( )方法来判断动作值是否为 MotionEvent. ACTION_MOVE 再进行处理。

4. onTrackBallEvent

onTrackBallEvent 方法用于手机中轨迹球的事件处理，所有的 View 控件同样全部实现了该方法，可以在 Activity 中重写该方法，也可以在各个 View 控件的实现类中重写。该方法的语法格式为：

```
public boolean onTrackballEvent (MotionEvent event)
```

参数 event 为手机轨迹球事件封装对象，其中封装了触发事件的详细信息，同样包括事件的类型、触发时间等，一般情况下，该对象会在用户操控轨迹球时被创建。

该方法的返回值与前面介绍的各个回调方法的返回值机制完全相同。另外，在模拟器运行状态下，可以通过 F6 键打开模拟器的轨迹球，然后便可以通过鼠标的移动来模拟轨迹球事件。

5. onFocusChanged

onFocusChanged 方法只能在 View 控件中重写，该方法是焦点改变的回调方法，当某个控件重写了该方法后，当焦点发生变化时，会自动调用该方法来处理焦点改变的事件。该方法的语法格式为：

```
protected void onFocusChanged (boolean gainFocus, int direction, Rect previouslyFocusedRect)
```

参数 gainFocus 表示触发该事件的 View 控件是否获得了焦点，当该控件获得焦点时，gainFocus 等于 true，否则等于 false。

参数 direction 表示焦点移动的方向，用数值表示。

参数 previouslyFocusedRect 表示在触发事件的 View 控件的坐标系中，前一个获得焦点的矩形区域，即表示焦点是从哪里来的，如果不可用则为 null。

焦点描述了按键事件（或者是屏幕事件等）的承受者，每次按键事件都发生在拥有焦点的 View 控件上。在应用程序中，可以对焦点进行控制，例如从一个 View 控件移动另一个 View 控件。表 3-18 列出了常用的与焦点有关的方法。

表 3-18　常用的与焦点有关的方法

方　法	说　明
setFocusable( )	设置 View 是否可以拥有焦点
isFocusable( )	监测此 View 是否可以拥有焦点
setNextFocusDownId( )	设置 View 的焦点向下移动后获得焦点 View 的 ID
hasFocus( )	返回 View 的父控件是否获得了焦点
requestFocus( )	尝试让此 View 获得焦点
isFocusableTouchMode( )	设置 View 是否可以在触摸模式下获得焦点，在默认情况下是不可以获得的

6. 接口 OnClickListener

该接口处理的是单击事件，在触控模式下，是在某个 View 控件上按下并抬起的组合动作，而在键盘模式下，是某个 View 控件获得焦点后单击确定键或者按下轨迹球事件。对应的回调方法为 onClick（View v），该接口需要实现 onClick 方法，参数 v 便为事件发生的事件源。

7. 接口 OnLongClickListener

接口 OnLongClickListener 与接口 OnClickListener 原理基本相同，只是该接口为 View 控件长按事件的捕捉接口，即当长时间按下某个 View 控件时触发的事件。对应的回调方法为 onLongClick（View v），参数 v 为事件源控件，当长时间按下此控件时才会触发该方法。该方法的返回值为一个 boolean 类型的变量，当返回 true 时，表示已经完整地处理了这个事件，并不希望其他的回调方法再次进行处理；当返回 false 时，表示并没有完全处理完该事件，希望其他方法继续对其进行处理。

8. 接口 OnFocusChangeListener

接口 OnFocusChangeListener 用来处理控件焦点发生改变的事件，如果注册了该接口，当某个控件失去焦点或者获得焦点时都会触发该接口中的回调方法 onFocusChange（View v，Boolean hasFocus），参数 v 便为触发该事件的事件源，参数 hasFocus 表示 v 的新状态，即 v 是否获得焦点。

9. 接口 OnKeyListener

接口 OnKeyListener 是对手机键盘进行监听的接口，通过对某个 View 控件注册该监听，当 View 控件获得焦点并有键盘事件时，便会触发该接口中的回调方法 onKey（View v，int keyCode，KeyEvent event），参数 v 为事件的事件源控件，参数 keyCode 为手机键盘的键盘码，参数 event 为键盘事件封装类的对象，其中包含了事件的详细信息，例如发生的事件、事件的类型等。

10. 接口 OnTouchListener

接口 OnTouchListener 是用来处理手机屏幕事件的监听接口，当为 View 控件的范围内触摸按下、抬起或滑动等动作时都会触发该事件。该接口对应的回调方法为 onTouch（View v，MotionEvent event），参数 v 同样为事件源对象，参数 event 为事件封装类的对象，其中封装了触发事件的详细信息，同样包括事件的类型、触发时间等信息。

11. 接口 OnCreateContextMenuListener

接口 OnCreateContextMenuListener 是用来处理上下文菜单显示事件的监听接口，该方法是定义和注册上下文菜单的另一种方式，对应的回调方法为 onCreateContextMenu（ContextMenu menu，View v，ContextMenuInfo info），参数 menu 为事件的上下文菜单，参数 v 为事件源 View，当该 View 获得焦点时才可能接收该方法的事件响应。参数 info 中封装了有关上下文菜单额外的信息，这些信息取决于事件源 View。该方法会在某个 View 中显示上下文菜单时被调用，可以通过实现该方法来处理上下文菜单显示时的一些操作。

本章小结

本章主要从 Android 用户界面开发出发，讲述了开发过程中经常使用到的控件，包括菜单、常用基础控件、对话框与消息框。界面中控件的结构及位置等需要通过有效的界面布局控制，Android 中提供了五种界面布局格式，即线性布局、相对布局、表格布局、绝对布局和框架布局。界面中还有一种必要的操作处理——外部操作的响应，通过有效的事件机制完成。

习　题

3-1　Android 系统支持哪几种菜单类型？分别通过什么方式实现？

3-2　Android 中常用的控件有哪些？

3-3　Android 中有几种布局方式？每一种方式中常用的 XML 属性及相应的设置方法是什么？

3-4　设计一个简易的乘法器，分别输入两个乘数，单击计算按钮得到计算结果，并且单击 menu 时弹出菜单显示退出或者是清屏。

第4章 Android数据存储与交互

在开发应用程序时，数据交互是贯穿整个开发过程的主线，也是应用程序本身使用的主线，因此程序中数据交互问题成为开发与设计人员面临的最基本问题。无论是底层驱动应用的开发还是桌面应用的开发，更甚至到大型商用软件的开发与设计，均涉及数据交互问题。任何应用程序都必须解决这一问题，即数据必须以某种合理的方式保存，不能丢失并且能够有效简便地使用和进行更新处理。通常情况下，应用于桌面的操作系统一般会提供一种公共文件系统，系统中的应用程序可以使用这个文件系统来存储和读取文件，该文件也可以被其他具有权限的应用程序读取。但是，不同于一般的桌面操作系统，Android 系统采用了一种不同的机制，所有的应用程序的数据和文件均为本应用程序所私有，但是它同时也提供了一种以标准方式供应用程序将私有数据开放共享给其他应用程序的机制。

Android 系统提供的数据存储方式主要有五种：

（1）共享优先数据存储——SharedPreferences

（2）数据库存储——SQLite Database

（3）文件存储——Files

（4）网络存储——Networks

（5）内容提供器——Content Providers

本章将详细讲述这五种数据存储方式，并通过实例实现各种存储方式。

4.1 共享优先数据存储

SharedPreferences 是一种简单的、轻量级的用于保存应用程序基本数据的类，该类通过用键值对（Name-Value Pair）的方式把简单数据类型（boolean、int、float、long 和 string）存储在应用程序的私有目录下（data/data/包名/shared_prefs/）自定义的 XML 文件中，即数据存储为 XML 文件格式。使用 SharedPreferences 进行数据存储有一个很好的优点就是它完全屏蔽了对文件系统的操作过程。

使用 SharedPreferences 进行数据的存储，首先需要获取一个 SharedPreferences 对象，获取该对象使用方法 getSharedPreferences( )，此方法是 Context 类提供的公共方法。

getSharedPreferences( )语法格式：

```
SharedPreferences  getSharedPreferences(String name, int mode)
```

参数 name 定义 SharedPreferences 的名称，这个名称与在 Android 文件系统中保存的文件同名，只要具有相同的 SharedPreferences 名称的键值对内容，都会保存在同一个文件中。

参数 mode 定义访问模式，SharedPreferences 提供了三种支持的基本访问模式，分别为 MODE_PRIVATE、MODE_WORLD_READABLE 和 MODE_WORLD_WRITEABLE。

MODE_PRIVATE 值为 0（0x00000000），称为私有模式，仅有创建程序才有权限对其进行读取或写入。

MODE_WORLD_READABLE 值为 1（0x00000001），称为全局读模式，不仅创建程序可以对其进行读取或写入，其他应用程序也具有读取操作的权限，但没有写入操作的权限。

MODE_WORLD_WRITEABLE 值为 2（0x00000002），称为全局写模式，创建程序和其他程序都可以对其进行写入操作，但没有读取的权限。

SharedPreferences 除了三种基本的访问模式，还支持全局读模式和全局写模式的叠加，即读写模式：

```
int nMode = MODE_WORLD_READABLE + MODE_WORLD_WRITEABLE;
```

通过使用 getSharedPreferences( ) 方法，可以获取进行数据存储的一个 SharedPreferences 对象，接下来就可以对数据进行修改。数据的修改是通过 SharedPreferences. Editor 类完成的，例如数据的清空、删除和添加等，修改完成后需要调用 commit( ) 函数保存修改内容。表 4-1 列出了 SharedPreferences. Editor 类常用的方法。

表 4-1　SharedPreferences. Editor 类常用的方法

方　法	说　明
clear( )	清除所有值
commit( )	保存
putBoolean(String key, boolean value)	保存一个 boolean 值
putFloat(String key, float value)	保存一个 float 值
putInt(String key, int value)	保存一个 int 值
putLong(String key, long value)	保存一个 long 值
putString(String key, long value)	保存一个 long 值
remove(String key)	删除该键对应的值
apply( )	保存（无返回值）

使用 SharedPreferences 读取已经保存好的数据，在 getSharedPreferences( ) 获取到 SharedPreferences 对象后，使用 SharedPreferences 类中定义的 getType( ) 方法读取相应类型的键值对。SharedPreferences 类定义的 getType( ) 方法以及其他常用方法可参考表 4-2。

表 4-2　SharedPreferences 类常用的方法

方　法	说　明
contains(String key)	判断是否包含相应的键值
edit( )	返回 SharedPreferences 的 Editor 接口
getAll( )	返回所有配置信息 Map
getBoolean(String key, boolean defValue)	获取一个 boolean 键值
getFloat(String key, float defValue)	获取一个 float 键值
getInt(String key, int defValue)	获取一个 int 键值
getLong(String key, long defValue)	获取一个 long 键值
getString(String key, String defValue)	获取一个 String 键值
registerOnSharedPreferenceChangeListener (SharedPreferences. OnSharedPreferenceChangeListener listener)	注册键值改变监听器
unregisterOnSharedPreferenceChangeListener (SharedPreferences. OnSharedPreferenceChangeListener listener)	注销键值改变监听器

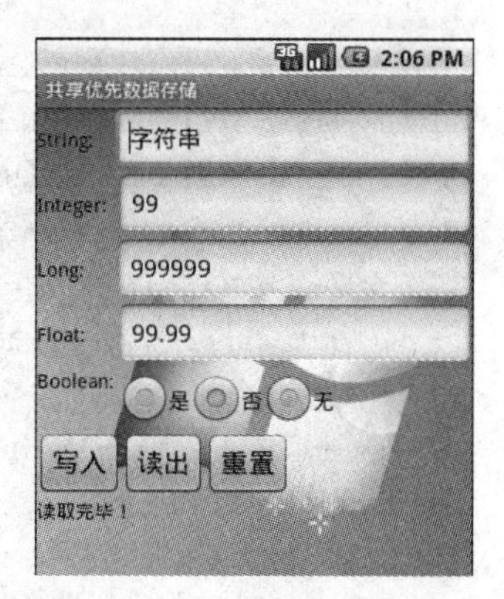

图 4-1　Chp04_SharedPreferences 用户界面

下面使用 SharedPreferences 进行数据存储，代码工程为 Chp04_SharedPreferences，其运行后的用户界面如图 4-1 所示。工程完成三个基本功能，即数据写入、数据读出和界面重置。使用 SharedPreferences 进行保存的数据有五种基本类型的数据（String、int、long、float 和 boolean），数据保存的 XML 文件即 SharedPreferences 文件，命名为“SharedFileName”。

SharedPreferences 文件保存在/data/data/com. SharedPreferences/shared _ prefs 目录下，即为 SharedFileName. xml。其中 com. SharedPreferences 为工程 Chp04_SharedPreferences 的包名称，可以在 DDMS 中使用 File Explore 查看到，如图 4-2 所示。在 File Explore 中可以看到 SharedFileName. xml 文件的四个基本属性：文件大小（Size）、最近更新日期（Date）和时间（Time）、文件权限（Permissions）以及附加信息（Info）。其中文件权限为“-rw-rw-rw-”，在 Linux 系统中，文件权限分别描述了创建者、同组用户和其他用户对文件的操作限制。x 表示可执行，r 表示可读，w 表示可写，d 表示目录，-表示普通文件。因此，“-rw-rw-rw”表示 SharedFileName. xml 可以被创建者、同组用户和其他用户进行读取和写入操作，但不可执行。“-rw-rw-rw”权限来自 SharedPreferences 访问模式设置为 MODE_WORLD_READABLE + MODE_WORLD_WRITEABLE 的结果。如果设置为 MODE_PRIVATE 则文件权限将成为“-rw-rw ---”，表示仅有创建者和同组用户具有读写文件的权限。

Name	Size	Date	Time	Permissions	Info
data		2011-12-31	16:30	drwxrwx--x	
anr		2012-01-28	15:58	drwxrwx--x	
app		2012-01-31	14:15	drwxrwx--x	
app-private		2011-12-31	16:28	drwxrwx--x	
backup		2011-12-31	16:30	drwx------	
dalvik-cache		2012-01-31	14:15	drwxrwx--x	
data		2012-01-31	13:09	drwxrwx--x	
android.tts		2011-12-31	16:30	drwxr-xr-x	
com.BasicControl		2012-01-29	03:13	drwxr-xr-x	
com.SharedPreferences		2012-01-31	13:20	drwxr-xr-x	
lib		2012-01-31	13:09	drwxr-xr-x	
shared_prefs		2012-01-31	13:25	drwxrwx--x	
SharedFileName.xml	274	2012-01-31	13:25	-rw-rw-rw-	
com.android.alarmclock		2011-12-31	16:30	drwxr-xr-x	

图 4-2　File Explore 中查看 SharedFileName. xml

Chp04_SharedPreferences 工程中，界面功能的代码如下所示。

```
1     public class MainAct extends Activity {
2         public String m_strSharedFileName = "SharedFileName";
3         public static int m_nMode = Context.MODE_WORLD_READABLE
4                                            + Context.MODE_WORLD_WRITEABLE;
```

```
5      public String m_strStingName = "StingName";
6      public String m_nIntName = "IntName";
7      public String m_lLongName = "LongName";
8      public String m_fFloatName = "FloatName";
9      public String m_bBooleanName = "BooleanName";
10     /** Called when the activity is first created. */
11     @Override
12     public void onCreate(Bundle savedInstanceState) {
13         super.onCreate(savedInstanceState);
14         setContentView(R.layout.main);
15         final EditText etxtStrValue = (EditText) findViewById(R.id.etxtStrMainAct);
16         final EditText etxtIntegerValue = (EditText) findViewById(R.id.etxtIntMainAct);
17         final EditText etxtLongValue = (EditText) findViewById(R.id.etxtLongMainAct);
18         final EditText etxtFloatValue = (EditText) findViewById(R.id.etxtFloatMainAct);
19         final RadioButton radTure = (RadioButton ) findViewById(R.id.rbtnYes);
20         final RadioButton radFalse = (RadioButton ) findViewById(R.id.rbtnNo);
21         final RadioButton radNull = (RadioButton ) findViewById(R.id.rbtnNull);
22         final Button btnWrite = (Button) findViewById(R.id.btnWrite);
23         final Button btnClear = (Button) findViewById(R.id.btnClear);
24         final Button btnRead = (Button) findViewById(R.id.btnRead);
25         final TextView txtResultMainAct = (TextView) findViewById(R.id.txtResultMainAct);
26         //写入
27         btnWrite.setOnClickListener(new Button.OnClickListener() {
28             public void onClick(View v) {
29             SharedPreferences clsSharedPreferences =getSharedPreferences(m_strSharedFileName,
30                                                     m_nMode);
31             int nIntValue = Integer.parseInt(etxtIntegerValue.getText().toString());
32             long lLongValue = Long.parseLong(etxtLongValue.getText().toString());
33             float fFloatValue = Float.parseFloat(etxtFloatValue.getText().toString());
34             boolean bBooleanValue = true;
35             if (radTure.isChecked())
36                 {bBooleanValue = true;}
37             else if (radFalse.isChecked())
38                 {bBooleanValue = false; }
39             clsSharedPreferences.edit()
40                 .putString(m_strStingName, etxtStrValue.getText().toString())
41                 .putInt(m_nIntName, nIntValue)
42                 .putLong(m_lLongName, lLongValue)
43                 .putFloat(m_fFloatName, fFloatValue)
44                 .putBoolean(m_bBooleanName, bBooleanValue)
45                 .commit();//一定要提交才生效
46                 txtResultMainAct.setText("保存成功!");
47                 }
```

```
48          });
49          //读出
50          btnRead.setOnClickListener(new Button.OnClickListener() {
51              public void onClick(View v){
52              SharedPreferences clsSharedPreferences =getSharedPreferences(m_strSharedFileName,
53                                                      m_nMode);
54                  String strStringValue =clsSharedPreferences.getString(m_strStingName, "");
55                  etxtStrValue.setText(strStringValue);
56                  int nIntValue =clsSharedPreferences.getInt(m_nIntName, 0);
57                  etxtIntegerValue.setText(String.valueOf(nIntValue));
58                  long lLongValue =clsSharedPreferences.getLong(m_lLongName, 0);
59                  etxtLongValue.setText(String.valueOf(lLongValue));
60                  float fFloatValue =clsSharedPreferences.getFloat(m_fFloatName, 0);
61                  etxtFloatValue.setText(String.valueOf(fFloatValue));
62                  boolean bBlooleanValue =clsSharedPreferences.getBoolean(m_bBooleanName, true);
63                  if (bBlooleanValue)
64                  {radTure.setChecked(true);
65                     radFalse.setChecked(false);
66                     radNull.setChecked(false); }
67                  else { radTure.setChecked(false);
68                     radFalse.setChecked(true);
69                     radNull.setChecked(false); }
70                  txtResultMainAct.setText("读取完毕!");
71                  }
72                  });
73              //重置
74              btnClear.setOnClickListener(new Button.OnClickListener() {
75              public void onClick(View v){
76                  etxtStrValue.setText("");
77                  etxtIntegerValue.setText("");
78                  etxtLongValue.setText("");
79                  etxtFloatValue.setText("");
80                  radTure.setChecked(false);
81                  radFalse.setChecked(false);
82                  radNull.setChecked(true);
83                  txtResultMainAct.setText("重置完毕!");
84                  }
85              });
86      }
87  }
```

代码中设置 SharedPreferences 的访问方式为“读+写”模式，需要保存或者修改的键值对为五类基本类型数据，其键值名称在代码 5~9 行声明。在 Activity 的 onCreate()方法中将数据写入操作、读取操作和界面重置操作绑定到三个相应的按钮上。26~48 行进行写入操作，使用 SharedPreferences. edit( )类的 putType( )方法，最后需要调用 commit( )保存，否则写入操作将无效。49 ~ 72 行完成键值对读取，仅使用 SharedPreferences. getType( )方法即可完成。74 ~ 85 行的重置操作仅仅是界面控件文本的清除，并未进行数据的本质交互，其实这个操作是很有必要的，很多情况下需要重置界面数据来完成错误输入后的一键清理和界面数据的保密。另外每个操作都设置了一个文本框用于及时呈现操作处理结果给用户，这对界面及操作的友好性显得非常重要。

在使用 SharedPreferences 进行数据存储和访问时，还存在一种常见的情形，那就是一个应用程序可否使用另一个应用程序的 SharedPreferences 文件，即 SharedPreferences 文件可否共享。答案是肯定的，例如应用程序 Chp04_SharedPreferencesOther（包名为 com. SharedPreferencesOther）读取或存储数据到 Chp04_SharedPreferences（包名为 com. SharedPreferences）中的 SharedPreferences 文件 SharedFileName. xml。这种情形需要使用 createPackageContext( )方法获取创建 SharedPreferences 文件的 Context 对象。

createPackageContext( )方法的语法为：

```
Context createPackageContext (String packageName, int flags)
```

参数 packageName 是目的包名称，也就是要得到Context 的包名，例如“com. SharedPreferences”；参数 flags 为标志位，可以是 CONTEXT_INCLUDE_CODE 和 CONTEXT_IGNORE_SECURITY 中的一个，CONTEXT_INCLUDE_CODE 代码包含标志，标志可以执行包里面的代码；CONTEXT_IGNORE_SECURIT 表示忽略安全警告，如果不加这个标志的话，有些功能是用不了的，会出现安全警告。createPackageContext( )方法在找不到包名的时候会报 NameNotFoundException 异常，在使用时需要捕获此异常。

例如，在其他应用程序中使用 Chp04_SharedPreferences（包名为 com. SharedPreferences）中的 SharedPreferences 文件 SharedFileName. xml，只需在获取 SharedPreferences 对象时，使用 createPackageContext( )方法返回的 Context. getSharedPreferences 方法进行获取即可。

显然，访问其他应用程序的 SharedPreferences 文件首先在创建 SharedPreferences 文件时需要将 SharedPreferences 的访问模式设置为全局读或全局写，其次需要明确共享 SharedPreferences 文件的包名称和 SharedPreferences 的名称，以通过 Context 获得 SharedPreferences 对象，另外还需要确切知道键值对的名称和键值类型，用以正确读取数据。下面代码即为 Chp04_SharedPreferencesOther 应用程序使用 Chp04_SharedPreferences 应用程序 SharedPreferences 文件时替换 getSharedPreferences( )方法的代码。

```
1     Context ctxt = null;
2     try {
3         ctxt = createPackageContext("com. SharedPreferences", CONTEXT_IGNORE_SECURITY);
4     }
5     catch (NameNotFoundException e) {
6             // TODO Auto-generated catch block
7             e. printStackTrace();
8         }
9         SharedPreferences clsSharedPreferences = ctxt. getSharedPreferences("SharedFileName",
10                m_nMode);
```

4.2 数据库存储

4.2.1 嵌入式数据库

随着数据存储的快速发展，数据库应用的范围更加深入和具体，那些仅适用于 PC、体积庞大、延时较长的数据库技术已不能满足针对性较强的嵌入式系统开发的需求，而且随着嵌入式系统的内存和各种永久存储介质容量的不断增加，嵌入式系统内数据处理量会不断增加，如何处理大量的数据对数据库的发展提出了更高的要求，嵌入式数据库系统应运而生。

嵌入式数据库系统是指应用于嵌入式系统的数据库系统，亦称为嵌入式实时数据库系统。嵌入式实时数据库系统以目前成熟的数据库技术为基础，针对嵌入式设备的具体特点，实现对移动设备和嵌入式设备上的数据存储、组织和管理。嵌入式数据库的名称来自其独特的运行模式，这种数据库嵌入到了应用程序进程中，消除了与客户机服务器配置相关的开销。嵌入式数据库实际上是轻量级的，在运行时，它们需要较少的内存。它们是使用精简代码编写的，对于嵌入式设备，其速度更快，效果更理想。嵌入式的运行模式允许嵌入式数据库通过 SQL 来轻松管理应用程序数据，而不依靠原始的文本文件，另外，嵌入式数据库还提供零配置运行模式。

嵌入式数据库系统具有以下特征：

（1）体积小；

（2）具有可靠性；

（3）具有可定制性；

（4）支持 SQL 查询语言；

（5）提供接口函数；

（6）具有实时性；

（7）有一定的底层控制能力；

（8）标准化发展。

嵌入式数据库系统在智能家电、无线通信、金融领域和导航定位系统领域有着广泛的应用。目前技术比较成熟的嵌入式数据库系统产品有很多，例如 Berkeley DB、Empress（商业数据库）、eXtremeDB、mSQL、Firebird 嵌入服务器和 SQLite 等。

Berkeley DB 是一个开放源代码的内嵌式数据库管理系统，能够为应用程序提供高性能的数据管理服务。使用时只需要调用一些简单的 API 就可以完成对数据的访问和管理，不使用 SQL 语言。Berkeley DB 为许多编程语言提供了实用的 API，包括 C、C ++、Java、Perl、Tcl、Python 和 PHP 等，所有同数据库相关的操作都由 Berkeley DB 函数库负责统一完成。Berkeley DB 具有很好的兼容性，可以运行于几乎所有的 UNIX 和 Linux 系统及其变种系统、Windows 操作系统以及多种嵌入式实时操作系统中。Berkeley DB 可扩展性同样很强，Database library 本身是很精简的，少于 300KB 的文本空间，但它能够管理规模高达 256TB 的数据库。它支持高并发度，成千上万个用户可同时操纵同一个数据库，Berkeley DB 能以足够小的空间占用量运行于有严格约束的嵌入式系统。Berkeley DB 在嵌入式应用中比关系数据库和面向对象数据库要好，有以下两点原因：

（1）数据库、程序库和应用程序在相同的地址空间中运行，所以数据库操作不需要进程间的通信。在一台机器的不同进程间或在网络中不同机器间进行进程通信所花费的开销，要远远大于函数调用的开销。

（2）Berkeley DB 对所有操作都使用一组 API，因此不需要对某种查询语言进行解析，也不用生成执行计划，大大提高了运行效率。

Empress 商业数据库具有微型内核结构，Empress 高度单元化，可根据需要选择需要的单元，从而缩小产品中 Empress 数据库所占用的资源，特别适合紧凑性的设计。Empress 提供了内核级的 CAPI 称为 MR，使运行速度最大化，用 MR 编写的应用程序在执行时不需要解析，在 MR 中还包括优秀的加锁控制、内存管理和基于记录数量的选择功能。Empress 可嵌入程序，该特性使应用程序和数据库工作于统一地址空间，增强了系统的稳定性，提高了系统的效率。另外 Empress 具有确定的响应时间，可以使数据的响应时间相对一致，使用者可以设定一个超时限制，如果在规定时间内没有完成插入修改等操作，系统会报错。Empress 支持多种硬件平台和软件平台，支持 SCSI、RAID、IDE、RAM、CD-RW、DVD-ROM 和 CF 等存储介质。Empress 拥有高度灵活的 SQL 接口，另外还支持 Unicode 码。

eXtremeDB 内存嵌入式实时数据库以其高性能、低开销、稳定可靠的极速实时数据管理能力在嵌入式数据管理领域及服务器实时数据管理领域占据一席之地。首先，它是一种内存数据库，eXtremeDB 将数据以程序直接使用的格式保存在主内存之中，不仅剔除了文件 I/O 的开销，也剔除了文件系统数据库所需的缓冲和 Cache 机制。其结果是每个交易只需一微秒甚至更短的极限速度，相比于类磁盘数据库而言，速度成百上千倍地提高。作为内存数据库，eXtremeDB 不仅性能高，而且数据存储的效率也非常高。为了提高性能并方便程序使用，数据在 eXtremeDB 中不做任何压缩，100MB 的空间可以保存高达 70MB 以上的有效数据，这是其他数据库所不可想象的。其次，它是一种混合数据库，eXtremeDB 不仅可以建立完全运行在主内存的内存数据库，更可以建立磁盘/内存混合介质的数据库。在 eXtremeDB，把这种建立在磁盘、内存或磁盘 + 内存的运行模式称为 eXtremeDB Fusion 融合数据库。eXtremeDB Fusion 兼顾数据管理的实时性与安全性要求，是实时数据管理的跨越式进步。第三，它是嵌入式数据库，eXtremeDB 内核以链接库的形式包含在应用程序之中，其开销只有 50 ~ 130KB。无论在嵌入式系统还是在实时系统之中，eXtrcmcDB 都天然地嵌入在应用程序之中，在最终用户毫不知情的情况下工作。eXtremeDB 的这种天然嵌入性对实时数据管理至关重要——各个进程都直接访问 eXtremeDB 数据库，避免了进程间通信，从而剔除了进程间通信的开销和不确定性。同时，eXtremeDB 独特的数据格式方便程序直接使用，剔除了数据复制及数据翻译的开销，缩短了应用程序的代码执行路径。第四，它具有应用定制的 API，应用程序对 eXtremeDB 数据库的操作接口是根据应用数据库设计而自动产生，不仅提升了性能，也剔除了通用接口所必不可少的动态内存分配，从而提高了应用系统的可靠性。定制过程简单方便，由高级语言定制 eXtremeDB 数据库中的表格、字段、数据类型、事件触发、访问方法等应用特征，通过 eXtremeDB 预编译器自动产生访问该数据库的 C/C ++ API。第五，它具有可预测的数据管理，eXtremeDB 独特的体系结构，保证了数据管理的可预测性。eXtremeDB 不仅更快、更小，而且更确定。在 80 系列双核 CPU 服务器上，eXtremeDB 在 1TB 内存里保存 15B 条记录；无论记录数多少，eXtremeDB 可以在八十分之一微秒的时间内提取一条记录。

mSQL（mini SQL）是一个单用户数据库管理系统，由于它的短小精悍，使其开发的应用系统特别受到互联网用户青睐。mSQL 是一种小型的关系数据库，性能不是太好，对 SQL 语言的支持也不够完全，但在一些网络数据库应用中是足够了。由于 mSQL 较简单，在运行简单的 SQL 语句时速度比 MySQL 略快，而 MySQL 在线程和索引上下了功夫，运行复杂的 SQL 语句时比 mSQL、PostgreSQL 等都要快一些。mSQL 的技术特点主要体现在安全性方面，mSQL 通过 ACL 文件设定各主机上各用户的访问权限，默认是全部可读/写。mSQL 缺乏 ANSI-SQL 的大多数特征，它仅仅实现了一个最少的 API，没有事务和参考完整性。mSQL 与 Lite（一种类似 C 的脚本语言）紧密结合，可以得到一个称为 W3-mSQL 的一个网站集成包，它是 JDBC、ODBC、Perl 和 PHP 的 API。

Firebird 嵌入服务器版本衍生自 Interbase，虽然它的体积比 Interbase 缩小了几十倍，但功能并无阉割。为了体现 Firebird 短小精悍的特色，在增加了超级服务器版本之后，又增加了嵌入版本。Firebird 嵌入服务器版本数据库文件与 Firebird 网络版本完全兼容，差别仅在于连接方式不同，可以实现零成本迁移。数据库文件仅受操作系统的限制，且支持将一个数据库分割成不同文件，突破了操作系统最大文件的限制，提高了 I/O 吞吐量。它完全支持 SQL92 标准，支持大部分 SQL-99 标准功能并且具有丰富的开发工具支持，绝大部分基于 Interbase 的组件，可以直接使用于 Firebird。Firebird 嵌入服务器版本支持事务、存储过程、触发器等关系数据库的所有特性并且可以自己编写扩展函数（UDF）。

SQLite 是一款轻型的数据库，是遵守 ACID 的关联式数据库管理系统，它的设计目标是嵌入式的，而且目前已经在很多嵌入式产品中得到应用，它占用资源非常少，在嵌入式设备中，可能只需要几百 K 的内存就够了。它能够支持 Windows/Linux/UNIX 等主流的操作系统，同时能够跟很多程序语言相结合，比如 Tcl、C#、PHP、Java 等，还有 ODBC 接口，同样比起 MySQL、PostgreSQL 这两款开源世界著名的数据库管理系统来讲，它的处理速度比它们都快。SQLite 第一个 Alpha 版本诞生于 2000 年 5 月，至今已经有十几个年头，SQLite 也已经发布了一个较新的版本 SQLite 3。

SQLite 具有三级模式的结构体系，即用户模式、逻辑模式和存储模式。相对于传统数据库，SQLite 具有更好的实时性、系统开销小、底层控制能力强，能够高效地利用嵌入式系统的有限资源，提高数据存储速度，增强系统的安全性。另外，SQLite 还具有如下特点：

（1）无需配置，无服务器，访问简单

使用 SQLite 前不需要安装设置，不需要管理员去管理，系统崩溃后可自动恢复。访问数据库的程序直接从磁盘上的数据库文件读写，没有中间的服务器进程。一个 SQLite 数据库是一个单独的普通磁盘文件，能够被定位在路径层次的任何地方。如果 SQLite 能读写磁盘文件则也能访问数据库。大多数 SQL 数据库引擎趋向于把数据存储为一个大的文件集合，通常这些文件在一个标准的定位中，只有数据库引擎本身能访问它。

（2）支持标准 SQL

SQLite 内嵌的 SQL 支持大部分 SQL92，支持视图、触发器和嵌套 SQL，还具有事务处理功能，自动维护事务完整性和原子性等特性，支持实体完整性和参照完整性，充分满足嵌入式应用开发的需求。

（3）具有精简性，支持可变长度的记录

SQLite 非常小，整个 SQLite 库小于 225KB，甚至可以压缩到 170KB。一般的 SQL 数据库引擎在表中为每一个记录分配一个固定的磁盘空间数，SQLite 只使用一个记录中实际存储信息的磁盘空间数，这会使数据库非常小，同时由于在磁盘上移动的信息很少，也使数据库很快。

（4）源代码开放，可靠性较好

SQLite 源代码是用 C 语言编写的，95% 有较好的注释，API 简单易用，并且有着 98% 以上的测试覆盖率。同时，官方还带有 TCL 的编译版本。

SQLite 数据库系统体系结构由四部分组成，即内核（Core）、SQL 编译器（SQL-Complier）后端（Backend）和附件（Accessories），如图 4-3 所示。

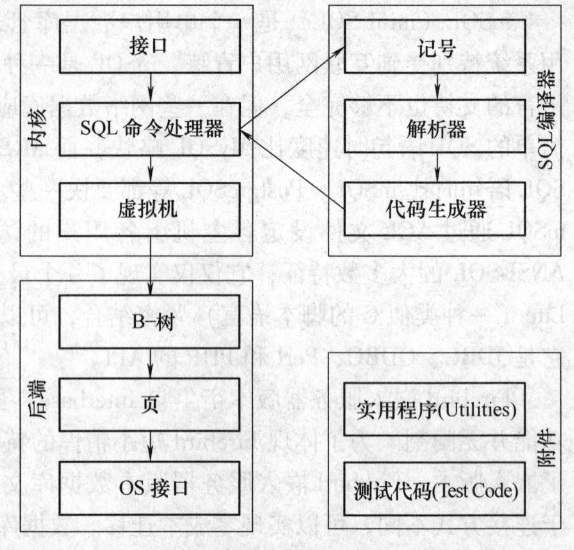

图 4-3　SQLite 数据库系统体系结构

1. 接口程序（Interface）

接口由 SQLite C-API 组成，不管是程序、脚本语言还是库文件，最终都是通过接口程序与 SQLite 交互的。SQLite 类库大部分的公共接口程序是由 main. c、legacy. c 和 vdbeapi. c 源文件中的功能执行的。但有些程序是分散在其他文件夹的，因为在其他文件夹里它们可以访问有文件作用域的数据结构。为了避免和其他软件在名字上有冲突，SQLite 类库中所有的外部符号都是以 sqlite3 为前缀来命名的，这些被用来做外部使用的符号是以 sqlite3_来命名的，这些符号用来形成 SQLite 的 API。

2. 分词器（Tokenizer）

当执行一个包含 SQL 语句的字符串时，接口程序要把这个字符串传递给 Tokenizer。Tokenizer 的任务是把原有字符串分成一个个标识符，并把这些标识符传递给语法分析器。Tokenizer 是在 C 文件夹 tokenize. c 中用手编译的。

3. 语法分析器（Parser）

语法分析器分析通过分词器产生的标识符语法的结构，并且得到一棵语法树。语法分析器同时也包含了重构语法树的优化器，因此能够得到一棵产生一个高效的字节编码程序的语法树。

4. 代码生成器（Code Generator）

代码生成器遍历语法树，并且生成一个等价的字节编码程序代码发生器。在语法分析器收集完符号并把之转换成完全的 SQL 语句时，它调用代码产生器来产生虚拟的机器代码，这些机器代码将按照 SQL 语句的要求来工作。

5. 虚拟机（Virtual Machine）

虚拟机（Virtual Machine）模块是一个内部字节编码语言的解释器，它通过执行字节编码语句来实现 SQL 语句的工作，它是数据库中数据的最终操作者，它把数据库看成表和索引的集合，而表和索引则是一系列的元组或者记录。

6. B/B + 树（B-Tree）

B/B + 树模块把每一个元组集组织进一个依次排好序的树状数据结构中，表和索引被分别置于单独的 B + 和 B 树中。该模块帮助虚拟机进行搜索，插入和删除树中的元组。它也帮助虚拟机创建新的树和删除旧的树。SQLite 数据库在磁盘里维护，使用源文件 btree. c 中的 B-Tree 执行。数据库中的每个表格和目录使用一个单独的 B-Tree。所有的 B-Tree 被存储在同样的磁盘文件里，文件格式的细节被记录在 btree. c 开头的备注里，B-Tree 子系统的接口程序被标题文件 btree. h 所定义。

7. 页面调度程序（Pager）

页面调度程序模块在原始文件的上层实现了一个面向页面的数据库文件抽象，它管理 B/B + 树使用的内存内缓存（数据库页的），另外也管理文件的锁定，并用日志来实现事务的 ACID 属性。B-Tree 模块要求信息来源于磁盘上固定规模的程序块，默认程序块的大小是 1024B，但是可以在 512 ~ 65536B 间变化，页面调度程序负责读、写和高速缓存这些程序块。页面调度程序还提供重新运算和提交抽象命令，它还管理关闭数据库文件夹。B-Tree 驱动器要求页面高速缓存器中的特别的页，当它想修改页或重新运行改变的时候，它会通报页面调度程序。为了保证所有的需求被快速、安全和有效地处理，页面调度程序处理所有的微小的细节。运行页面高速缓存的代码在专门的 C 源文件 pager. c 中。页面高速缓存的子系统的接口程序被目标文件 pager. h 所定义。

8. 操作系统接口（OS Interface）

操作系统接口模块提供了对应于不同本地操作系统的统一的交界面，为了在 POSIX 和 Win32 之间提供一些可移植性，SQLite 操作系统的接口程序使用一个提取层。OS 提取层的接口程序被定义在 os. h. 每个支持的操作系统有它自己的执行文件：UNIX 使用 os_unix. c，Windows 使用 os_win. c。每个具体的操作器具有它自己的标题文件：os_unix. h，os_win. h 等。

9. 工具（Utilities）

工具模块中包含各种各样的实用功能，还有一些如内存分配、字符串比较、Unicode 转换之类的公共服务也在工具模块中。这个模块就是一个包罗万象的工具箱，很多其他模块都需要调用和共享它。

10. 测试代码（Test Code）

测试模块中包含了无数的回归测试语句，用来检查数据库代码的每个细微角落。这个模块是 SQLite 性能如此可靠的原因之一。

4.2.2 Android SQLite 数据库

Android 作为目前主流的移动操作系统，完全符合 SQLite 占用资源少的优势，在 Android 平台上，集成了嵌入式关系型数据库 SQLite。Android 开发中使用 SQLite 数据库系统进行数据交互时，SQLite 数据库的建立和基本操作通过两种方式实现，一种是使用 SQLite 命令，一种是使用库类，下面分别就两种方法进行阐述。

sqlite3 是 SQLite 数据库自带的一个基于命令行的 SQL 命令执行工具，并可以显示命令的执行结果，sqlite3 工具被集成在 Android 系统中。首先需要在 cmd 中使用命令进入 Linux 命令行界面，其命令为：

```
adb shell
```

执行完指令后，新的一行出现一个“#”说明启动成功，此时就可以启动 sqlite3 工具，命令为：

```
sqlite3
```

执行后显示版本等信息，提示符也变为“sqlite >”，这时就可以进行数据库的建立及其他操纵，在进行数据库建立及其他数据库基本操作之前，先来介绍 sqlite3 提供的常用特殊指令，这些特殊指令均带有点前缀，可以通过使用“. help”命令查看，结果如下面所示，其各特殊指令的功能参考如表 4-3 所示。

```
sqlite > . help
. help
. bail ON |OFF              Stop after hitting an error.  Default OFF
. databases                 List names and files of attached databases
. dump ? TABLE? ...         Dump the database in an SQL text format
. echo ON |OFF                Turn command echo on or off
. exit                      Exit this program
. explain ON |OFF             Turn output mode suitable for EXPLAIN on or off.
. header(s) ON |OFF           Turn display of headers on or off
. help                        Show this message
. import FILE TABLE           Import data from FILE into TABLE
. indices TABLE               Show names of all indices on TABLE
. load FILE ? ENTRY?          Load an extension library
. mode MODE ? TABLE?             Set output mode where MODE is one of:
                                 csv       Comma-separated values
                                 column    Left-aligned columns.  (See . width)
                                 html      HTML < table > code
                                 insert    SQL insert statements for TABLE
                                 line      One value per line
                                 list      Values delimited by . separator string
                                 tabs      Tab-separated values
                                 tcl       TCL list elements
. nullvalue STRING          Print STRING in place of NULL values
. output FILENAME               Send output to FILENAME
. output stdout           Send output to the screen
. prompt MAIN CONTINUE          Replace the standard prompts
. quit                    Exit this program
. read FILENAME           Execute SQL in FILENAME
. schema ? TABLE?           Show the CREATE statements
. separator STRING          Change separator used by output mode and . import
. show                      Show the current values for various settings
. tables ? PATTERN?         List names of tables matching a LIKE pattern
. timeout MS                Try opening locked tables for MS milliseconds
. timer ON |OFF             Turn the CPU timer measurement on or off
. width NUM NUM ...         Set column widths for "column" mode
sqlite >
```

表 4-3 sqlite3 提供的特殊指令

指 令	功 能
. bail ON I OFF	遇到错误时停止，默认为 OFF
. databases	显示数据库名称和文件位置
. dump ? TABLE? ...	将数据库以 SQL 文本形式导出
. echo ON I OFF	开启和关闭回显
. exit	退出
. explain ON I OFF	开启或关闭适当输出模式，如果开启模式将更改为 column，并自动设置宽度
. header（s）ON I OFF	开启或关闭标题显示
. help	显示帮助信息
. import FILE TABLE	将数据从文件导入表中
. indices TABLE	显示表中所有的列名
. load FILE ? ENTRY?	导入扩展库
. mode MODE ? TABLE?	设置输入格式
. nullvalue STRING	打印时使用 STRING 代替 NULL
. output FILENAME	将输入保存到文件
. output stdout	将输入显示在屏幕上
. prompt MAIN CONTINUE	替换标准提示符
. quit	退出
. read FILENAME	在文件中执行 SQL 语句
. schema ? TABLE?	显示表的创建语句
. separator STRING	更改输入和导入的分隔符
. show	显示当前设置变量值
. tables ? PATTERN?	显示符合匹配模式的表名
. timeout MS	尝试打开被锁定的表 MS
. timer ON I OFF	开启或关闭 CPU 计时器
. width NUM NUM ...	设置 column 模式的宽度

下面开始使用 sqlite3 命令行工具，实现数据库的创建、查询、插入、修改和删除等基本操作。为了讲述方便，在这里首先建立工程 Chp04_SQLite，其包名称为 com. SQLite。

在 Android 系统中，每个应用程序的数据库都保存在各自的/data/data/ < package name >/databases 目录下，在 Linux 控制台中可以查看应用程序/data/data/ < package name > 目录下是否存在 databases 目录，进入目录/data/data/ < package name > 使用指令：

```
# cd /data/data/ <package_name>
```

例如，cd /data/data/com. SQLite，查看该目录下的所有文件，使用 ls 命令：

```
# ls
```

若文件列表中没有 databases 文件夹，则需要新创建该文件夹，使用命令：

```
# mkdir databases
```

文件夹中如有多余的文件，可根据需要删除，删除文件的命令：

```
# rm <file_name>
```

命令执行成功后，再次使用 ls 命令可以看到文件列表中已经包含 databases 文件夹。使用 cd 命令进入 databases 文件夹，即可以在文件中创建需要的数据库文件。在 SQLite 数据库中，每个数据库保存在一个独立的文件中，使用 sqlite3 工具后加数据库文件名的方式打开数据库文件，如果数据库文件不存在，sqlite3 工具则自动创建该数据库，因此创建和打开数据库均可以使用命令：

```
# sqlite3 <database_name.db>
```

尽管提供了数据库名，例如，sqlite3 mySQLite1. db，但如果这个数据库并不存在，SQLite 并不会真正地创建它。SQLite 会等到真正地向其中增加了数据库对象之后才创建它，比如在其中创建了表或视图。SQLite 采用这种策略的原因是考虑到用户在将数据库写到外部文件之前可能对数据库做一些永久性的设置，如页的大小等。有些设置，如页大小、字符集（UTF-8 或 UTF-16）等，一旦数据库创建之后就不能再修改了。这个中间期是能够修改这些设置的唯一机会。因此，采用默认设置，要将数据库写到磁盘，仅需要在其中创建一个表。

```
create table UserInfo (User_No integer primary key autoincrement, User_Name text not null,Sex
text);
```

User_No 和 User_Name 是字段，亦称为列名，primary key 设置此字段为主键，autoincrement 设置字段值自动增长 +1，not null 指定字段不可为空，integer、text 是 SQLite 支持的数据类型，SQLite 支持的基本数据类型有以下几类：

VARCHAR/NVARCHAR(15)/TEXT/INTEGER/FLOAT/BOOLEAN/CLOB/BLOB/TIMESTAMP/NUMERIC(10, 5)/VARYING CHARACTER (24) /NATIONAL VARYING CHARACTER(16)

查看表是否创建成功可以使用“. tables”特殊命令，如需查看其他信息可以使用表 4-3 中的其他特殊命令。至此，数据库 mySQLite1. db 创建成功，可以定位到应用程序 /data/data/<package name>/databases 目录，mySQLite1. db 文件已经包含在其中了，如图 4-4 所示。

Threads | Heap | Allocation Tracker | File Explorer

Name	Size	Date	Time	Permissions	Info
data		2011-12-31	16:30	drwxrwx--x	
anr		2012-01-28	15:58	drwxrwx--x	
app		2012-02-02	03:51	drwxrwx--x	
app-private		2011-12-31	16:28	drwxrwx--x	
backup		2011-12-31	16:30	drwx------	
dalvik-cache		2012-02-02	03:51	drwxrwx--x	
data		2012-02-02	02:38	drwxrwx--x	
android.tts		2011-12-31	16:30	drwxr-xr-x	
com.BasicControl		2012-01-29	03:13	drwxr-xr-x	
com.HR_Menu		2012-02-01	16:56	drwxr-xr-x	
com.SQLite		2012-02-02	02:49	drwxr-xr-x	
databases		2012-02-02	04:13	drwxrwxrwx	
mySQLite1.db	4096	2012-02-02	04:13	-rw-r--r--	

图 4-4　创建 mySQLite1. db 数据库

数据库创建完成后，数据库的其他基本操作，例如向表中插入数据、删除数据、修改数据，添加新表、新列等操作均是通过 SQLite 支持的 SQL 语句完成的。SQLite 虽然很精简小巧，但是支持的 SQL 语句不逊色于其他开源数据库，其支持的 SQL 语句如表 4-4 所示。

表 4-4 SQLite 支持的 SQL 语句

SQLite-SQL	说 明
ATTACH DATABASE	将一个已经存在的数据库添加到当前数据库链接
BEGIN TRANSACTION	开启事务
COMMENT	注释不是 SQL 命令，但会出现在 SQL 查询中
COMMIT TRANSACTION	提交事务
COPY	复制（SQLite3 已经删除该命令）
CREATE INDEX	创建索引
CREATE TABLE	创建表
CREATE TRIGGER	创建触发器
CREATE VIEW	创建视图
DELETE	删除
DETACH DATABASE	拆分已经存在的数据库
DROP INDEX	删除索引
DROP TABLE	删除表
DROP TRIGGER	删除触发器
DROP VIEW	删除视图
END TRANSACTION	结束事务
EXPLAIN	非标准的扩展功能
EXPRESSION	略
UPDATE	更新
SELECT	查询
INSERT	插入
ON CONFLICT CLAUSE	定义了解决约束冲突的算法（非独立 SQL 命令）
PRAGMA	用于修改 SQLite 库或者查新 SQLite 库内部数据的特殊命令
REPLACE	重命名
ROLLBACK TRANSACTION	事务回滚

例如，向表 UserInfo 中插入两条数据后查询表中数据：

```
sqlite > insert into UserInfo values (null,'张三', '男');
sqlite > insert into UserInfo values(null, '李四', '女');
sqlite >select *  from UserInfo;
```

结果如下：

1 | 张三 | 男

2 | 李四 | 女

在 Android 应用程序开发过程中，使用 SQLite 命令行操作数据库一般用于调试和测试，在实际开发中并非常用，最常用的操作 SQLite 数据库的方法是使用库类 API。下面介绍 Android 系统中创建和操作 SQLite 数据库的另外一种方式——使用库类 API。

SQLiteDatabase 类是 Android 系统提供的用于管理和操作 SQLite 数据库的 API，一个 SQLiteDatabase 对象实例相当于一个 SQLite 数据库。该类提供了创建、删除、执行 SQL 语句和其他常用数据库操作的方法。

在 Android 中创建和打开一个数据库使用 SQLiteDatabase 类的静态方法 openOrCreateDatabase()来实现，openOrCreateDatabase()方法会自动去检测数据库是否存在，如果存在则打开，如果不存在则创建这个数据库，创建成功则返回一个 SQLiteDatabase 对象，否则抛出异常 FileNotFoundException。

openOrCreateDatabase( )语法格式:

```
static SQLiteDatabase openOrCreateDatabase(String path, CursorFactory factory);
```

方法第一个参数 path 是数据库路径，路径必须是数据库的全路径。方法第二个参数 factory 是指定返回一个 Cursor 子类的 factory，如果没有则指定为 null，即使用默认 factory。例如建立数据库 mySQLite2. db。

```
1    SQLiteDatabase dbSQLiteDatabase = null;
2    dbSQLiteDatabase = SQLiteDatabase.openOrCreateDatabase(
3                        "/data/data/com.SQLite/databases/mySQLite2.db", null);
```

一个数据库中可以包含多个表，每一条数据都保存在一个指定的表中，要创建表可以通过 SQLiteDatabase 类的 execSQL（String sql）方法来执行一条 SQL 语句。execSQL( )方法能够执行大部分的 SQL 语句。例如创建表 UserInfo，并添加字段 User_Id、User_Name、User_Sex。

```
1    String sqlCreateTable = "CREATE TABLE UserInfo " +
2    "(User_Id INTEGER PRIMARY KEY,User_Name TEXT NOT NULL,User_Sex TEXT)";
3    dbSQLiteDatabase.execSQL(sqlCreateTable);//执行 SQL 语句
```

向表中插入数据存在两种方法，一种可以调用 SQLiteDatabase 的 insert( )方法来添加数据，insert( )方法的语法格式为:

```
long insert(String table, String nullColumnHack, ContentValues values)
```

参数 table 是表名称，参数 nullColumnHack 空列的默认值，参数 values 的类型是 ContentValues 类型，ContentValues 其实就是一个封装了列名称和列值的 Map，通过 ContentValues 的 put( )方法就可以把数据放到 ContentValues 中，然后插入到表中去，put( )方法的语法格式为:

```
put(String key,Type value)
```

参数 key 值是字段名称，value 值是字段的值。

插入数据的另一种方法是使用 SQL 语句，使用 SQLiteDatabase 类的 execSQL( )方法实现，类似于创建新表。

例如，向表 UserInfo 添加两行记录:

```
1    //使用 insert()方法
2    ContentValues  cvContentValues = new ContentValues();
3    cvContentValues.put("User_Id", 1);
4    cvContentValues.put("User_Name", "张三");
5    cvContentValues.put("User_Sex", "男");
6    m_dbSQLiteDatabase.insert("UserInfo", null, cvContentValues);
7    //使用 SQL 语句
8    String  sqlInsertData = "INSERT INTO UserInfo " +
9                "(User_Id, User_Name, User_Sex) values (2, '李四', '女')";
10    m_dbSQLiteDatabase.execSQL(sqlInsertData);
```

删除数据也有两种方法，一种是调用 SQLiteDatabase 类的 delete( )方法，另一种是使用 execSQL( )执行 SQL 语句完成。delete( )方法的语法格式为:

```
int delete(String table, String whereClause, String[] whereArgs);
```

参数 table 是表名，参数 whereClause 是删除条件，参数 whereArgs 是删除条件值数组。

```
1	//delete()
2	String strWhereClause = "User_Id = ?";//删除条件
3	String[] strArrayWhereArg = {String.valueOf(1)};//删除条件参数
4	m_dbSQLiteDatabase.delete("UserInfo", strWhereClause, strArrayWhereArg);
5	//SQL 语句
6	String  sqlDeleteData = "DELETE FROM UserInfo WHERE User_Id = 2 ";
7	m_dbSQLiteDatabase.execSQL(sqlDeleteData);
```

在 Android 中查询数据是通过 Cursor 类来实现的，使用 SQLiteDatabase.query()方法时，会得到一个 Cursor 对象，Cursor 指向的就是每一条数据。它提供了很多有关查询的方法，如表 4-5 所示。

表 4-5　Cursor 类提供的常用方法

方　法	说　明
move()	以当前的位置为参考，将 Cursor 移动到指定的位置，成功返回 true，失败返回 false
moveToPosition(int position)	将 Cursor 移动到指定的位置，成功返回 true，失败返回 false
moveToNext()	将 Cursor 向前移动一个位置，成功返回 true，失败返回 false
moveToLast()	将 Cursor 向后移动一个位置，成功返回 true，失败返回 false
movetoFirst()	将 Cursor 移动到第一行，成功返回 true，失败返回 false
isBeforeFirst()	返回 Cursor 是否指向第一项数据之前
isAfterLast()	返回 Cursor 是否指向最后一项数据之后
isClosed()	返回 Cursor 是否关闭
isFirst()	返回 Cursor 是否指向第一项数据
isLast()	返回 Cursor 是否指向最后一项数据
isNull(int columnindex)	返回指定位置的值是否为 null
getCount()	返回总的数据项数
getInt(int columnindex)	返回当前行中指定的索引数据

SQLiteDatabase.query()方法有多种重载方法，其中常用的重载方法语法格式为：

```
Cursor android.database.sqlite.SQLiteDatabase.query(String table, String[] columns, String se-
lection, String[] selectionArgs, String groupBy, String having, String orderBy)
```

方法中各参数的意义如表 4-6 所示。

表 4-6　query()方法参数意义

参　数	意　义
String table	表名称
String [] columns	返回的属性列名称
String selection	查询条件
String [] selectionArgs	如果在查询条件中使用的问号，则需要定义替换符的具体内容
String groupBy	分组方式
String having	定义组的过滤器
String orderBy	排序方式

例如，查询表 UserInfo 中所有记录，并将名字和性别提取出来用于显示。

```
1	String strMsg = "查到的数据:";
2	Cursor resultCursor =  m_dbSQLiteDatabase.query("UserInfo",
```

```
3                new String[] {"User_Id", "User_Name", "User_Sex"}, null, null, null, null, null);
4    if(resultCursor.moveToFirst()){
5        for(int i=0;i<resultCursor.getCount();i++){
6            strMsg+=resultCursor.getString(1)+resultCursor.getString(2);//获取名字+性别
7        }
8    }
```

如果添加了数据后发现数据有误，这时需要修改这个数据，可以使用 SQLiteDatabase 类的 updata( )方法来更新一条数据。updata( )方法的语法格式为：

```
int update(String table, ContentValues values, String whereClause, String[] whereArgs)
```

例如，将表 User_Name 中 User_Id 字段值为 1 的记录的 User_Name 字段修改为 "WangWu"，其代码为：

```
1    ContentValues updateValues = new ContentValues();
2    updateValues.put("User_Name", "WangWu");
3    String strWhereClause = "User_Id = ?";
4    String[] strArrayWhereArg = {String.valueOf(1)};
5    m_dbSQLiteDatabase.update("UserInfo", updateValues, strWhereClause, strArrayWhereArg);
```

数据库的基本操作，除了数据库创建与打开，表的新建、插入、删除、查询和修改外，基本的数据库操作还包括表的删除、数据库的关闭和数据库的删除。表的删除通过调用 SQLiteDatabase 类中 execSQL( )方法执行，数据库的关闭通过调用 SQLiteDatabase 类的 close ( )方法，数据库的删除直接使用 Activity 的 deleteDatabase（string dbName）即可。

下面代码实现了数据库的建立与打开，表的新建，插入数据，删除数据，查询数据，修改数据和关闭数据库，其运行后的效果如图 4-5 所示。

```
1    public class MainAct extends Activity {
2        private  SQLiteDatabase m_dbSQLiteDatabase = null;
3        private  TextView m_txtResultDisplay = null;
4            /** Called when the activity is first created. */
5            @Override
6            public void onCreate(Bundle savedInstanceState) {
7                super.onCreate(savedInstanceState);
8                setContentView(R.layout.main);
9                m_txtResultDisplay = (TextView)findViewById(R.id.txtResultDisplay);
10               //创建或者打开数据库
11               Button btnCreateOpenDB = (Button)findViewById(R.id.btnCreateOpenDB);
12               btnCreateOpenDB.setOnClickListener(new OnClickListener(){
13               @Override
14               public void onClick(View v) {
15                   // TODO Auto-generated method stub
16                   m_dbSQLiteDatabase = OpenOrCreateDB();
17                   if(m_dbSQLiteDatabase == null)
18                       m_txtResultDisplay.setText("数据库创建/打开失败!");
19                   else
20                       m_txtResultDisplay.setText("数据库创建/打开成功!");
```

```
21                }
22            });
23            //创建表
24            Button btnCreateTable = (Button) findViewById(R.id.btnCreateTable);
25            btnCreateTable.setOnClickListener(new OnClickListener(){
26              @Override
27              public void onClick(View v) {
28                    // TODO Auto-generated method stub
29                    Boolean bCreateTable = CreateTable();
30                    if(bCreateTable = = false)
31                        m_txtResultDisplay.setText("创建表失败!");
32                    else
33                        m_txtResultDisplay.setText("创建表成功!");
34              }
35            });
36            //插入数据
37            Button btnInsertData = (Button) findViewById(R.id.btnInsertData);
38            btnInsertData.setOnClickListener(new OnClickListener(){
39              @Override
40              public void onClick(View v) {
41                    // TODO Auto-generated method stub
42                    Boolean bInsertData = InsertData();
43                    if(bInsertData = = false)
44                        m_txtResultDisplay.setText("插入数据失败!");
45                    else
46                        m_txtResultDisplay.setText("插入数据成功!");
47              }
48            });
49            //删除数据
50            Button btnDeleteData = (Button) findViewById(R.id.btnDeleteData);
51            btnDeleteData.setOnClickListener(new OnClickListener(){
52              @Override
53              public void onClick(View v) {
54                    // TODO Auto-generated method stub
55                    Boolean bDeleteData = DeleteData();
56                    if(bDeleteData = = false)
57                        m_txtResultDisplay.setText("删除数据失败!");
58                    else
59                        m_txtResultDisplay.setText("删除数据成功!");
60              }
61            });
62            //查询数据
63            Button btnQueryData = (Button) findViewById(R.id.btnQueryData);
64            btnQueryData.setOnClickListener(new OnClickListener(){
65              @Override
66              public void onClick(View v) {
```

```
67                    // TODO Auto-generated method stub
68                    QueryData();
69                }
70            });
71            //修改数据
72            Button btnUpdateData = (Button)findViewById(R.id.btnUpdateData);
73            btnUpdateData.setOnClickListener(new OnClickListener(){
74            @Override
75            public void onClick(View v) {
76                    // TODO Auto-generated method stub
77                    Boolean bUpdateData = UpdateData();
78                    if(bUpdateData = = false)
79                        m_txtResultDisplay.setText("修改数据失败!");
80                    else
81                        m_txtResultDisplay.setText("修改数据成功!");
82                }
83            });
84            //关闭数据库
85              Button btnCloseDB = (Button)findViewById(R.id.btnCloseDB);
86              btnCloseDB.setOnClickListener(new OnClickListener(){
87              @Override
88              public void onClick(View v) {
89                  // TODO Auto-generated method stub
90                  CloseDB();
91                  m_txtResultDisplay.setText("数据库已关闭!");
92          }
93        });
94      }
95      //创建或者打开数据库
96      private  SQLiteDatabase  OpenOrCreateDB()
97      {
98        SQLiteDatabase dbSQLiteDatabase = null;
99        try{
100               dbSQLiteDatabase = SQLiteDatabase.openOrCreateDatabase(
101                 "/data/data/com.SQLite/databases/mySQLite2.db", null);
102         }
103         catch(Exception ex){
104               return null;
105         }
106         return dbSQLiteDatabase;
107     }
108     //创建表
109     private boolean CreateTable()
110     {
111       String sqlCreateTable  = "CREATE TABLE UserInfo " +
112     "(User_Id INTEGER PRIMARY KEY,User_Name TEXT NOT NULL," +
```

```
113     "User_Sex TEXT)";
114     if(m_dbSQLiteDatabase = = null) return false;
115     m_dbSQLiteDatabase. execSQL(sqlCreateTable);
116     return true;
117     }
118     //插入数据
119     private boolean InsertData()
120     {
121         if(m_dbSQLiteDatabase = = null) return false;
122         //insert()
123         ContentValues  cvContentValues = new ContentValues();
124         cvContentValues. put("User_Id", 1);
125         cvContentValues. put("User_Name", "ZhangS");
126         cvContentValues. put("User_Sex", "Male");
127         m_dbSQLiteDatabase. insert("UserInfo", null, cvContentValues);
128         //SQL 语句
129         String  sqlInsertData = "INSERT INTO UserInfo " +
130           "(User_Id, User_Name, User_Sex) values (2, 'LiSi', 'Female')";
131         m_dbSQLiteDatabase. execSQL(sqlInsertData);
132         return true;
133       }
134      //删除数据
135       private boolean DeleteData()
136       {
137      if(m_dbSQLiteDatabase = = null) return false;
138         //delete()
139         String strWhereClause = "User_Id = ?";//删除条件
140         String[] strArrayWhereArg = {String. valueOf(1)};//删除条件参数
141         m_dbSQLiteDatabase. delete("UserInfo", strWhereClause, strArrayWhereArg);
142         //SQL 语句
143         String  sqlDeleteData = "DELETE FROM UserInfo WHERE User_Id = 2 ";
144         m_dbSQLiteDatabase. execSQL(sqlDeleteData);
145         return true;
146       }
147       //查询数据
148         private boolean QueryData()
149       {
150             if(m_dbSQLiteDatabase = = null) return false;
151             String strMsg = "查到的数据:";
152             Cursor resultCursor =  m_dbSQLiteDatabase. query("UserInfo",
153                         new String[] {"User_Id", "User_Name", "User_Sex"}, null,
154                         null, null, null, null);
155             if(resultCursor. moveToFirst()){
156               for(int i = 0;i < resultCursor. getCount();i + +){
157                   strMsg + = resultCursor. getString(1) + resultCursor. getString(2);
158             }
```

```
159                }
160                m_txtResultDisplay.setText(strMsg);
161                return true;
162          }
163          //修改数据
164            private boolean UpdateData()
165          {
166                if(m_dbSQLiteDatabase==null)return false;
167                ContentValues updateValues=new ContentValues();
168                updateValues.put("User_Name", "WangWu");
169                String strWhereClause="User_Id=?";
170                String[] strArrayWhereArg={String.valueOf(1)};
171                m_dbSQLiteDatabase.update("UserInfo", updateValues,
172                          strWhereClause, strArrayWhereArg);
173                return true;
174          }
175          //关闭数据库
176          private boolean CloseDB()
177          {
178                if(m_dbSQLiteDatabase==null)return false;
179                m_dbSQLiteDatabase.close();
180                return true;
181          }
182    }
```

Android 除了提供 SQLiteDatabase 类操作 SQLite 数据库的同时，还为 SQLiteDatabase 类提供了一个辅助类，即 SQLiteOpenHelper 类。SQLiteOpenHelper 类主要用来管理数据库的创建和版本更新。SQLiteOpenHelper 是一个抽象类，一般的用法是定义一个继承该类的类，并且实现其三个抽象方法。

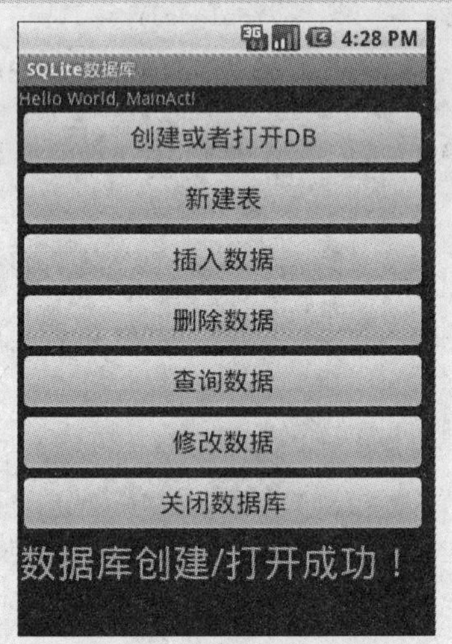

图 4-5　SQLite 数据库基本操作

onCreate（SQLiteDatabase db）：在数据库第一次生成的时候会调用这个方法，一般用于生成数据库表。

onUpgrade（SQLiteDatabase db，int oldVersion，int newVersion）：当数据库需要升级的时候，Android 系统会主动地调用这个方法。一般在这个方法里边删除数据表，并建立新的数据表，当然是否还需要做其他的操作，完全取决于应用的需求。

onOpen（SQLiteDatabase）：当打开数据库时的回调函数，一般不会用到。

另外，SQLiteOpenHelper 类还有两个方法经常用到，分别为 getWritableDatabase( ) 和 getReadableDatabase( )，其功能是创建或者打开读写或者只读数据库。当在程序当中调用这两个方法的时候，如果当时没有数据库，那么 Android 系统就会自动生成一个数据库。

4.3　文件存储

Android 文件系统是基于 Linux 的文件系统，其文件存储和访问有三种方式。首先，应用程序能够创建仅能够用于自身访问的私有文件，这类文件存放在应用程序自己的目录内，即/data/data/ <package_name>/files 目录，这类存储称为内部存储。其次，Android 系统提供了对 SD 卡等外部设备的访问方法，这类文件存储方式称为外部存储。另外，Android 系统还可以访问保存在资源目录中的原始文件以及 XML 文件，此类文件一般保存在/res/raw 目录和/res/xml 目录下。

4.3.1　内部存储

Android 系统进行内部文件存储主要用到两个基本方法，openFileOutput( )方法和 openFileInput( )方法。openFileOutput( )方法的主要功能是为写入数据做准备而打开应用程序的私有文件，若需要打开的文件不存在，则创建这个私有文件。openFileInput( )方法的功能是为读取数据做准备而打开应用程序私有文件。openFileOutput( )方法和 openFileInput( )方法的语法格式为：

```
FileOutputStream openFileOutput(String name, int mode);
FileInputStream openFileInput (String name);
```

参数 name 是文件名,文件名中不能包含分隔符“/”,新建或者需要打开的文件存放在/data/data/ <package_name>/files 目录。参数 mode 是文件操作模式，系统支持四种基本文件操作模式，分别为 MODE_PRIVATE、MODE_APPEND、MODE_WORLD_READABLE 和 MODE_WORLD_WRITEABLE，各个模式的意义分别为：

MODE_PRIVATE——私有模式或者缺陷模式。文件仅能够被文件创建程序进行访问或者具有相同 UID 的程序进行访问，其他应用程序无权访问。另外，在该模式下，写入的内容会覆盖原文件的内容。

MODE_APPEND——追加模式。此模式下如果文件已经存在，则在文件的结尾处添加新数据，不会覆盖以前的数据。

MODE_WORLD_READABLE——全局读模式。允许任何程序读取私有文件。

MODE_WORLD_WRITEABLE——全局写模式。允许任何程序写入私有文件。

除了四种基本模式外，还可以通过模式相加的方法设置全局读写模式，即“读模式 + 写模式”MODE_WORLD_READABLE + MODE_WORLD_WRITEABLE。

下面代码使用 openFileOutput( )方法和 openFileInput( )方法进行文件的创建和读写：

```
1    public class MainAct extends Activity {
2      private EditText m_etxtFileContent = null;
3      private TextView m_txtInfoDisplay = null;
4      private String m_strFileName = "myFile.txt";
5        @Override
6        public void onCreate(Bundle savedInstanceState) {
7            super.onCreate(savedInstanceState);
```

```
8           setContentView(R.layout.main);
9           m_etxtFileContent = (EditText)findViewById(R.id.etxtFileContent);
10          m_txtInfoDisplay = (TextView)findViewById(R.id.txtInfoDisplay);
11          Button btnWrite = (Button)findViewById(R.id.btnWrite);
12          Button btnRead = (Button)findViewById(R.id.btnRead);
13          btnWrite.setOnClickListener(new OnClickListener(){
14          public void onClick(View v) {
15              WriteData();
16              m_txtInfoDisplay.setText("未进行读出操作");
17            }
18          });
19          btnRead.setOnClickListener(new OnClickListener(){
20          public void onClick(View v) {
21              ReadData();
22            }
23          });
24      }
25      //写入
26      private void WriteData()
27      {
28      String strFileContent = m_etxtFileContent.getText().toString();
29      FileOutputStream  file = null;
30      try{
31          file = openFileOutput(m_strFileName,Context.MODE_APPEND);
32          file.write(strFileContent.getBytes());
33          file.flush();
34          file.close();
35      }
36      catch (Exception e) {
37          return;
38      }
39      }
40      //读出
41      private void ReadData()
42      {
43      FileInputStream  file = null;
44      try{
45          file = openFileInput(m_strFileName);
46          if(file.available() = =0)return;
47          byte[] byteFileContent = new byte[file.available()];
48          while(file.read(byteFileContent)! = -1);
49          m_txtInfoDisplay.setText(new String(byteFileContent));
50      }
51      catch (Exception e) {
52          return;
```

```
53    }
54        }
55  }
```

上面代码中使用到了四个控件，EditText 用于输入文件保存的内容，TextView 用于读取文件内容的显示，两个按钮分别用于文件写入和文件读出。

代码 25 ~ 39 行使用 openFileOutput（）方法进行文件的创建和写入操作，操作模式为 MODE_APPEND 追加模式，调用 write()函数是将数据写入文件，使用此函数写入的数据首先要放入数据缓冲区中，缓冲区满后才主动写入到文件中，所以在调用 close()函数关闭 FileOutputStream 前一般需要使用 flush()函数将所有缓冲区中的数据写入到文件。使用 openFileOutput()方法时会报异常未处理错误，此方法一般需要捕获 FileNotFoundException 异常，使用 write()、flush()和 close()方法时同样会报异常未处理错误，使用时一般需要捕获 IOException 异常，所示代码中使用了 try/catch 用于捕获出现的异常。

代码 40 ~ 54 行使用 openFileInput()方法进行文件读取操作，同样在实际使用过程中会遇到错误提示，因为文件操作可能会遇到各种问题而最终导致操作失败，因此代码应该使用 try/catch 捕获可能产生的异常。

代码运行界面如图 4-6 所示，文件创建的目录如图 4-7 所示。

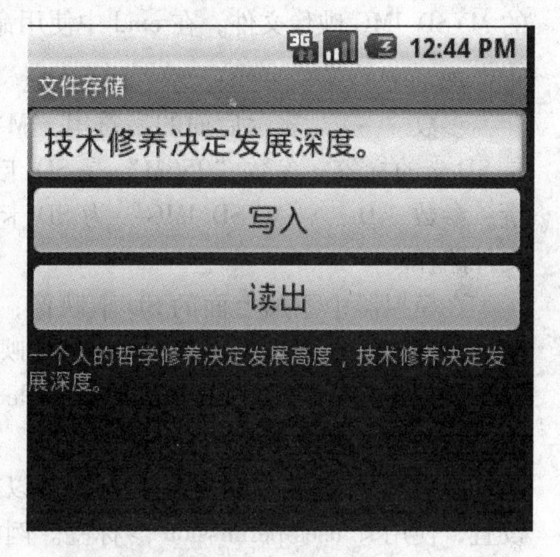

图 4-6　内部存储

Threads | Heap | Allocation Tracker | File Explorer

Name	Size	Date	Time	Permissions
data		2012-02-03	14:04	drwxrwx--x
anr		2012-02-03	14:03	drwxrwx--x
app		2012-02-03	14:33	drwxrwx--x
app-private		2012-02-03	14:02	drwxrwx--x
backup		2012-02-03	14:04	drwx------
dalvik-cache		2012-02-03	14:33	drwxrwx--x
data		2012-02-03	14:04	drwxrwx--x
android.tts		2012-02-03	14:03	drwxr-xr-x
com.File		2012-02-03	14:31	drwxr-xr-x
files		2012-02-03	14:32	drwxrwx--x
myFile.txt	27	2012-02-03	14:32	-rw-rw----

图 4-7　内部存储文件位置

4.3.2　SD 卡存储

SD 卡（Secure Digital Memory Card）亦称为安全数码卡，是一种基于半导体快闪记忆器的新一代记忆设备，它被广泛地应于便携式装置上。SD 卡由日本松下、东芝及美国 SanDisk

公司于 1999 年 8 月共同开发研制。大小犹如一张邮票的 SD 记忆卡，重量只有 2g，但却拥有高记忆容量、快速数据传输率、极大的移动灵活性以及很好的安全性。

Android 系统对 SD 卡的支持解决了内部存储空间小与存储文件大的矛盾，作为外部存储的主要设备，Android 系统中提供了很多方法用于支持 SD 卡的便捷访问。Android 系统使用 SD 卡进行文件外部存储不同于内部存储，其不同在于使用 SD 卡不能设置文件访问权限，不能设置访问模式。

在模拟器重添加 SD 卡，需要事先创建 SD 卡的映像文件，例如，创建一个 256MB 大小的 MYSD. IMG 映像文件，在 cmd 中使用命令：

```
mksdcard -lMYSD 256M D:\MYSD.IMG
```

参数“-l”表示后面的字符串“MYSD”是 SD 卡的标签；参数“256M”是 SD 卡的容量；参数“D: \ MYSD. IMG”为 SD 卡映像文件路径。

在模拟器中添加上面的 SD 卡映像，最简单的方法就是在创建模拟器时添加 SD 映像文件，如图 4-8 所示，在 SD Card 标签 File 选项中添加创建的映像文件。

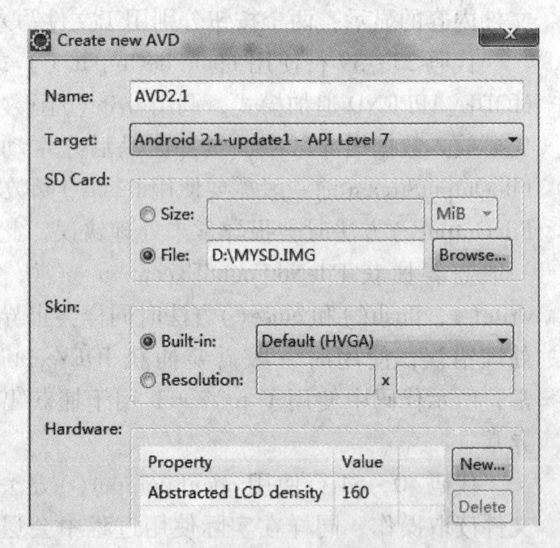

图 4-8 在创建模拟器时添加 SD 映像文件

使用 SD 卡进行文件存储，首先可以在 AndroidManifset. xml 文件进行访问 SD 卡的权限设置，使用 < uses-permission >标签。例如：

```
1 <!-- 在 SDCard 中创建与删除文件权限 -->
2 <uses-permission
3     Android:name="android.permission.MOUNT_UNMOUNT_FILESYSTEMS"/>
4 <!-- 往 SDCard 写入数据权限 -->
5 <uses-permission
6 Android:name="android.permission.WRITE_EXTERNAL_STORAGE"/>
```

下面代码实现了 SD 卡中文件的创建和读取。

```
1  public class MainAct extends Activity {
2      /** Called when the activity is first created. */
3    private EditText m_etxtFileContent = null;
4    private TextView m_txtInfoDisplay = null;
5    private String m_strFileName = "myFile.txt";
6        @Override
7        public void onCreate(Bundle savedInstanceState) {
8            super.onCreate(savedInstanceState);
9            setContentView(R.layout.main);
10           m_etxtFileContent = (EditText)findViewById(R.id.etxtFileContent);
11           m_txtInfoDisplay = (TextView)findViewById(R.id.txtInfoDisplay);
12           Button btnWrite = (Button)findViewById(R.id.btnWrite);
13           Button btnRead = (Button)findViewById(R.id.btnRead);
```

```
14            btnWrite.setOnClickListener(new OnClickListener(){
15            public void onClick(View v) {
16                WriteSD();
17                m_txtInfoDisplay.setText("未进行读出操作");
18              }
19            });
20            btnRead.setOnClickListener(new OnClickListener(){
21              public void onClick(View v) {
22                ReadSD();
23              }
24            });
25        }
26        //写入
27        private void WriteSD()
28        {
29        if (Environment.getExternalStorageState()
30                .equals(Environment.MEDIA_MOUNTED)) {
31            File file=new File(Environment.getExternalStorageDirectory(),
32                    m_strFileName);
33            try {
34                FileOutputStream outPutStream=new FileOutputStream(file);
35                outPutStream.write(m_etxtFileContent.getText().toString().getBytes());
36                outPutStream.close();
37                Toast.makeText(this, "写入文件成功",Toast.LENGTH_LONG).show();
38            } catch (Exception e) {
39                Toast.makeText(this, "写入文件失败",Toast.LENGTH_SHORT).show();
40            }
41        }
42        else {
43            // 此时 SDcard 不存在或者不能进行读写操作的
44            Toast.makeText(this,"SDcard 不存在或者不能进行读写操作",
45                    Toast.LENGTH_SHORT).show();
46        }
47    }
48    //读出
49    private void ReadSD(){
50        if (Environment.getExternalStorageState().equals(
51                Environment.MEDIA_MOUNTED)) {
52            File file=new File(Environment.getExternalStorageDirectory(),
53                    m_strFileName);
54            try {
55                FileInputStream inputStream=new FileInputStream(file);
56                byte[] byteFileContent=new byte[inputStream.available()];
57                inputStream.read(byteFileContent);
58                m_txtInfoDisplay.setText(new String(byteFileContent));
59                Toast.makeText(this, "读取成功",Toast.LENGTH_LONG).show();
```

```
60              } catch (Exception e) {
61                  Toast.makeText(this, "读取失败",Toast.LENGTH_SHORT).show();
62              }
63          } else {
64              // 此时 SDcard 不存在或者不能进行读写操作的
65              Toast.makeText(this,"此时 SDcard 不存在或者不能进行读写操作",
66                      Toast.LENGTH_SHORT).show();
67          }
68      }
69  }
```

代码中 26～47 行实现了 SD 卡内容的创建和写入操作，其中用到了 Environment 类，获取一些环境配置等信息，此类常用的静态方法如表 4-7 所示。

表 4-7　Environment 类常用的静态方法

静态方法	说明
getDataDirectory( )	获取到 Android 中的 data 数据目录
getDownloadCacheDirectory( )	获取到下载的缓存目录
getExternalStorageDirectory( )	获取到外部存储的目录，一般指 SDcard
getExternalStorageState(File path)	获取外部设置的当前状态，一般指 SDcard，Android 系统中对于外部设置的状态，比较常用的 MEDIA_MOUNTED（SDcard 存在并且可以进行读写）MEDIA_MOUNTED_READ_ONLY（SDcard 存在，只可以进行读操作）
getRootDirectory( )	获取到 Android Root 路径
isExternalStorageEmulated(File path)	返回 Boolean 值，判断外部设置是否有效
isExternalStorageRemovable( )	返回 Boolean 值，判断外部设置是否可以移除

代码运行界面如图 4-9 所示，创建文件的位置如图 4-10 所示。

图 4-9　使用 SD 卡进行文件读取

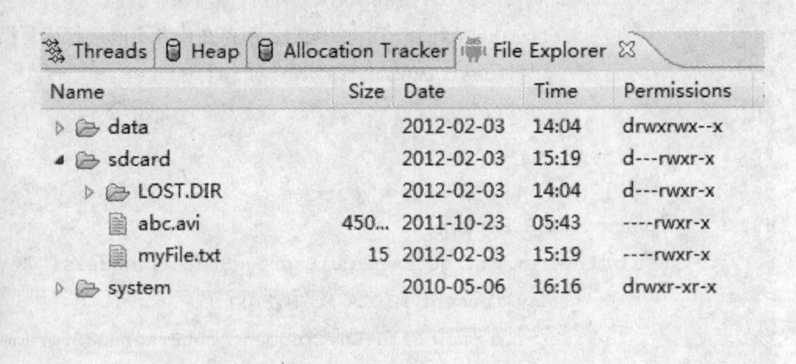

图 4-10　SD 创建文件位置

4.3.3　资源文件访问

Android 系统对资源文件的存储只要是对原始格式文件和 XML 文件的访问，原始格式文件可以是任何格式的文件，例如视频格式文件、音频格式文件、图像文件和数据文件等，在应用程序编译和打包时，/res/raw 目录下的所有文件都会保留原有格式不变。/res/xml 目录下的 XML 文件，一般用来保存格式化的数据，在应用程序编译和打包时会将 XML 文件转换为高效的二进制格式，应用程序运行时会以特殊的方式进行访问。

读取原始格式文件，一般通过调用 getResource( ) 函数获得资源对象 Resources，然后通过调用资源对象的 openRawResource( ) 函数，以二进制流的形式打开指定的原始格式文件，在读取文件结束后，调用 close( ) 函数关闭文件流。

读取 XML 格式文件，一般通过调用资源对象 Resources 的 getXml( ) 函数，获取到 XML 解析器 XmlPullParser，XmlPullParser 是 Android 平台标准的 XML 解析器，XmlPullParser 类实现了操作 XML 文件常用的方法。

```
1      public class MianAct extends Activity {
2        private Resources m_resResources = null;
3          /** Called when the activity is first created. */
4          @Override
5          public void onCreate(Bundle savedInstanceState) {
6              super.onCreate(savedInstanceState);
7              setContentView(R.layout.main);
8              Button btnRaw = (Button)findViewById(R.id.btnRaw);
9              Button btnXml = (Button)findViewById(R.id.btnXml);
10             final TextView txtRaw = (TextView)findViewById(R.id.txtRaw);
11             final TextView txtXml = (TextView)findViewById(R.id.txtXml);
12             m_resResources = this.getResources();
13           //访问原始格式文件
14             btnRaw.setOnClickListener(new OnClickListener(){
15             @Override
16             public void onClick(View v) {
17                 // TODO Auto-generated method stub
18                 String strFileContent = "Raw:";
19                 try{
20                     InputStream inStream = null;
21                     inStream = m_resResources.openRawResource(R.raw.myfreeformat);
22                     byte[] byteArray = new byte[inStream.available()];
23                     while (inStream.read(byteArray) != -1);
24                     strFileContent += new String(byteArray,"utf-8");
25                 }
26                 catch(Exception ex){
27                 }
28                 txtRaw.setText(strFileContent);
29             }
30         });
31         //访问 XML 文件
32         btnXml.setOnClickListener(new OnClickListener(){
33             @Override
34             public void onClick(View v) {
```

```
35                // TODO Auto-generated method stub
36                XmlPullParser xmlParser = m_resResources.getXml(R.xml.myxml);
37                String strFileContent = "Xml:";
38                try {
39                    while (xmlParser.next() != XmlPullParser.END_DOCUMENT) {
40                        if ((xmlParser.getName() != null) && xmlParser.getName()
41                                .equals("poemsentence")) {
42                            int count = xmlParser.getAttributeCount();
43                            for (int i = 0; i < count; i++) {
44                            String strName = xmlParser.getAttributeName(i);
45                            String strValue = xmlParser.getAttributeValue(i);
46                            if ((strName != null) && strName.equals("value")) {
47                                strFileContent += strValue;
48                                }
49                            }
50                        }
51                    }
52                } catch (Exception e) {
53                }
54                txtXml.setText(strFileContent);
55            }
56        });
57    }
58 }
```

代码 20 ~ 24 行读取资源文件目录 raw 内的原始格式文件 myfreeformat.txt，使用 Resources 类的 openRawResource( ) 方法，第 24 行表示字符串的编码方式以 UTF-8 方式编码。代码 36 ~ 55 行访问目录 xml 下的 XML 文件 myxml.xml，使用方法 getXml( ) 获取 XmlPullParser 对象后，调用其方法 getName( ) 获取 XML 节点元素的名字，通过比较后，对符合条件的节点元素使用方法使用 getAttributeCount( ) 函数获取元素的属性数量，并分别调用 getAttributeName( ) 获取属性名称，调用 getAttributeValue( ) 获取相应的值。

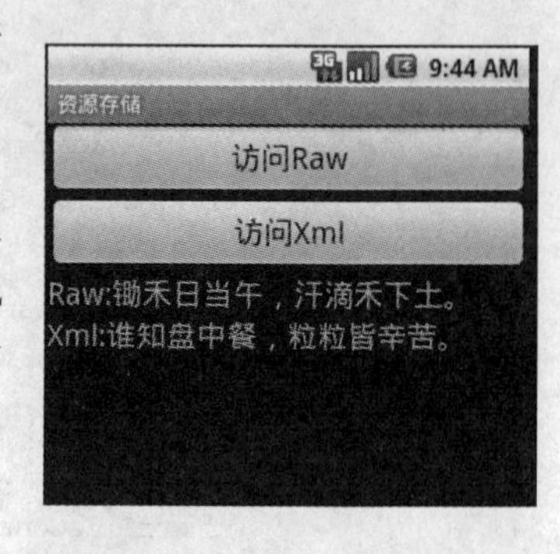

图 4-11　访问资源文件

运行结果如图 4-11 所示。

4.4　内容提供器

内容提供器（Content Provider）是 Android 系统中基本组件之一，用来存储和获取数据并使这些数据可以被所有的应用程序访问，它为存储和读取数据提供了一种通用的接口机制，由于在 Android 系统中没有一个公共的内存区域供多个应用共享存储数据，所以它是应用程序之间共享数据的唯一方法。

4.4.1 内容解析器

内容解析器（Content Resolver）用来处理对内容提供器的访问，用户不能直接调用 Content Provider 的接口函数，而是通过 Content Resolver 来完成。Content Resolver 调用 Content Provider 是通过通用资源标志符确定需要访问的数据集。通用资源标志符（Uniform Resource Identifier，URI）用来定位任何远程或本地的可用资源，Content Provider 使用的 URI 语法格式为：

```
content://<authority>/<data_path>/<id>
```

content：//是通用前缀，表示该 URI 用于 Content Provider 定位资源，已经由 Android 所规定无需修改。

<authority>是授权者名称，用来确定具体由哪一个 Content Provider 提供资源，一般<authority>都由类的小写全称组成，以保证唯一性。

<data_path>是数据路径，用来确定请求的是哪个数据集，如果 Content Provider 仅提供一个数据集，数据路径则是可以省略的；如果 Content Provider 仅提供多个数据集，数据路径则必须指明具体是哪一个数据集，数据集的数据路径可以写成多段格式。

<id>是数据编号，用来唯一确定数据集中的一条记录，Content Provider 将其存储的数据以数据表的形式提供给访问者，在数据表中每一行为一条记录，每一列为具有特定类型和意义的数据。每一条数据记录都包括一个"_ID"数值字段，该字段唯一标识一条数据。如果请求的数据并不只限于一条数据，则<id>可以省略。

当外部应用需要对 Content Provider 中的数据进行添加、删除、修改和查询操作时，可以使用 ContentResolver 类来完成，要获取 ContentResolver 对象，可以使用 Activity 提供的 getContentResolver()方法。ContentResolver 类提供了与 Content Provider 类相同签名的四个方法。

```
public Uri insert(Uri uri, ContentValues values);
public int delete(Uri uri, String selection, String[] selectionArgs);
public int update(Uri uri, ContentValues values, String selection, String[] selectionArgs);
public Cursor query(Uri uri, String[] projection, String selection, String[] selectionArgs,
String sortOrder);
```

4.4.2 内容提供者

Android 系统为一些常见的数据类型，诸如音乐、视频、图像、手机通信录联系人信息等，内置了一系列的 Content Provider，这些都位于 android. provider 包下，持有特定的权限，可以在应用程序中访问这些 Content Provider。所以，如果需要公开共享应用程序的数据，可有两种实现方式，一种是创建自定义的 Content Provider，这时需要继承 Content Provider 类，另一种将数据写到已存在的 Content Provider 中，这种情况需要公开的数据和已存在的 Content Provider 数据结构一致，当然前提是获取写该 Content Provider 的权限。

创建 Content Provider，首先需要继承 Content Provider 类，并重载类的六个方法，六个方法分别为：

```
public int delete(Uri uri, String selection, String[] selectionArgs)
public Uri insert(Uri uri, ContentValues values)
public Cursor query(Uri uri, String[] projection, String selection,String[] selection-
```

```
                Args, String sortOrder)
public int update(Uri uri, ContentValues values, String selection,String[] selectionArgs)
public boolean onCreate()
public String getType(Uri uri)
```

各方法的功能如表 4-8 所示。

表 4-8 Content Provider 重载方法

方法	功能
delete()	删除数据集
insert()	添加数据集
qurey()	查询数据集
update()	更新数据集
onCreate()	初始化底层数据集和建立数据连接等工作
getType()	返回指定 URI 的 MIME 数据类型，如果 URI 是单条数据，则返回的 MIME 数据类型应以 vnd. android. cursor. item 开头；如果 URI 是多条数据，则返回的 MIME 数据类型应以 vnd. android. cursor. dir/开头

其次，因为 URI 代表了要操作的数据，所以需要解析 URI，并从 URI 中获取数据。Android 系统提供了两个用于操作 URI 的工具类，分别为 UriMatcher 和 ContentUris 。

UriMatcher 类用于匹配 URI，使用时首先需要注册 URI 路径，注册方法如下：

```
uriMatcher = new UriMatcher(UriMatcher.NO_MATCH);
//如果匹配 content:// com.mycontentprovider / mydb 路径,返回匹配码为 1
uriMatcher.addURI("com.mycontentprovider", "mydb", 1);
uriMatcher.addURI("com.mycontentprovider", "mydb/#", 2);
```

常量 UriMatcher. NO_MATCH 表示不匹配任何路径的返回码，UriMatcher. NO_MATCH 的值为-1。addURI()函数用来添加新的匹配项，如果匹配，则返回匹配码，其语法格式为：

```
public void  addURI  (String authority, String path, int code)
```

参数 authority 表示匹配的授权者名称，参数 path 表示数据路径，#代表任何数字，参数 code 表示返回代码。

ContentUris 类用于获取 URI 路径后面的 ID 部分，它有两个比较实用的方法，即 withAppendedId(uri, id) 和 parseId(uri)。方法 withAppendedId(uri, id) 用于为路径加上 ID 部分，例如：

```
Uri uri = Uri.parse("content:// com.mycontentprovider / mydb");
Uri resultUri = ContentUris.withAppendedId(uri, 10);

//则生成后的 Uri 为:content:// com.mycontentprovider/mydb/10
```

parseId(uri) 方法用于从路径中获取 ID 部分，例如：

```
Uri uri = Uri.parse("content:// com.mycontentprovider/mydb/10");
long personid = ContentUris.parseId(uri);
```

则上面代码获取的结果为 10。

最后，在完成 Content Provider 类的代码实现后，需要在 AndroidManifest. xml 文件中进行注册，注册 Content Provider 使用 <provider> 标签，例如：

```
<application android:icon="@drawable/icon" android:label="@string/app_name">
  <provider android:name=".MyContent Provider"
        android:authorities="com.mycontentprovider"/>
 </application>
```

内容提供器创建完成后，使用内容提供器需要通过前面提到的内容解析器，每个 Android 组件都具有一个 ContentResolver 对象，获取 ContentResolver 对象的方法是调用 getContentResolver( ) 函数，获取到 ContentResolver 对象之后就可以使用其提供的查询、添加、删除和更新方法操作数据了。下面通过编写代码具体实现 Content Provider 的创建和在其他应用程序中具体使用创建的 Content Provider。

下面代码创建了一个 Content Provider 组件，供其他应用程序进行数据交换。

```
1    public class MyContent Provider extends Content Provider{
2      public static final String STR_MINESINGLE="vnd.android.cursor.item/vnd.mydb" ;
3      public static final String STR_MINEMULTIPLE="vnd.android.cursor.dir/vnd.mydb";
4      public static final String STR_AUTHORITY="com.mycontentprovider";
5      public static final String STR_PATHSINGLE="mydb/#";
6      public static final String STR_PATHMULTIPLE="mydb";
7      public static final String STR_URI="content://com.mycontentprovider/mydb";
8      public static final Uri    URI_CONTENT=Uri.parse(STR_URI);
9      public static final String TBCOL_ID="_id";
10     public static final String TBCOL_USERNAME="user_name";
11     public static final String TBCOL_USERSEX="user_sex";
12     private static final String DB_NAME="mydb.db";
13     private static final String DB_TABLE="TbUserInfo";
14     private static final int  DB_VERSION=1;
15     private SQLiteDatabase   dbSQLiteDatabase;
16     private DBOpenHelper   dbOpenHelper;
17     private static final int INT_MULTIPLE=1;
18     private static final int INT_SINGLE=2;
19     private static final UriMatcher uriMatcher;
20     static {
21         uriMatcher=new UriMatcher(UriMatcher.NO_MATCH);
22         uriMatcher.addURI(STR_AUTHORITY, STR_PATHMULTIPLE, INT_MULTIPLE);
23         uriMatcher.addURI(STR_AUTHORITY, STR_PATHSINGLE, INT_SINGLE);
24     }
25     @Override
26     public String getType(Uri uri) {
27         switch(uriMatcher.match(uri)){
28             case INT_MULTIPLE:
29                 return STR_MINEMULTIPLE;
30             case INT_SINGLE:
31                 return STR_MINESINGLE;
32             default:
```

```
33                throw new IllegalArgumentException("Unkown uri:" + uri);
34            }
35        }
36        @Override
37        public int delete(Uri uri, String selection, String[] selectionArgs) {
38            int count = 0;
39            switch(uriMatcher.match(uri)){
40                case INT_MULTIPLE:
41                    count = dbSQLiteDatabase.delete(DB_TABLE, selection, selectionArgs);
42                    break;
43                case INT_SINGLE:
44                    String segment = uri.getPathSegments().get(1);
45                    count = dbSQLiteDatabase.delete(DB_TABLE, TBCOL_ID + " = "
46                            + segment, selectionArgs);
47                    break;
48                default:
49                    throw new IllegalArgumentException("Unsupported URI:" + uri);
50            }
51            getContext().getContentResolver().notifyChange(uri, null);
52            return count;
53        }
54
55        @Override
56        public Uri insert(Uri uri, ContentValues values) {
57            long id = dbSQLiteDatabase.insert(DB_TABLE, null, values);
58            if ( id > 0 ){
59                Uri newUri = ContentUris.withAppendedId(URI_CONTENT, id);
60                getContext().getContentResolver().notifyChange(newUri, null);
61                return newUri;
62            }
63            throw new SQLException("Failed to insert row into " + uri);
64        }
65        @Override
66        public boolean onCreate() {
67            Context context = getContext();
68            dbOpenHelper = new DBOpenHelper(context, DB_NAME, null, DB_VERSION);
69            dbSQLiteDatabase = dbOpenHelper.getWritableDatabase();
70
71            if (dbSQLiteDatabase == null)
72                return false;
73            else
74                return true;
75        }
76        @Override
77        public Cursor query(Uri uri, String[] projection, String selection,
78                String[] selectionArgs, String sortOrder) {
```

```
79          SQLiteQueryBuilder qb = new SQLiteQueryBuilder();
80          qb.setTables(DB_TABLE);
81          switch(uriMatcher.match(uri)){
82              case INT_SINGLE:
83                  qb.appendWhere(TBCOL_ID + " = " + uri.getPathSegments().get(1));
84                  break;
85              default:
86                  break;
87          }
88          Cursor cursor = qb.query(dbSQLiteDatabase,
89                  projection,
90                  selection,
91                  selectionArgs,
92                  null,
93                  null,
94                  sortOrder);
95          cursor.setNotificationUri(getContext().getContentResolver(), uri);
96          return cursor;
97      }
98
99      @Override
100         public int update(Uri uri, ContentValues values, String selection,
101                 String[] selectionArgs) {
102             int count;
103             switch(uriMatcher.match(uri)){
104                 case INT_MULTIPLE:
105                     count = dbSQLiteDatabase.update(DB_TABLE, values,
106                             selection, selectionArgs);
107                     break;
108                 case INT_SINGLE:
109                     String segment = uri.getPathSegments().get(1);
110                 count = dbSQLiteDatabase.update(DB_TABLE, values, TBCOL_ID + " = "
111                         + segment, selectionArgs);
112                 break;
113             default:
114                 throw new IllegalArgumentException("Unknow URI:" + uri);
115         }
116         getContext().getContentResolver().notifyChange(uri, null);
117         return count;
118     }
119       private static class DBOpenHelper extends SQLiteOpenHelper {
120               public DBOpenHelper(Context context, String name,
121                       CursorFactory factory, int version) {
122                   super(context, name, factory, version);
123               }
124               private static final String DB_CREATE = "create table " +
```

```
125                 DB_TABLE + " (" + TBCOL_ID + " integer primary key autoincrement, " +
126                 TBCOL_USERNAME + " text not null, " + TBCOL_USERSEX + " text);";
127             @Override
128             public void onCreate(SQLiteDatabase _db) {
129                 _db.execSQL(DB_CREATE);
130             }
131             @Override
132             public void onUpgrade(SQLiteDatabase _db, int _oldVersion,
133                     int _newVersion) {
134                 _db.execSQL("DROP TABLE IF EXISTS " + DB_TABLE);
135                 onCreate(_db);
136             }
137         }
138     }
```

代码创建了一个 Content Provider，并重写了相应的方法对应相应的操作，另外代码实现了 SQLite 数据库 mydb.db 的建立，并创建表 TbUserInfo，表中有三个字段分别为用户 ID（_id）、用户姓名（user_name）和用户性别（user_sex）。使用创建的 Content Provider 的代码如下所示：

```
1    public class MainAct extends Activity {
2      private static final String STR_URI = "content://com.mycontentprovider/mydb";
3      private static final Uri    URI_CONTENT = Uri.parse(STR_URI);
4      private static final String TBCOL_ID = "_id";
5      private static final String TBCOL_USERNAME =  "user_name";
6      private static final String TBCOL_USERSEX = "user_sex";
7
8      private EditText m_etxtName = null;
9      private EditText m_etxtID = null;
10     private TextView m_txtResult = null;
11     private RadioButton m_radMale = null;
12     private RadioButton m_radFemale = null;
13     private Button m_btnAdd = null;
14     private Button m_btnQueryAll = null;
15     private Button m_btnClearDisp = null;
16     private Button m_btnDelAll = null;
17     private Button m_btnQueryById = null;
18     private Button m_btnDelById = null;
19     private Button m_btnUpdateById = null;
20     private CheckBox m_ckbKeyOp = null;
21     private CheckBox m_ckbNonKeyOp = null;
22     private ContentResolver m_crContentResolver = null;
23
24         @Override
```

```
25        public void onCreate(Bundle savedInstanceState) {
26            super.onCreate(savedInstanceState);
27            setContentView(R.layout.main);
28            m_crContentResolver = this.getContentResolver();
29            m_radMale = (RadioButton)findViewById(R.id.radMale);
30            m_radFemale = (RadioButton)findViewById(R.id.radFemale);
31            m_etxtName = (EditText)findViewById(R.id.etxtName);
32            m_etxtID = (EditText)findViewById(R.id.etxtID);
33            m_txtResult = (TextView)findViewById(R.id.txtResult);
34            m_btnAdd = (Button)findViewById(R.id.btnAdd);
35            m_btnQueryAll = (Button)findViewById(R.id.btnQueryAll);
36            m_btnClearDisp = (Button)findViewById(R.id.btnClearDisp);
37            m_btnDelAll = (Button)findViewById(R.id.btnDelAll);
38            m_btnQueryById = (Button)findViewById(R.id.btnQueryById);
39            m_btnDelById = (Button)findViewById(R.id.btnDelById);
40            m_btnUpdateById = (Button)findViewById(R.id.btnUpdateById);
41            m_ckbKeyOp = (CheckBox)findViewById(R.id.ckbKeyOp);
42            m_ckbNonKeyOp = (CheckBox)findViewById(R.id.ckbNonKeyOp);
43
44        m_btnAdd.setEnabled(false);
45        m_btnQueryAll.setEnabled(false);
46        m_btnDelAll.setEnabled(false);
47        m_btnQueryById.setEnabled(false);
48        m_btnDelById.setEnabled(false);
49        m_btnUpdateById.setEnabled(false);
50
51        m_btnAdd.setOnClickListener(new OnClickListener(){
52          @Override
53          public void onClick(View v) {
54            if(m_etxtName.getText().toString().hashCode() = =0)
55            {
56                Toast.makeText(getApplicationContext(), "姓名不能为空!",
57                        Toast.LENGTH_LONG).show();
58                return;
59            }
60            ContentValues cvContentValues = new ContentValues();
61            cvContentValues.put(TBCOL_USERNAME, m_etxtName.getText().toString());
62            if(m_radMale.isChecked())
63            {cvContentValues.put(TBCOL_USERSEX, m_radMale.getText().toString());}
64            else if(m_radFemale.isChecked())
65            {cvContentValues.put(TBCOL_USERSEX, m_radFemale.getText().toString());}
66            m_crContentResolver.insert(URI_CONTENT, cvContentValues);
```

```
67            m_txtResult.setText("添加成功!");
68            }
69        });
70        m_btnQueryAll.setOnClickListener(new OnClickListener(){
71          @Override
72          public void onClick(View v) {
73            Cursor csCursor=m_crContentResolver.query(URI_CONTENT,
74            new String[] {TBCOL_ID, TBCOL_USERNAME, TBCOL_USERSEX},
75                    null, null, null);
76            if (csCursor = = null){
77                    m_txtResult.setText("未返回任何数据!");
78                    return;
79            }
80            if (csCursor.getCount() = =0){
81                    m_txtResult.setText("未返回任何数据!");
82                    return;
83            }
84            String strResult="查找到:" + String.valueOf(csCursor.getCount())
85                 + "条记录\n";
86            if (csCursor.moveToFirst()){
87                    do{
88                        strResult + = "ID:" + csCursor.getInt(csCursor
89                               .getColumnIndex(TBCOL_ID)) + ",";
90                        strResult + = "姓名:" + csCursor.getString(csCursor
91                               .getColumnIndex(TBCOL_USERNAME)) + ",";
92                        strResult + = "性别:" + csCursor.getString(csCursor
93                               .getColumnIndex(TBCOL_USERSEX)) + "\n";
94                    }while(csCursor.moveToNext());
95                }
96                m_txtResult.setText(strResult);
97            }
98        });
99        m_btnClearDisp.setOnClickListener(new OnClickListener(){
100               @Override
101               public void onClick(View v) {
102                   m_etxtID.setText("");
103                   m_etxtName.setText("");
104                   m_txtResult.setText("");
105               }
106           });
107           m_btnDelAll.setOnClickListener(new OnClickListener(){
108               @Override
109               public void onClick(View v) {
```

```
110                 m_crContentResolver.delete(URI_CONTENT, null, null);
111                 m_txtResult.setText("已完成全部删除");
112             }
113         });
114         m_btnQueryById.setOnClickListener(new OnClickListener(){
115             @Override
116             public void onClick(View v) {
117                 if(m_etxtID.getText().toString().hashCode() = =0)
118                 {
119                     Toast.makeText(getApplicationContext(), "ID 不能为空!",
120                             Toast.LENGTH_LONG).show();
121                         return;
122                 }
123                 Uri uri =Uri.parse(STR_URI + "/" + m_etxtID.getText().toString());
124                 Cursor csCursor =m_crContentResolver.query(uri,
125                 new String[] {TBCOL_ID, TBCOL_USERNAME, TBCOL_USERSEX},
126                          null, null, null);
127                 if (csCursor = = null){
128                         m_txtResult.setText("未返回任何数据!");
129                         return;
130                 }
131                 if (csCursor.getCount()  = = 0){
132                         m_txtResult.setText("未返回任何数据!");
133                         return;
134                 }
135                 String strResult ="";
136                 if (csCursor.moveToFirst()){
137                         strResult += "ID:" + csCursor.getInt(csCursor
138                             .getColumnIndex(TBCOL_ID)) + ",";
139                         strResult += "姓名:" + csCursor.getString(csCursor
140                             .getColumnIndex(TBCOL_USERNAME)) + ",";
141                         strResult += "性别:" + csCursor.getString(csCursor
142                             .getColumnIndex(TBCOL_USERSEX)) + "\n";
143                 }
144                 m_txtResult.setText(strResult);
145                 }
146             });
147             m_btnDelById.setOnClickListener(new OnClickListener() {
148                 @Override
149                 public void onClick(View v) {
150                     if(m_etxtID.getText().toString().hashCode() = =0 )
151                     {
152                         Toast.makeText(getApplicationContext(), "ID 不能为空!",
```

```
153                         Toast.LENGTH_LONG).show();
154                     return;
155                 }
156                 Uri uri = Uri.parse(STR_URI + "/" + m_etxtID.getText().toString());
157                 int nResult = m_crContentResolver.delete(uri, null, null);
158                 if(nResult > 0)m_txtResult.setText("删除成功!");
159                 else m_txtResult.setText("删除失败!");
160             }
161         });
162         m_btnUpdateById.setOnClickListener(new OnClickListener() {
163             @Override
164             public void onClick(View v) {
165             if(m_etxtID.getText().toString().hashCode() = = 0 )
166             {
167                 Toast.makeText(getApplicationContext(), "ID不能为空!",
168                         Toast.LENGTH_LONG).show();
169                 return;
170             }
171             ContentValues cvContentValues = new ContentValues();
172             cvContentValues.put(TBCOL_USERNAME, m_etxtName.getText().toString());
173             if(m_radMale.isChecked())
174             {cvContentValues.put(TBCOL_USERSEX, m_radMale.getText().toString());}
175             else if(m_radFemale.isChecked())
176             {cvContentValues.put(TBCOL_USERSEX, m_radFemale.getText().toString());}
177             Uri uri = Uri.parse(STR_URI + "/" + m_etxtID.getText().toString());
178             int nResult = m_crContentResolver.update(uri, cvContentValues, null, null);
179                 if(nResult > 0)m_txtResult.setText("更新成功!");
180                 else m_txtResult.setText("更新失败!");
181             }
182         });
183         m_ckbKeyOp.setOnCheckedChangeListener(new OnCheckedChangeListener(){
184                 @Override
185                 public void onCheckedChanged(CompoundButton buttonView,
186                     boolean isChecked) {
187                 if(isChecked)
188                 {
189                     m_ckbNonKeyOp.setChecked(false);
190                     m_etxtID.setEnabled(true);
191                     m_btnAdd.setEnabled(false);
192                     m_btnQueryAll.setEnabled(false);
193                     m_btnDelAll.setEnabled(false);
194                     m_btnQueryById.setEnabled(true);
195                     m_btnDelById.setEnabled(true);
```

```
196                      m_btnUpdateById.setEnabled(true);
197                  }
198              }
199          });
200      m_ckbNonKeyOp.setOnCheckedChangeListener(new OnCheckedChangeListener(){
201              @Override
202              public void onCheckedChanged(CompoundButton buttonView,
203                      boolean is Checked) {
204                  if(is Checked)
205                  {
206                      m_ckbKeyOp.setChecked(false);
207                      m_etxtID.setEnabled(false);
208                      m_btnAdd.setEnabled(true);
209                      m_btnQueryAll.setEnabled(true);
210                      m_btnDelAll.setEnabled(true);
211                  m_btnQueryById.setEnabled(false);
212                      m_btnDelById.setEnabled(false);
213                      m_btnUpdateById.setEnabled(false);
214                  }
215              }
216          });
217      }
218  }
```

代码 2 ~ 22 行定义了一些常用的常量和界面控件，代码 5 1 ~ 69 行实现添加按钮单击响应事件，其功能是完成数据库表中一行数据的添加。代码 70 ~ 98 行完成查询按钮的单击响应事件，完成查询数据表中所有数据并显示。代码 98 ~ 113 行完成界面数据的清理，功能中不涉及数据库的实质性操作。代码 114 ~ 146 行为根据数据表主键（_id）字段查询相应数据并显示查询结果，响应查询按钮单击。代码 147 ~ 161 行完成删除指定_id 的一行数据，代码 162 ~ 182 完成指定_id 的一行数据的修改。代码 183 ~ 217 行完成两个复选框的互斥响应，一个响应需要指定主键的操作，例如根据 ID 查询、删除和修改，另一个相应无需指定 ID 字段信息的操作，例如查询全部数据、删除全部数据等。程序运行效果如图 4-12所示。

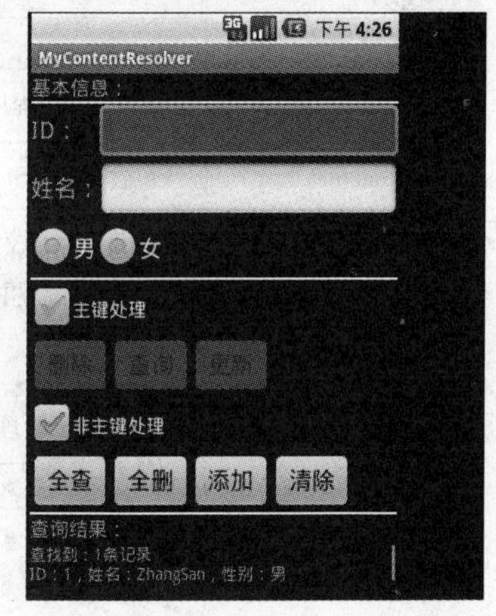

图 4-12　MainAct 用户界面

4.5　网络存储

Android 数据交互方式中使用网络访问和获取数据是其中一种重要的方式，Android 的网络存储最重要的方式是支持 HTTP 协议，编写的 Android 网络应用就相当于一个浏览器。Android 的网络应用将在第 6 章中具体涉及，本节仅简要概述 Android 如何使用网络发送和接收数据。

Android 发送请求和获取网络数据有很多方式，下面代码是 Android 发送请求和获取网络数据的一种常用方法，代码中获取到网络数据流 InputStream，使用获取到的 InputStream 进行数据操作就变得十分的简便了。

```
String path = "http://www.android.com/images/opensourceproject.gif";
URL url = new URL(path);
HttpURLConnection conn = (HttpURLConnection)url.openConnection();
conn.setConnectTimeout(6* 1000);
InputStream inStream = conn.getInputStream();
```

另外，由于 Android 的应用程序是使用 Java 来开发的，所以网络应用使用的也是 J2SE 的包。Android 中常用的网络包如表 4-9 所示。

表 4-9　Android 中常用网络包

包	描　述
java. net	提供与联网有关的类，包括流和数据包(Datagram)sockets、Internet 协议和常见 HTTP 处理。该包是一个多功能网络资源。有经验的 Java 开发人员可以立即使用这个熟悉的包创建应用程序
java. io	虽然没有提供显式的联网功能，但是仍然非常重要。该包中的类由其他 Java 包中提供的 socket 和连接使用。它们还用于与本地文件（在与网络进行交互时会经常出现）的交互
java. nio	包含表示特定数据类型的缓冲区的类。适合用于两个基于 Java 语言的端点之间的通信
org. apache. *	表示许多为 HTTP 通信提供精确控制和功能的包。可以将 Apache 视为流行的开源 Web 服务器
android. net	除核心 java. net. *类以外，包含额外的网络访问 socket。该包包括 URI 类，后者频繁用于 Android 应用程序开发，而不仅仅是传统的联网方面
android. net. http	包含处理 SSL 证书的类
android. net. wifi	包含在 Android 平台上管理有关 WiFi（802. 11 无线 Ethernet）所有方面的类。并不是所有设备都配备了 WiFi 功能，特别是 Android 在 Motorola 和 LG 等手机制造商的“翻盖手机”领域获得了成功
android. telephony. gsm	包含用于管理和发送 SMS（文本）消息的类。一段时间后，可能会引入额外的包来为非 GSM 网络提供类似的功能，比如 CDMA 或 android. telephony. cdma 等网络

本 章 小 结

本章主要讲述了 Android 数据交互方面的内容，系统中数据交互主要通过五种方式实现，共享优先数据机制、SQLite 数据库、File 文件机制、内容提供器控件和网络存储。其中在应用程序中最常用也是最有效的数据交互方式是使用 SQLite 数据库。

习　题

4-1　Android 平台采用的是哪种类型的数据库？它的文件存储和访问有哪几种方式？

4-2　Android 提供了哪些类来操作数据库？可以对数据库进行哪些操作？这些操作又是如何实现的？

4-3　什么是内容提供器（Content Provider）？Content Provider 使用的 URI 语法格式及每一部分的含义是什么？

4-4　编写一个电话通信录，使之可以输入姓名、号码、地址、邮箱，并可以向前、向后浏览数据记录，也可以添加、修改、删除数据。

第 5 章 Android 后台服务与事件广播

Android 系统通过提供的 Service 组件实现不直接与用户进行交互的后台服务，适合于进程内服务及进程间服务。Android 系统的事件广播消息的内容包括应用程序的数据信息或者是系统信息，例如电池电量变化、网络连接变化、接收到的短信及系统设置提示信息。事件广播机制依靠 BroadCastReciver 组件实现，可以接收到指定的广播消息。常驻程序 AppWidget 又称为“窗口小部件”，是在 HomeScreen 上显示的小部件，实现了桌面（Launcher）上显示控件的机能，并能够响应用户的单击操作。开发时常用 AppWidgetProvider 类和 AppWidgetProviderInfo 类实现。

5.1 Service 进程服务

Service 组件是 Android 系统中提供的四大组件之一，同样也是 Android 系统提供的后台运行服务。Service 组件与 Activity 组件不同，它并不能同用户直接交互，是一种无界面的后台应用。

5.1.1 Service 组件生命周期

相比 Activity 组件生命周期，Service 组件的生命周期要简单很多，在其整个生命周期中只继承了 onCreate( )、onStart( ) 和 onDestroy( ) 三个事件回调方法，分别用于创建、启动和销毁 Service。当第一次启动 Service 时，先后调用了 onCreate( ) 和 onStart( ) 两个方法，当停止 Service 时，则执行 onDestroy( ) 方法。如果 Service 已经启动了，当再次启动 Service 时，不会再执行 onCreate( )方法，而是直接执行 onStart( )方法。另外，在启动 Service 时，根据 onStartCommand( )的返回值不同，有两个附加的模式，一种模式为 START_STICKY，用于显式启动和停止 Service；另一种为 START_NOT_STICKY 或 START_REDELIVER_INTENT，用于有命令需要处理时才运行的模式。

Service 的运行方式有两种，一种通过调用 Context. startService( )方法启动，另一种是通过调用 Context. bindService( )方法启动。调用 Context. startService( )方法启动，调用者与服务之间没有关联，即使调用者退出了，服务仍然运行。这种方式在服务未被创建时，系统会先调用服务的 onCreate( )方法，接着调用 onStart( )方法。如果调用 startService( )方法前服务已经被创建，多次调用 startService( )方法并不会导致多次创建服务，但会导致多次调用 onStart ( )方法。采用 startService( )方法启动的服务，只能调用 Context. stopService( )方法结束服务，服务结束时会调用 onDestroy( )方法。使用 Context. bindService( )方法启用服务，调用者与服务绑定在了一起，调用者一旦退出，服务也就终止。这种方式启动服务时会回调 onBind( ) 方法，该方法在调用者与服务绑定时被调用，当调用者与服务已经绑定，多次调用 Context. bindService( )方法并不会导致该方法被多次调用。采用 Context. bindService( )方法启动服务时只能调用 onUnbind( )方法解除调用者与服务之间的绑定，服务结束时会调用 onDestroy( )方法。在选择 Service 运行方式时，同一个 Service 可以同时混合使用，Service 两种启动方式的生命周期如图 5-1所示。

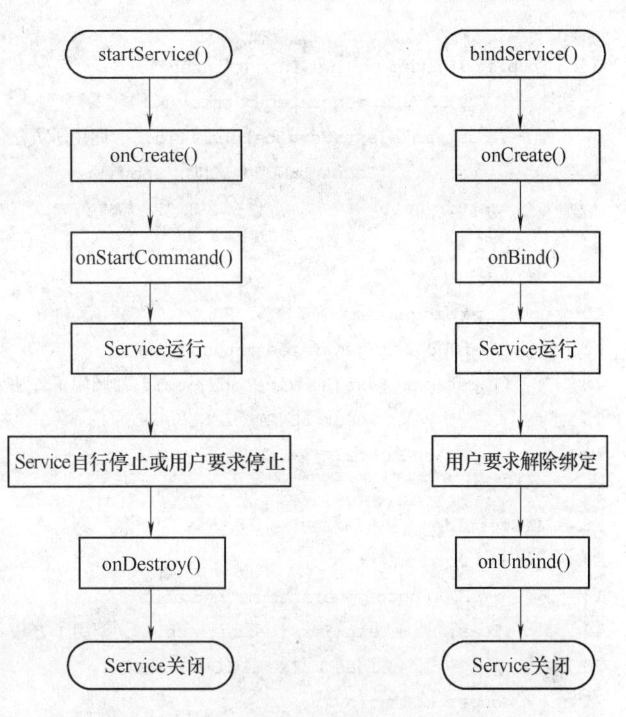

图 5-1　Service 两种启动方式的生命周期

Android 系统在处理拥有 Service 的进程优先级时，选择了较高优先级处理。一般情况下，Service 优先级要比 Activity 优先级要高，所以在系统资源紧张时，Service 也不轻易被系统终止回收，Android 系统会尽量保持拥有 Service 的进程运行。由于 Service 运行于后台，没有与用户可交互的界面接口，一般可以认为 Service 是永久运行于系统后台的组件进程。当 Service 正在调用 onCreate( )、onStartCommand( )或者 onDestory( )方法时，用于当前 Service 的进程则变为前台进程以避免被系统终止；当 Service 已经被启动时，拥有它的进程比用户可见的进程优先级低一些，但比不可见的进程要高，这就意味着 Service 同样不会被系统终止；如果 Service 已经被绑定，那么拥有 Service 的进程则拥有最高的优先级，可以认为 Service 是可见的；如果 Service 可以使用 startForeground（int，Notification）方法来将其设置为前台进程，那么系统就认为是对用户可见的，并不会在内存不足时终止此进程。

5.1.2　Service 服务

Service 服务可分为两种类型，一种为本地服务（Local Service），另一种为远程服务(Remote Service)。本地服务用于程序内部，通常用于实现应用程序自身中一些耗时的任务处理，比如查询升级信息。远程服务主要用于系统内部的应用程序之间，可被其他应用程序复用，比如天气预报服务。

定义 Service 服务类时，必须继承 Service 基类，在其最小代码集中至少要复写 onBind(Intent intent）方法，此方法是 Service 被绑定后调用的方法，方法返回相应的 Service 对象。为了能完成实际的功能，Service 类中一般需要复写其他的事件回调方法，例如 onCreate( )、onStart( )、onDestroy( )等。onCreate( )方法用于创建 Service，在使用时可以在内部完成必要的初始化等前期服务处理。onStart( )方法在 Service 启动时调用，例如在调用 startService( )方法启动 Service 后，系统会调用 onStart( )方法，并通过 Intent 传递参数。onDestroy( )方法用于销毁 Service，并释放所有占用的资源，销毁后的 Service 不能被程序可见，程序不能再继续使用，直到 Service 再次启动后方能为程序可见。

下面代码实现了一个简单的本地服务 Service 类：

```
1    public class ServiceLocal extends Service{
2      @Override
```

```
3     public IBinder onBind(Intent intent) {
4         // TODO Auto-generated method stub
5         Toast.makeText(ServiceLocal.this, "调用了方法 onBind()",
6                     Toast.LENGTH_LONG).show();
7         return null;
8     }
9     @Override
10    public void onCreate() {
11        // TODO Auto-generated method stub
12        Toast.makeText(ServiceLocal.this, "调用了方法 onCreate()",
13                    Toast.LENGTH_LONG).show();
14        super.onCreate();
15    }
16    @Override
17    public void onDestroy() {
18        // TODO Auto-generated method stub
19        Toast.makeText(ServiceLocal.this, "调用了方法 onDestroy()",
20                    Toast.LENGTH_LONG).show();
21        super.onDestroy();
22    }
23    @Override
24    public void onStart(Intent intent, int startId) {
25        // TODO Auto-generated method stub
26        Toast.makeText(ServiceLocal.this, "调用了方法 onStart()",
27                    Toast.LENGTH_LONG).show();
28        super.onStart(intent, startId);
29    }
30 }
```

类似于 Activity 的使用方法，在完成 Service 类后并在程序使用 Service 前需要注册 Service，注册在 AndroidManifest.xml 文件中。没有注册的 Service 不能被系统所见，系统无法使用非注册的 Service。注册时，使用 <service> 标签，其中的 android:name 为 Service 类的定义名称，需要同定义的 Service 类名称保持一致。Service 的启动方式也和 Activity 类似，有显式启动和隐式启动两种方式，显式启动直接传递给 Intent 相应的 Service 类，而隐式启动需要在注册 Service 时，提供 Intent-filter 的 action 属性，启动机制完全类似于 Activity 的启动机制。

```
<service android:name=".ServiceLocal">
    <intent-filter>
      <action android:name="com.ServiceLocal.SERVICE_LOCAL"/>
    </intent-filter>
</service>
```

注册完成后，就可以正常使用定义的 Service 服务，例如通过 Activity 测试上面定义的本地服务的启动、停止、绑定和解绑操作。

```
1 public class MainAct extends Activity {
2     private String m_strActName = "com.ServiceLocal.SERVICE_LOCAL";
3     /** Called when the activity is first created. */
4     @Override
5     public void onCreate(Bundle savedInstanceState) {
6         super.onCreate(savedInstanceState);
7         setContentView(R.layout.main);
8         Button btnStart = (Button)findViewById(R.id.btnStart);
9         Button btnStop = (Button)findViewById(R.id.btnStop);
10        Button btnBind = (Button)findViewById(R.id.btnBind);
11        Button btnUnbind = (Button)findViewById(R.id.btnUnbind);
12        //启动按钮响应
13        btnStart.setOnClickListener(new OnClickListener(){
14        @Override
15        public void onClick(View v) {
16            // TODO Auto-generated method stub
17            Intent inttIntent = new Intent();
18            inttIntent.setAction(m_strActName);
19            startService(inttIntent);
20        }
21        });
22        //停止按钮响应
23        btnStop.setOnClickListener(new OnClickListener(){
24        @Override
25        public void onClick(View v) {
26            // TODO Auto-generated method stub
27            Intent inttIntent = new Intent();
28            inttIntent.setAction(m_strActName);
29            stopService(inttIntent);
30        }
31        });
32        //绑定按钮响应
33        btnBind.setOnClickListener(new OnClickListener(){
34        @Override
35        public void onClick(View v) {
36            // TODO Auto-generated method stub
37            Intent inttIntent = new Intent();
38            inttIntent.setAction(m_strActName);
39            bindService(inttIntent, connService, Service.BIND_AUTO_CREATE);
40        }
41        });
42    //绑定按钮响应
43        btnUnbind.setOnClickListener(new OnClickListener(){
44        @Override
45        public void onClick(View v) {
46            // TODO Auto-generated method stub
```

```
47            Intent inttIntent = new Intent();
48            inttIntent.setAction(m_strActName);
49            unbindService(connService);
50        }
51        });
52    }
53    // 连接内部类
54    private ServiceConnection connService = new ServiceConnection() {
55        @Override
56        public void onServiceConnected(ComponentName name, IBinder service) {
57            // TODO Auto-generated method stub
58        }
59        @Override
60        public void onServiceDisconnected(ComponentName name) {
61            // TODO Auto-generated method stub
62        }
63        };
64    }
```

上面代码中主要运用了四个常用的操作 Service 的方法，方法的功能如表 5-1 所示。

表 5-1　常用操作 Service 的方法

方　法	功 能 说 明
startService(Intent)	启动 Service
stopService(Intent)	停止 Service
bindSerivce(Intent service, ServiceConnection conn, int flags)	绑定 Service 参数说明： service：通过该参数（也就是 Intent）可以启动指定的 Service conn：该参数是一个 ServiceConnection 对象，这个对象用于监听访问者与 Service 之间的连接情况，当访问者与 Service 连接成功时，将回调 ServiceConnection 对象的 onServiceConnected（ComponentName name，Ibinder service）方法；如果断开，将回调 onServiceDisConnected（ComponentName name）方法 flags：指定绑定时是否自动创建 Service
unbindService(conn)	解除绑定

运行上面的程序，其主界面如图 5-2 所示，单击相应的按钮可现实相应的调用方法的提示信息。

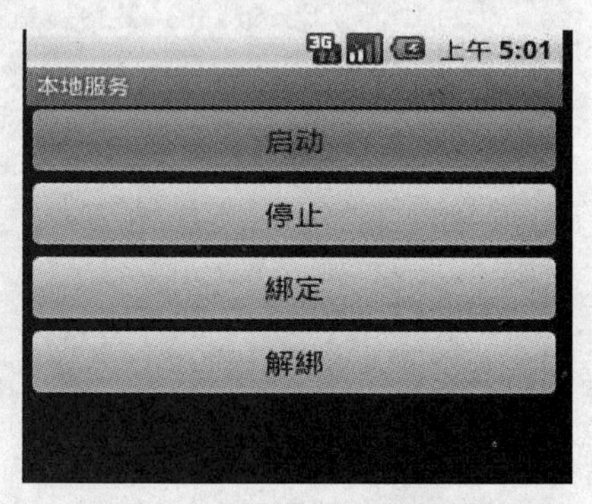

图 5-2　本地服务主要操作

远程服务亦称为跨进程服务，服务和使用服务不在同一个进程中。使用远程服务不同于使用本地服务，一般先使用 AIDL 定义服务接口，在 Service 类中实现定义接口的方法及属性，通过 AIDL 接口实现使用服务的组件调用服务。

AIDL（Android Interface Definition Language）为 Android 接口描述语言，由于 Android 系统中的进程之间不

能共享内存，因此需要提供一些机制在不同进程之间进行数据通信。为了使其他的应用程序也可以访问特定应用程序提供的服务，Android 系统采用了远程过程调用（Remote Procedure Call，RPC）方式来实现。与很多其他的基于 RPC 的解决方案一样，Android 使用一种接口定义语言（Interface Definition Language，IDL）来作为公开服务的接口。Android 应用程序组件中的 Activity、Broadcast 和 Content Provider 都可以进行跨进程访问，Android 应用程序组件 Service 同样实现了跨进程访问。因此，将这种可以跨进程访问的服务称为 AIDL 服务。

建立 AIDL 服务要比建立本地服务复杂一些，一般通过五个具体步骤实现：

首先，在 Eclipse 中建立的 Android 工程的 Java 包目录中建立一个扩展名为 aidl 的文件，该文件的语法类似于 Java 代码。

```
1  package com.ServiceRemote;
2  interface DisplayMsg{
3    void setMsg(String msg);
4    String displayMsg();
5  }
```

其次，如果 AIDL 文件的内容是正确的，保存文件时 ADT 会自动生成一个 Java 接口文件，扩展名为“.java”。

```
1   /* This file is auto-generated.  DO NOT MODIFY. * /
2   package com.ServiceRemote;
3   public interface DisplayMsg extends android.os.IInterface
4   {
5   /* *  Local-side IPC implementation stub class. * /
6   public static abstract class Stub extends
7                   android.os.Binder implements com.ServiceRemote.DisplayMsg
8   {
9   private static final java.lang.String DESCRIPTOR = "com.ServiceRemote.DisplayMsg";
10  /* *  Construct the stub at attach it to the interface. * /
11  public Stub()
12  {
13  this.attachInterface(this, DESCRIPTOR);
14  }
15  /* *
16      * Cast an IBinder object into an com.ServiceRemote.DisplayMsg interface,
17      * generating a proxy if needed.
18      * /
19  public static com.ServiceRemote.DisplayMsg asInterface(android.os.IBinder obj)
20  {
21  if ((obj = =null)) {
22  return null;
23  }
24  android.os.IInterface iin = (android.os.IInterface)obj.queryLocalInterface(DESCRIPTOR);
25  if (((iin! =null)&&(iin instanceof com.ServiceRemote.DisplayMsg))) {
26  return ((com.ServiceRemote.DisplayMsg)iin);
27  }
28  return new com.ServiceRemote.DisplayMsg.Stub.Proxy(obj);
```

```
29 }
30 public android.os.IBinder asBinder()
31 {
32 return this;
33 }
34 @Override public boolean onTransact(int code, android.os.Parcel data,
35             android.os.Parcel reply, int flags) throws android.os.RemoteException
36 {
37 switch (code)
38 {
39 case INTERFACE_TRANSACTION:
40 {
41 reply.writeString(DESCRIPTOR);
42 return true;
43 }
44 case TRANSACTION_setMsg:
45 {
46 data.enforceInterface(DESCRIPTOR);
47 java.lang.String _arg0;
48 _arg0 = data.readString();
49 this.setMsg(_arg0);
50 reply.writeNoException();
51 return true;
52 }
53 case TRANSACTION_displayMsg:
54 {
55 data.enforceInterface(DESCRIPTOR);
56 java.lang.String _result = this.displayMsg();
57 reply.writeNoException();
58 reply.writeString(_result);
59 return true;
60 }
61 }
62 return super.onTransact(code, data, reply, flags);
63 }
64 private static class Proxy implements com.ServiceRemote.DisplayMsg
65 {
66 private android.os.IBinder mRemote;
67 Proxy(android.os.IBinder remote)
68 {
69 mRemote = remote;
70 }
71 public android.os.IBinder asBinder()
72 {
73 return mRemote;
74 }
```

```
75    public java.lang.String getInterfaceDescriptor()
76    {
77    return DESCRIPTOR;
78    }
79    public void setMsg(java.lang.String msg) throws android.os.RemoteException
80    {
81    android.os.Parcel _data = android.os.Parcel.obtain();
82    android.os.Parcel _reply = android.os.Parcel.obtain();
83    try {
84    _data.writeInterfaceToken(DESCRIPTOR);
85    _data.writeString(msg);
86    mRemote.transact(Stub.TRANSACTION_setMsg, _data, _reply, 0);
87    _reply.readException();
88    }
89    finally {
90    _reply.recycle();
91    _data.recycle();
92    }
93    }
94    public java.lang.String displayMsg() throws android.os.RemoteException
95    {
96    android.os.Parcel _data = android.os.Parcel.obtain();
97    android.os.Parcel _reply = android.os.Parcel.obtain();
98    java.lang.String _result;
99    try {
100   _data.writeInterfaceToken(DESCRIPTOR);
101   mRemote.transact(Stub.TRANSACTION_displayMsg, _data, _reply, 0);
102   _reply.readException();
103   _result = _reply.readString();
104   }
105   finally {
106   _reply.recycle();
107   _data.recycle();
108   }
109   return _result;
110   }
111   }
112   static final int TRANSACTION_setMsg =
113                     (android.os.IBinder.FIRST_CALL_TRANSACTION + 0);
114   static final int TRANSACTION_displayMsg =
115                     (android.os.IBinder.FIRST_CALL_TRANSACTION + 1);
116   }
117   public void setMsg(java.lang.String msg) throws android.os.RemoteException;
118   public java.lang.String displayMsg() throws android.os.RemoteException;
119   }
```

第三，建立服务类，与本地服务类的建立相同，需要继承 Service 基类。

```
1  public class ServiceRemote extends Service{
2        private Stub m_clsDisplayMsgCls = new DisplayMsgCls();
3        @Override
4        public IBinder onBind(Intent intent) {
5              // TODO Auto-generated method stub
6              return m_clsDisplayMsgCls;
7        }
8  }
```

第四，实现由 AIDL 文件生成的 Java 接口。

```
1  public class DisplayMsgCls extends DisplayMsg.Stub{
2        private String m_strMsg = "";
3    @Override
4    public String displayMsg() throws RemoteException {
5          // TODO Auto-generated method stub
6          return "需要显示的信息为:" + m_strMsg;
7  }
8  @Override
9  public void setMsg(String msg) throws RemoteException {
10           // TODO Auto-generated method stub
11           this.m_strMsg = msg;
12       }
13 }
```

第五，在 AndroidManifest.xml 文件中配置 AIDL 服务，其中 <action> 标签中 android:name 的属性值就是客户端要引用该服务的 ID，也就是 Intent 类的参数值，与本地服务相同。

```
1    <service android:name=".ServiceRemote">
2        <intent-filter>
3              <action android:name="com.ServiceLocal.SERVICE_REMOTE"/>
4        </intent-filter>
5   </service>
```

远程服务建立完成后，可以通过 Activity 测试此服务，下面代码实现了上述远程服务的连接测试，其运行效果如图 5-3 所示。

```
1  public class MainAct extends Activity {
2      private DisplayMsg m_infDisplayMsg = null;
3      private String m_strActName = "com.ServiceLocal.SERVICE_REMOTE";
4        /** Called when the activity is first created. */
5        @Override
6        public void onCreate(Bundle savedInstanceState) {
7            super.onCreate(savedInstanceState);
8            setContentView(R.layout.main);
9            Button btnConn = (Button)findViewById(R.id.btnConn);
10            btnConn.setOnClickListener(new OnClickListener(){
11              @Override
12              public void onClick(View v) {
```

```
13              // TODO Auto-generated method stub
14              Intent inttService = new Intent();
15              inttService.setAction(m_strActName);
16              bindService(inttService, connService, Service.BIND_AUTO_CREATE);
17          }
18        });
19    }
20    private ServiceConnection connService = new ServiceConnection() {
21      @Override
22      public void onServiceConnected(ComponentName name, IBinder service) {
23        // TODO Auto-generated method stub
24        m_infDisplayMsg = DisplayMsg.Stub.asInterface(service);
25        if (m_infDisplayMsg ! = null)
26              try {
27                  //方法调用
28                  m_infDisplayMsg.setMsg("你好!");
29                  String strMsg = m_infDisplayMsg.displayMsg();
30                  //显示方法调用返回值
31                  Toast.makeText(MainAct.this, strMsg,
32                              Toast.LENGTH_LONG).show();
33              }catch (RemoteException e) {
34                     e.printStackTrace();
35              }
36        }
37        @Override
38        public void onServiceDisconnected(ComponentName name) {
39              // TODO Auto-generated method stub
40        }
41      };
42  }
```

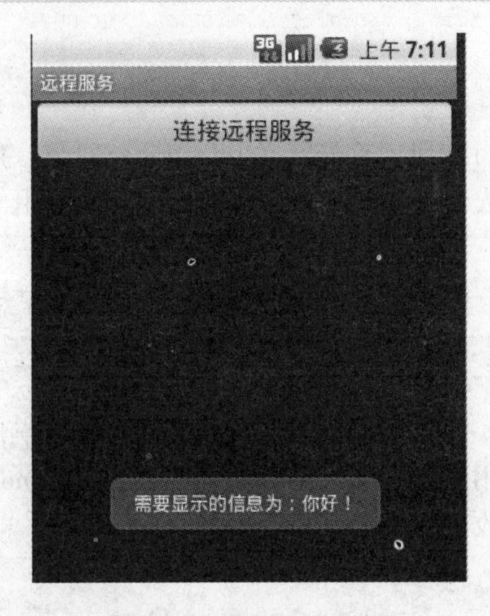

图 5-3 远程服务测试效果

5.2 BroadCastReciver 广播

广播（BroadCast）是 Android 系统中广泛使用的运用在应用程序间传递信息的一种机制。广播接收器（BroadCast Receiver）则是用于接收并处理这些广播通知的组件，是 Android 基本组件之一。它和事件处理机制类似，只不过事件的处理机制是程序组件级别的，而广播处理机制是系统级别的。

Android 系统中可以接收的广播有两种，一种是普通广播（Normal broadcasts），另一种是有序广播（Ordered broadcasts）。普通广播一般通过 Context. sendBroadcast( )发送，它是完全异步的，广播接收者的运行也是没有顺序的，几乎同时运行。有序广播通过 Context. sendOrderedBroadcast( )方法发送，按照接收者的优先级顺序接收广播，优先级别在 intent-filter 中的 priority 中声明。

使用广播接收器接收广播通知，需要首先定义一个广播接收器，定义广播接收器类需要继承 BroadcastReceiver 基类来实现，并且必须重写其中的 onReceive( )方法，此方法用于响应相应的广播事件处理。

```
1 public class myBroadcastReceiver extends BroadcastReceiver {
2   @Override
3   public void onReceive(Context context, Intent intent) {
4         // TODO Auto-generated method stub
5   }
6 }
```

完成广播接收器的定义后，在使用前必须进行注册。Android 系统提供两种注册广播接收器的方法，即静态注册和动态注册。静态注册是在 AndroidManifest. xml 文件中注册，通过添加 <receive>标签并在标签内用 <intent-filter>标签设置过滤器完成。

```
1  <receiver android:name = ". mySelfBroadcastReceiver" >
2  <intent-filter >
3      <action android:name = "com. BroadcastSelf. MYSELF_ACTION" />
4  </intent-filter >
5  </receiver >
```

动态注册是在代码中进行注册，首先定义 IntentFilter 对象，然后在需要注册的地方使用 Context. registerReceive( )方法注册广播接收器，取消注册使用 Context. unregisterReceive( )方法。

```
1 IntentFilter filter = new IntentFilter("com. BroadcastSelf. MYSELF_ACTION ");
2 mySelfBroadcastReceiver clsReceiver = new mySelfBroadcastReceiver ();
3 registerReceiver(clsReceiver, filter);
```

广播接收器注册完成后，使用广播接收器进行广播事件的响应同样也有两种方式，一种是主动广播，另外一种是使用系统标准广播。主动广播使用 sendBroadcast( )方法发送广播，使用注册过的广播接收器接收广播并处理相应事件。

```
1 Intent myIntent = new Intent();
2 myIntent. setAction("com. BroadcastSelf. MYSELF_ACTION");
```

```
3    myIntent.putExtra("strBroadcastMsg", "发送内容:123456;");
4  sendBroadcast(myIntent);
```

使用系统标准广播是指使用 Android 系统中定义的标准广播 Action，这种方式不需要使用 sendBroadcast( )方法进行广播发送，而是直接使用已经完成注册的广播接收器进行接收。常用的系统广播 Action 如表 5-2 所示。

表 5-2 Android 系统中定义的标准广播 Action

标准广播 Action	说 明
ACTION_TIME_TICK	时间改变，每分钟发送一次广播
ACTION_TIME_CHANGED	时间重新设置
ACTION_TIMEZONE_CHANGED	时区改变
ACTION_BOOT_COMPLETED	系统启动完成
ACTION_PACKAGE_ADDED	添加包
ACTION_PACKAGE_CHANGED	改变包
ACTION_PACKAGE_REMOVED	删除包
ACTION_PACKAGE_RESTARTED	重启包
ACTION_PACKAGE_DATA_CLEARED	清理包中数据
ACTION_UID_REMOVED	用户 ID 被删除
ACTION_BATTERY_CHANGED	电量改变
ACTION_POWER_CONNECTED	电源连接
ACTION_POWER_DISCONNECTED	电源断开
ACTION_SHUTDOWN	系统关闭

5.3 AppWidget 常驻程序

AppWidget 是 Application Widget 的缩写，是桌面组件的一部分，也称为常驻程序，是在 HomeScreen 上显示的小部件。AppWidget 的开发不同于普通的 Android 应用，AppWidget 是运行在别的进程中的程序，其使用 RemoteViews 更新 UI。一旦系统发生变更，很容易引起 AppWidget 的更新。其支持的组件有限，事件类型也很少，所以一般用于更新周期较长，事件比较简单的用于桌面显示的组件。对于桌面应用而言，AppWidget 所支持的组件类型有限，原因是 AppWidget 使用 RemoteViews 更新信息，而 RemoteViews 支持的布局样式类型有三种，分别为 RrameLayout、LinearLayout 和 RelativeLayout，支持的组件类型也有限，一般包含 AnalogClock、Button 、Chronometer、ImageButton 、ImageView、ProgressBar 、TextView 等几种常用桌面组件。

5.3.1 AppWidget 框架

AppWidget 是 Android 1.5 平台添加的应用程序框架，可以在 Android 操作系统的主窗体上放置常驻程序。在主窗体上长按触控面板，弹出的界面如图 5-4 所示，选择“窗口小部件”后即可显示所有已安装好的常驻程序列表。

使用 AppWidget 框架开发常驻程序时，经常用到的类有四个，分别是 AppWidgetProvider 、AppWidgetProviderInfo、AppWidgetManager 和 RemoteViews。

AppWidgetProvider 类定义了 App Widget 的基本声明周期函数，继承自 BroadcastReceiver 类，所以 AppWidget 的实质是一种广播机制。基本周期函数包括 onDeleted( )、onDisabled( )、onEnabled( )、onReceive( )和 onUpdate( )。onDeleted( )方法在 AppWidget 被删除时调用，其语法格式为：

```
public void onDeleted(Context context, int[] appWidgetIds)
```


图 5-4 添加应用到主屏幕

参数 context 为 AppWidget 运行的上下文环境，参数 appWidgetIds 为需要删除的 appWidgetIds 列表。

onDisabled( )方法的调用时机为最后一个 AppWidget 实例被删除时，此方法的语法格式为：

```
public void onDisabled(Context context)
```

onEnabled( )方法是在 AppWidget 首次被创建时调用，其语法格式为：

```
public void onEnabled(Context context)
```

onReceive( )方法在每一次广播此方法都被调用，并且先于其他回调方法，此方法的语法格式为：

```
public void onReceive(Context context, Intent intent)
```

参数 intent 为被接收的广播 intent。

onUpdate( )方法用于更新 AppWidget，此方法在 AppWidgetProviderInfo 中的属性 uodatePeriodMills 定义的时间间隔被调用，当用户添加并返回 AppWidget 时也被调用，其语法格式为：

```
public void onUpdate(Context context, AppWidgetManager appWidgetManager, int[] appWidgetIds)
```

AppWidgetProviderInfo 类为 AppWidget 提供元数据，包括布局、更新频率等数据，该对象定义在 XML 文件当中。AppWidgetManager 类负责管理 AppWidget 和向 AppWidgetProvider 发送消息。RemoteViews 类是一个可以运行在其他进程中的类，是构造 AppWidget 的基础元素，目前 Android 平台上的 RemoteViews 支持的布局仅有 RrameLayout、LinearLayout 和 RelativeLayout 三种，并且不支持自定义的类。

5.3.2 AppWidget 创建

创建 AppWidget 较为复杂，首先需要为 AppWidget 实现 Layout 布局样式，例如创建 app-widget. xml 布局文件，在布局文件中仅添加一个 ImageButton。

```
1   <? xml version = "1.0" encoding = "utf-8"? >
2   < LinearLayout xmlns:android = "http://schemas.android.com/apk/res/android"
3       android:orientation = "vertical"
4       android:layout_width = "fill_parent"
5       android:layout_height = "fill_parent"
6       >
7   < ImageButton
8       android:id = "@ + id/imgBtnImage"
9       android:layout_width = "wrap_content"
10      android:layout_height = "wrap_content"
11      android:src = "@ drawable/neulogo"
12      android:background = "#00000000"/ >
13  < /LinearLayout >
```

其次，实现 AppWidgetProvider 类，自定义 AppWidgetProvider 类需要继承 AppWidgetProvider 基类，类似于第 2 章测试 Activity 生命周期时的程序代码，下面代码实现了一个简单的自定义 AppWidgetProvider 类，并复写了包含的周期回调函数。

```
1 public class MyAppWidgetProvider extends AppWidgetProvider{
2     @ Override
3     public void onDeleted(Context context, int[] appWidgetIds) {
4           // TODO Auto-generated method stub
5           super.onDeleted(context, appWidgetIds);
6           System.out.println("onDeleted");
7 }
8 @ Override
9 public void onDisabled(Context context) {
10            // TODO Auto-generated method stub
11            super.onDisabled(context);
12            System.out.println("onDisabled");
13 }
14 @ Override
15 public void onEnabled(Context context) {
16            // TODO Auto-generated method stub
17            super.onEnabled(context);
18            System.out.println("onEnabled");
19 }
20 @ Override
21 public void onReceive(Context context, Intent intent) {
22            // TODO Auto-generated method stub
23            super.onReceive(context, intent);
24            System.out.println("onReceive");
```

```
25 }
26 @Override
27 public void onUpdate(Context context, AppWidgetManager appWidgetManager,
28         int[] appWidgetIds) {
29         // TODO Auto-generated method stub
30         super.onUpdate(context, appWidgetManager, appWidgetIds);
31         System.out.println("onUpdate");
32   }
33 }
```

第三，定义 AppWidgetProviderinfo 对象，需要在 res/xml 文件夹当中定义一个 XMI 文件，用于存放 AppWidget 的 provider 配置信息，例如定义名为 appwidgetconfig. xml 文件。

```
1 <?xml version="1.0" encoding="utf-8"?>
2 <appwidget-provider
3   xmlns:android="http://schemas.android.com/apk/res/android"
4   android:minWidth="260dp"
5   android:minHeight="200dp"
6   android:updatePeriodMillis="1000000"
7   android:initialLayout="@layout/appwidget">
8 </appwidget-provider>
```

属性 android：minWidth 定义 AppWidget 组件最小宽度，属性 android：minHeight 定义 AppWidget 组件最小高度，属性 android：updatePeriodMillis 定义更新的事件周期，属性 android：initialLayout 定义 AppWidget 的布局文件。

最后，配置清单文件，使用 <receiver> 标签注册 AppWidgetProvider，并在内部使用 <intent-filter> 元素的 <action> 子标签指定 AppWidget 接收的动作，使用 <meta-data> 标签指定 AppWidget 的布局文件。

```
1 <receiver android:name=".MyAppWidgetProvider" >
2   <intent-filter>
3       <action android:name="android.appwidget.action.APPWIDGET_UPDATE"/>
4   </intent-filter>
5   <meta-data android:name="android.appwidget.provider"
6         android:resource="@xml/appwidgetconfig"/>
7 </receiver>
```

至此，AppWidget 创建成功，添加窗口小部件后其效果图如图 5-5 所示。

图 5-5　添加窗口小部件

本章小结

本章主要讲述了 Android 后台服务、事件广播和常驻程序。后台服务有系统提供的 Service 组件实现，可分为本地服务和远程服务。事件广播机制只要依靠 BroadCast Reciver 组件实现。常驻程序 AppWidget 又称为窗口小部件，是在 HomeScreen 上显示的小部件，开发时常用 AppWidgetProvider 类和 AppWidgetProviderInfo 类实现。

习 题

5-1 试阐述 Service 两种启动方式的生命周期。

5-2 Service 分为哪两种类型？常用的操作 Service 的方法有哪些？

5-3 创建一个基于 Broadcast Receiver 的应用，使其当在模拟器中设置/日期和时间中调整一下系统时间时，自动启动该应用程序发出提示信息。

5-4 使用 AppWidget 框架开发常驻程序时，经常用到的类有哪些？创建一个个性化的 AppWidget。

第6章　媒介与网络

以娱乐为主的Android、iPhone等移动设备中多媒体开发占有很大比重，随着移动设备硬件性能的提高和外部存储设备容量的增加，需要开发功能完善、高质量的多媒体处理软件。Android平台为多种常见媒体类型提供了内建的编码/解码支持，因而可以通过程序实现音频、视频播放等操作。Android中提供了相应类来获取图像文件信息，进行图像的平移、旋转及缩放等操作，保存为指定格式图像文件。本章主要学习编写专业的绘图或控制图形动画的应用程序以及如何使用内置浏览器、获取网络资源。

6.1　Android音频与视频

音频和视频的播放要调用底层硬件，实现播放、暂停、停止、快进和快退等操作，在硬件层基础上是框架层，框架层音频和视频播放采用C和C++，比较复杂。本章主要介绍如何采用应用层API开发音频和视频播放应用程序。

6.1.1　Android音频/视频播放状态

音频及视频的播放会用到MediaPlayer类，该类提供了播放、暂停、重复、停止播放等方法。它是播放媒体文件最为常用的类。该类位于android. media包中，除了基本操作之外，还可以提供用于铃声管理、脸部识别以及音频路由控制的各种类。MediaPlayer类被设计用来播放大容量的音频文件以及同样可支持播放操作（停止、开始、暂停等）和查找操作的流媒体。其还可支持与媒体操作相关的监听器。Socket通信是在双方建立起连接后就可以直接进行数据的传输，在连接时可实现信息的主动推送，而不需要每次由客户端向服务器发送请求。Socket是一种抽象层，应用程序通过它来发送和接收数据，使用Socket可以将应用程序添加到网络中，与处于同一网络中的其他应用程序进行通信。简单来说，Socket提供了程序内部与外界通信的端口，并为通信双方提供了数据传输通道。

对播放音频/视频文件和流的控制是通过一个状态机来管理的。使状态发生转移的方法如表6-1所示。

表6-1　Mediaplayer类中常用方法

方　法	说　明
static MediaPlayer create (Context context, Uri uri)	静态方法，通过Uri创建一个多媒体播放器
int getCurrentPosition()	返回int，得到当前播放位置
int getDuration()	返回int，得到文件的时间
int getVideoHeight()	返回int，得到视频的高度
int getVideoWidth()	返回int，得到视频的宽度
boolean isLooping()	返回boolean，是否循环播放

（续）

方　法	说　明
boolean isPlaying( )	返回 boolean，是否正在播放
void pause( )	无返回值，暂停
void prepareAsync( )	无返回值，准备异步
void release( )	无返回值，释放 MediaPlayer 对象
void reset( )	无返回值，重置 MediaPlayer 对象
void start( )	无返回值，开始播放
void stop( )	无返回值，停止播放
void seekTo（int msec）	无返回值，指定播放的位置（以毫秒为单位的时间）
void setAudioStreamType（int streamtype）	无返回值，指定流媒体的类型
void setDataSource（String path）	无返回值，设置多媒体数据来源【根据路径】
void setDisplay（SurfaceHolder sh）	无返回值，设置用 SurfaceHolder 来显示多媒体
void setLooping（boolean looping）	设置是否循环播放
void setOnBufferingUpdateListener（MediaPlayer. OnBufferingUpdateListener listener）	监听事件，网络流媒体的缓冲监听
void setOnCompletionListener（MediaPlayer. OnCompletionListener listener）	监听事件，网络流媒体播放结束监听
void setOnErrorListener（MediaPlayer. OnErrorListener listener）	监听事件，设置错误信息监听
void setOnVideoSizeChangedListener（MediaPlayer. OnVideoSizeChangedListener listener）	监听事件，视频尺寸监听
void etVolume（float leftVolume，float rightVolume）	无返回值，设置音量

图 6-1 显示了一个 MediaPlayer 对象被支持的播放控制操作驱动的生命周期和状态。椭圆代表 MediaPlayer 对象可能驻留的状态。弧线表示驱动 MediaPlayer 对象在各个状态之间迁移的播放控制操作。

这张状态转换图清晰地描述了 MediaPlayer 对象的各个状态，也列举了主要的方法的调用时序，每种方法只能在一些特定的状态下使用，如果使用时 MediaPlayer 对象的状态不正确，则会引发 IllegalStateException 异常 。Android 音频/视频有如下 10 种状态。

Idle 状态：当使用 new( )方法创建一个 MediaPlayer 对象或者调用了其 reset( )方法时，该 MediaPlayer 对象处于 Idle 状态。这两种方法的一个重要差别就是：如果在这个状态下调用了 getDuration( ) 等方法（相当于调用时机不正确），通过 reset( )方法进入 Idle 状态的话会触发 OnErrorListener. onError( ) ，并且 MediaPlayer 对象会进入 Error 状态；如果是新创建的 MediaPlayer 对象，则并不会触发 onError( )，也不会进入 Error 状态。

End 状态：通过 release( ) 方法可以进入 End 状态，只要 MediaPlayer 对象不再被使用，就应当尽快将其通过 release( ) 方法释放掉，以释放相关的软硬件组件资源，这其中有些资源是只有一份的（相当于临界资源）。如果 MediaPlayer 对象进入了 End 状态，则不会再进入任何其他状态了。

Initialized 状态：这个状态比较简单，MediaPlayer 对象调用 setDataSource( ) 法就进入 Initialized 状态，表示此时要播放的文件已经设置好了。

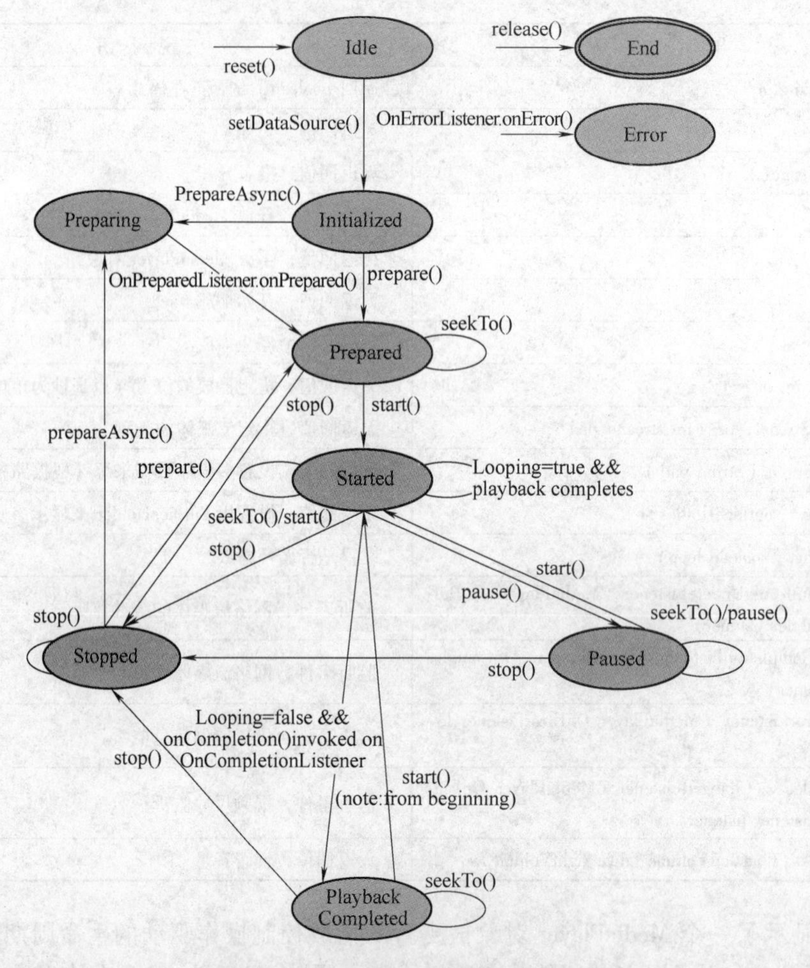

图 6-1　Android 音频/视频播放状态图

Prepared 状态：初始化完成之后还需要通过调用 prepare( )或 prepareAsync( )方法，这两个方法一个是同步的一个是异步的，只有进入 Prepared 状态，才表明 MediaPlayer 到目前为止都没有错误，可以进行文件播放。

Preparing 状态：这个状态比较好理解，主要是和 prepareAsync( )方法配合，如果异步准备完成，会触发 OnPreparedListener. onPrepared( )方法，进而进入 Prepared 状态。

Started 状态：显然，MediaPlayer 对象一旦准备好，就可以调用 start( ) 方法，这样 MediaPlayer 对象就处于 Started 状态，这表明 MediaPlayer 对象正在播放文件过程中。可以使用 isPlaying( )方法测试 MediaPlayer 对象是否处于 Started 状态。如果播放完毕，而又设置了循环播放，则 MediaPlayer 仍然会处于 Started 状态，类似地，如果在该状态下 MediaPlayer 对象调用了 seekTo( )方法或者 start( ) 方法均可以让 MediaPlayer 对象停留在 Started 状态。

Paused 状态：Started 状态下 MediaPlayer 对象调用 pause( ) 方法可以暂停 MediaPlayer 对象，从而进入 Paused 状态，MediaPlayer 对象暂停后再次调用 start( )方法则可以继续 MediaPlayer 对象的播放，转到 Started 状态，暂停状态时可以调用 seekTo( ) 方法，这是不会改变状态的。

Stop 状态：Started 或者 Paused 状态下均可调用 stop( )方法停止 MediaPlayer 对象，而处于 Stop 状态的 MediaPlayer 对象要想重新播放，需要通过 prepareAsync( )方法和 prepare( )方法回到先前的 Prepared 状态重新开始才可以。

PlaybackCompleted 状态：文件正常播放完毕，而又没有设置循环播放的话就进入该状态，并会触发 OnCompletionListener 的 onCompletion( ) 方法。此时可以调用 start( ) 方法重新从头播放文件，也可以调用 stop( )方法停止 MediaPlayer 对象，或者也可以调用 seekTo( )方法来重新定位播放位置。

Error 状态：如果由于某种原因 MediaPlayer 对象出现了错误，会触发 OnErrorListener. onError( ) 事件，此时 MediaPlayer 对象即进入 Error 状态，及时捕捉并妥善处理这些错误是很重要的，它可以及时释放相关的软硬件资源，改善用户体验。通过 setOnErrorListener (android. media. MediaPlayer. OnErrorListener) 可以设置该监听器。如果 MediaPlayer 对象进入了 Error 状态，可以通过调用 reset( )方法来恢复，使得 MediaPlayer 对象重新返回到 Idle 状态。

6.1.2 Android 音频播放

本小节给出了音频播放案例，前面介绍了音频/视频播放状态和方法，下面介绍软件播放音频文件。播放 MediaPlayer 中的音频和视频可以通过以下三种方式：从源文件播放，从文件系统播放和从流媒体播放。从源文件播放是指资源文件放在" /res/raw" 文件夹下，然后发布时被打包成 APK 一起安装在手机上。从文件系统播放是指在 Android 系统的外部存储设备（如 SD 卡）和内部设备上的文件播放。流媒体播放则指放在网络上的文件，也是流媒体等网络资源播放。下面分别进行详细介绍。

1. 从源文件播放

下面说明应用 MediaPlayer 对象播放一个音频文件的步骤：

（1）使用 new 的方式创建播放器对象：new MediaPlayer( )，也可以使用 MediaPlayer. create (this, R. raw. test) 函数，这时就不用调用 setDataSource ( ) 方法了。

（2）设置音频源：使用 setDataSource (Audio_ PATH) 方法。

（3）准备音频源：使用 prepare( )方法。

（4）播放音频：使用 start( )方法启动音频播放的方法，例如：mplayer. start( )方法。

（5）停止播放：如要暂停播放或停止播放，则调用 pause( )方法或 stop( )方法。

（6）释放资源：音频文件播放结束应该调用 release( )方法释放播放器占用的系统资源。

如果要重新播放音频文件，需要调用 reset( )方法返回空闲状态，再从第（2）步开始重复其他各步骤。

【例 6-1】 使用 mediaplayer 类从源文件中播放音频。

（1）新建名为 Mediaplayer 的工程，在项目的 res/raw 文件夹下面放入 Android 支持的文件，如本例中放入 jnstyle. mp3。

（2）在 res/drawablewen 文件夹下分别放入 icon. png，bg. png，play. png，pause. png，stop. png。

（3）在 res/values/string. xml 文件中添加如下代码：

```
1   <?xml version="1.0"encoding="utf-8"?>
2   <AbsoluteLayout
3       android:background="@drawable/bg"
4       android:id="@+id/widget32"
5       android:layout_width="fill_parent"
6       android:layout_height="fill_parent"
7       xmlns:android="http://schemas.android.com/apk/res/android">
8   <TextView
9       android:layout_x="300px"
10       android:layout_y="10px"
11       android:id="@+id/myTextView1"
12       android:layout_width="wrap_content"
13       android:layout_height="wrap_content"
14       android:text="@string/hello"
15       android:textColor="@drawable/yellow">
16    </TextView>
17    <ImageButton
18       android:id="@+id/myButton1"
19       android:layout_width="wrap_content"
20       android:layout_height="wrap_content"
21       android:src="@drawable/play"
22       android:layout_x="100px"
23       android:layout_y="60px">
24    </ImageButton>
25    <ImageButton
26       android:id="@+id/myButton3"
27       android:layout_width="wrap_content"
28       android:layout_height="wrap_content"
29       android:src="@drawable/pause"
30       android:layout_x="300px"
31       android:layout_y="180px">
32    </ImageButton>
33    <ImageButton
34       android:id="@+id/myButton2"
35       android:layout_width="wrap_content"
36       android:layout_height="wrap_content"
37       android:src="@drawable/stop"
38       android:layout_x="540px"
39       android:layout_y="800px">
40    </ImageButton>
41    </AbsoluteLayout>
```

（4）在 res/values/color.xml 文件中添加如下代码：

```
1   <?xml version="1.0" encoding="utf-8"?>
2   <resources>
3   <drawable name="yellow">#FF8000</drawable>
4   </resources>
```

（5）在 mediaplayer/mediaplayer. sample/MainActivity. java 文件中添加如下代码：

```
1   package mediaplaer. sample;
2   import android. app. Activity;
3   import android. content. ContextWrapper;
4   import android. media. MediaPlayer;
5   import android. os. Bundle;
6   import android. view. View;
7   import android. widget. ImageButton;
8   import android. widget. TextView;
9   public class MainActivity extends Activity
10          {
11                /* 声明一个 ImageButton,TextView,MediaPlayer 变量* /
12                private ImageButton mButton01, mButton02, mButton03;
13                private TextView mTextView01;
14                private MediaPlayer mMediaPlayer01;
15                /* 声明一个 Flag 作为确认音乐是否暂停的变量并预设为 false* /
16                private boolean bIsPaused = false;
17                /* *  Called when the activity is first created. * /
18                @ Override
19  public void onCreate(Bundle savedInstanceState)
20          {
21                super. onCreate(savedInstanceState);
22                setContentView(R. layout. main);
23                /* 透过 findViewById 建立 TextView 与 ImageView 对象* /
24                mButton01 = (ImageButton) findViewById(R. id. myButton1);
25                mButton02 = (ImageButton) findViewById(R. id. myButton2);
26                mButton03 = (ImageButton) findViewById(R. id. myButton3);
27                mTextView01 = (TextView) findViewById(R. id. myTextView1);
28                /*  onCreate 时建立 MediaPlayer 对象 * /
29                mMediaPlayer01 = new MediaPlayer();
30                /*  将音乐以 Import 的方式保存 res/raw/always. mp3 * /
31                mMediaPlayer01 = MediaPlayer. create(MainActivity. this, R. raw. jnstyle);
32                /*  执行播放音乐的按钮 * /
33                mButton01. setOnClickListener(new ImageButton. OnClickListener()
34                  {
35                      @ Override
36                      /* 重写 OnClick 事件* /
37                      public void onClick(View v)
38                            {
39                                // TODO Auto-generated method stub
40                                try
41                                    {
42                                        if (mMediaPlayer01 ! = null)
43                                              {
44                                                    mMediaPlayer01. stop();
45                                    }
```

```
46                  /* MediaPlayer 取得播放资源与 stop()之后
47                  * 要准备 Playback 的状态前一定要使用 MediaPlayer.prepare()* /
48                  mMediaPlayer01.prepare();
49                  /* 开始或反复播放* /
50                  mMediaPlayer01.start();
51                  /* 改变 TextView 为开始播放状态* /
52                  mTextView01.setText(R.string.start);
53                  }
54                  catch (Exception e)
55                  {
56                      // TODO Auto-generated catch block
57                      mTextView01.setText(e.toString());
58                      e.printStackTrace();
59                  }
60              }
61      });
62      /* 停止播放 * /
63      mButton02.setOnClickListener(new ImageButton.OnClickListener()
64      {
65          @Override
66          public void onClick(View arg0)
67              {
68                  // TODO Auto-generated method stub
69                  try
70                      {
71                          if (mMediaPlayer01 ! = null)
72                          {
73                              /* 停止播放* /
74                              mMediaPlayer01.stop();
75                              /* 改变 TextView 为停止播放状态* /
76                              mTextView01.setText(R.string.close);
77                          }
78                      }
79              catch (Exception e)
80                  {
81                      // TODO Auto-generated catch block
82                      mTextView01.setText(e.toString());
83                      e.printStackTrace();
84                  }
85              }
86      });
87      /* 暂停播放 * /
88      mButton03.setOnClickListener(new ImageButton.OnClickListener()
89          {
90              @Override
91              public void onClick(View arg0)
```

```
92                    {
93                    // TODO Auto-generated method stub
94                    try
95                    {
96                          if (mMediaPlayer01 ! = null)
97                               {
98                                    /* 是否为暂停状态 = 否* /
99                                         if(bIsPaused = = false)
100                                        {
101                                              /* 暂停播放* /
102                                             mMediaPlayer01. pause();
103                               /* 设定 Flag 为 true 表示 Player 状态为暂停* /
104                                              bIsPaused = true;
105                                              /* 改变 TextView 为暂停播放* /
106                                        mTextView01. setText(R. string. pause);
107                                        }
108                                   /* 是否为暂停状态 = 是* /
109                                   else if(bIsPaused = = true)
110                                   {
111                                         /* 重复播放状态* /
112                                         mMediaPlayer01. start();
113                              /* 设定 Flag 为 false 表示 Player 状态为非暂停状态* /
114                                                   bIsPaused = false;
115                                                   /* 改变 TextView 为开始播放* /
116                                                  mTextView01. setText(R. string. start);
117                                              }
118                                         }
119                                    }
120                        catch (Exception e)
121                        {
122                                   // TODO Auto-generated catch block
123                                   mTextView01. setText(e. toString());
124                                   e. printStackTrace();
125                        }
126                    }
127                    });
128                    /*  当 MediaPlayer. OnCompletionLister 会执行的 Listener * /
129    mMediaPlayer01. setOnCompletionListener(new MediaPlayer. OnCompletionListener()
130              {
131                  // @ Override
132                  /* 重写文件播放完毕事件* /
133                  public void onCompletion(MediaPlayer arg0)
134                        {
135                              try
136                                 {
137                                   /* 解除资源与 MediaPlayer 的指派关系 * /
```

```
138                                mMediaPlayer01.release();
139                                /* 改变 TextView 为播放结束* /
140                            mTextView01.setText(R.string.OnCompletionListener);
141                            }
142                        catch (Exception e)
143                          {
144                               mTextView01.setText(e.toString());
145                               e.printStackTrace();
146                          }
147                      }
148                  });
149                  /* 当 MediaPlayer.OnErrorListener 会执行的 Listener * /
150              mMediaPlayer01.setOnErrorListener(new MediaPlayer.OnErrorListener()
151                {
152                  @Override
153                  /* 重写错误处理事件* /
154                  public boolean onError(MediaPlayer arg0, int arg1, int arg2)
155                      {
156                          // TODO Auto-generated method stub
157                          try
158                          {
159                              /* 发生错误时解除资源与 MediaPlayer 的指派* /
160                              mMediaPlayer01.release();
161                              mTextView01.setText(R.string.OnErrorListener);
162                          }
163                  catch (Exception e)
164                      {
165                          mTextView01.setText(e.toString());
166                          e.printStackTrace();
167                      }
168                return false;
169                }
170            });
171        }
172        private ContextWrapper getContext()
173        {
174            // TODO Auto-generated method stub
175            return null;
176        }
177        @Override
178        /* 加暂停状态事件* /
179        protected void onPause()
180            {
181                // TODO Auto-generated method stub
182                try
183                {
```

```
184                          /* 暂停时解除资源与 MediaPlayer 的指派关系* /
185                          mMediaPlayer01.release();
186                      }
187                      catch (Exception e)
188                      {
189                          mTextView01.setText(e.toString());
190                          e.printStackTrace();
191                      }
192                      super.onPause();
193                  }
194          }
```

(6) 运行该程序结果如图 6-2 所示。

图 6-2 音频播放示例

2. 从文件系统播放

从文件系统播放音乐（如 SDCard/TFCard），需要使用 new 操作符创建 MediaPlayer 对象。获得 MediaPlayer 对象之后，需要依次调用 setDataSource( )方法和 prepare( )方法，以便设置数据源，让播放器完成准备工作，然后调用 start( )方法播放音乐。

程序代码如下：

```
1    MediaPlayer MyMediaPlayer = new MediaPlayer(); //实例化 MediaPlayer
2    String path  =  "/SDCard/she.mp3";                    //播放路径
3    MyMediaPlayer.setDataSource(path);                    //获取播放资源
4    MyMediaPlayer.prepare();                              //准备
5    MyMediaPlayer..start();                               //播放
```

3. 从网络播放

Android 平台支持在线播放媒体音乐，通过 progress download 的方式播放在线音频资源。Progress download 由底层的 OpenCore 多媒体库提供支持，应用开发者不必关心具体实现细节，只需要设置网络音频资源的地址，就可以完成在线播放的工作，极大地提高了开发效率。由于从网络下载播放音频资源需要较长的时间，在准备音频资源的时候，需要使用 prepareAsync( )方法，这个方法是异步执行的，不会阻塞程序的主进程。MediaPlayer 对象通过 MediaPlayer. OnPreparedListener 通知 MediaPlayer 的准备状态。

在特定的情况下，程序需要通过代理服务器访问在线资源，例如，通过 APN CMWAP 访问网络时，需要设置 CMWAP 的代理地址（如 10.0.0.172：80）。Android 平台提供了非常方便的解决方案，使开发者可以非常简单地通知 OpenCore 使用指定的代理来访问网络音频资源：在网络媒体的 URL 后，通过“x-http-proxy”指定代理服务器地址。另外，由于 Android 平台支持 Multi PDP，可以同时建立多个 APN 连接，所以开发者必须指明哪个连接端口需要使用代理服务器，通过“x-net-interface”指定连接端口（如何建立 CMWAP 数据连接，获取连接端口名，本文不再详细叙述，请读者查看相关的 Android 技术文章），下面的程序简单说明了通过代理服务器播放音频资源的步骤：

```
1  private void playFromNetwork() {
2          String path = "http://website/path/test.mp3";//设置资源路径
3          String CMWAP_HOST = "10.0.0.172";
4          String CMWAP_PORT = "80";                //设置 CMWAP 的代理地址
5          String SOCKET_INTERFACE = "cminnet0";    //假设 CMWAP 连接的端口名
6          String urlWithProxy = path + "? x-http-proxy =" + CMWAP_HOST + ":" + CMWAP_PORT + "&" +
"x-net-interface =" + SOCKET_INTERFACE ;//指定代理服务器地址
7          try {
8              MediaPlayer MyMediaPlayer = new MediaPlayer();//实例化 MediaPlayer
9              player.setDataSource(path);            //获取资源路径
10             player.setOnPreparedListener(new MediaPlayer.OnPreparedListener() {
11             public void onPrepared(MediaPlayer player) {
12             MyMediaPlayer.start();      //播放}
13             });
14         player.prepareAsync();                  //准备
15         } catch (Exception e) {
16         e.printStackTrace();}
17  }
```

6.1.3 Android 视频播放

视频播放同样用到 MediaPlayer 类，包括状态和状态管理都是一样的。由于采用 Medical-Player 类开发视频播放，用于视频播放的播放承载体必须是实现了表面视图处理接口的（SurfaceHolder）视图组件，即需要使用 SurfaceView 组件来显示播放的视频图像。因此，Android 也可以使用一个封装好的视频播放控件——VideoView 控件。下面介绍如何实现视频播放。

在 Android 系统中，经常使用 android. widget 包中的视频视图类 VideoView 播放视频文件。VideoView 类可以从不同的来源读取图像，计算和维护视频的画面尺寸，以使其适应于任何管理器，并提供一些诸如缩放、着色之类的显示选项。VideoView 中常用方法见表 6-2。

表 6-2 VideoView 对象中常用方法

方 法	说 明
VideoView（Context context）	创建一个默认属性的 VideoView 实例
boolean canPause( )	返回 boolean，判断是否能够暂停播放视频
int getBufferPercentage( )	返回 int，获得缓冲区的百分比
int getCurrentPosition( )	返回 int，获得当前的位置
int getDuration( )	返回 int，得到播放视频的总时间
boolean isPlaying( )	返回 boolean，是否正在播放视频
boolean onTouchEvent(MotionEvent ev)	返回 boolean，应用该方法来处理触屏事件
void seekTo(int msec)	无返回值，设置播放位置
void setMediaController(MediaController Controler)	无返回值，设置媒体控制器
void setOnCompletionListener(MediaPlayer. OnCompletion-Listener 1)	无返回值，注册在媒体文件播放完毕时调用的回调函数
void setOnPreparedListener(MediaPlayer. OnPreparedListener 1)	无返回值，注册在媒体文件加载完毕时调用的回调函数
void setVideoPath(String path)	无返回值，设置视频文件的路径
void setVideoURI(Uri uri)	无返回值，设置视频文件的统一资源标识符
void start( )	无返回值，开始播放视频文件
void stopPlayback( )	无返回值，回放视频文件

【例 6-2】 应用 VideoView 组件设计一个视频播放器。

其具体步骤如下所示：

（1）新建名为 MyVideoPlayer 的工程，在项目的 res/raw 文件夹下面放入 Android 支持的文件，如本例中放入 JNStyle. mp4。

（2）在 res/layout/activity. main 文件中添加如下代码：

```
1  <LinearLayout xmlns:android="http://schemas.android.com/apk/res/android"
2       android:orientation="vertical"
3       android:layout_width="fill_parent"
4       android:layout_height="fill_parent"
5  >
6  <VideoView //声明一个 videoview 组件
7       android:id="@+id/video"
8       android:layout_width="fill_parent"
9       android:layout_height="fill_parent"
10 />
11 </LinearLayout>
```

（3）在 MyVideo/com.example.myvideo/MainActivity.java 文件中添加如下代码：

```
1   package com.example.myvideo;
2   import java.io.File;
3   import android.os.Bundle;
4   import android.app.Activity;
5   import android.graphics.PixelFormat;
6   import android.widget.MediaController;
7   import android.widget.VideoView;

8   public class MainActivity extends Activity {
9   //声明视频视图 VideoView
10  private VideoView video;
11  //声明媒体控制组件 MediaController
12  private MediaController ctlr;
13  @Override
14  protected void onCreate(Bundle savedInstanceState) {
15  super.onCreate(savedInstanceState);
16  //设置窗口特征
17  getWindow().setFormat(PixelFormat.TRANSLUCENT);
18  //设置当前 Activity 布局
19  setContentView(R.layout.activity_main);
20  //实例化 File
21  File clip = new File("/sdcard/JNStyle.mp4");
22  //判断文件是否存在
23  if(clip.exists()){
24      //通过 findViewById 方法获得 VideoView 实例
25      video=(VideoView)findViewById(R.id.video);
26      //设置视频播放路径
27      video.setVideoPath(clip.getAbsolutePath());
28      //实例化 MediaController
29      ctlr=new MediaController(this);
30      //MediaController 和 MediaPlayer 相互关联
31      ctlr.setMediaPlayer(video);
32      //使 VideoView 获得焦点
33      video.requestFocus();
```

```
34          }
35      }
36  }
```

程序的运行结果如图 6-3 所示。

图 6-3 手机中显示的 VideoView 组件设计视频播放器

6.2 Android 图形绘制与特效

6.2.1 几何图形绘制类

在 Android 中涉及几何图形绘制的这些工具类都很形象。在绘制图形中，首先需要一张画布，这里就是 Android 中的 Canvas；其次还需要画笔，这里就是 Android 中的 Paint；再次需要不同的颜色，这里就是 Android 中的 Color。接下来如果要画线还需要连接路径，这里就是 Android 中的 Path。还可以借助工具直接画出各种图形如圆、椭圆、矩形等，这里就是 Android 中的 ShapeDrawable 类，当然它还有很多子类，例如，OvalShape（椭圆）、RectShape（矩形）等。

1. 画布（Canvas）

Canvas 就是我们所说的画布，位于 android. graphics 包中，提供了一些画各种图形的方法，例如，矩形、圆、椭圆等。该类的常用方法如表 6-3 所示。

表 6-3　Canvas 常用方法

方法名称	方法描述
void Canvas（）	创建一个空的画布，可以使用 setBitmap()方法来设置绘制具体的画布
void Canvas（Bitmap bitmap）	以 bitmap 对象创建一个画布，将内容都绘制在 bitmap 上，bitmap 不得为 null
void drawColor()	设置 Canvas 的背景颜色
setBitmap()	设置具体画布
boolean clipRect(RectF rect)	设置显示区域，即设置裁剪区
void rotate(float degress)	旋转画布
Skew()	设置偏移量
void drawText（String text，float x，float y，Paint paint）	以（x，y）为起始坐标，使用 paint 绘制文本
void drawPoint（float x，float y，Paint paint）	在坐标（x，y）上使用 paint 画点
void drawLine（float startX，float startY，float stopX，float stopY，Paint paint）	以（startX，startY）为起始坐标点，（stopX，stopY）为终止坐标点，使用 paint 画线
void drawCircle(float cx，float cy，float radius，Paint paint)	以（cx，cy）为原点，radius 为半径，使用 paint 画圆
void drawOval（RectF oval，Paint paint）	使用 paint 画矩形 oval 的内切椭圆
void DrawRect（RectF rect，Paint paint）	使用 paint 画矩形 rect
void drawRoundRect（RectF rect，float rx，float ry，Paint paint）	画圆角矩形
void clipRect(float left，float top，float right，float bottom)	剪辑矩形
boolean clipRegion（Region region）	剪辑区域

2. 画笔（Paint）

Paint 用来描述图形的颜色和风格，如线宽、颜色、字体等信息。Paint 位于 android. graphics 包中，该类的常用方法如表 6-4 所示。

表 6-4　Paint 常用方法

方法名称	方法描述
Paint()	构造方法，使用默认设置
void setColor（int color）	设置颜色
void setStrokeWidth（float width）	设置线宽
void setTextAlign（Paint. Align align）	设置文字对齐
void setTextSize（float textSize）	设置文字尺寸
shader setShader（Shader shader）	设置渐变
void setAlpha（int a）	设置 Alpha 值
void reset()	复位 Paint 默认设置

3. 颜色（Color）

Color 类定义了一些颜色变量和一些创建颜色的方法。颜色的定义一般使用 RGB 三原色定义。Color 位于 android. graphics 包中，其常用属性和方法如表 6-5 所示。

表 6-5 Color 常用属性和方法

属性名称	方法描述
BALCK	黑色
BLUE	蓝色
CYAN	青色
DKGRAY	深灰色
GRAY	灰色
GREEN	绿色
LTGRAY	浅灰色
MAGENTA	紫色
RED	红色
TRANSPARENT	透明
WHITE	白色
YELLOW	黄色

4. 点到点的连接路径（Path）

当想要画一个圆的时候，只需要制定圆心（点）和半径就可以了。那么，如果要画一个梯形呢？需要有点和连线。Path 一般用来画从一点到另一个点之间的连线。Path 位于 android. graphics 包中，其常用方法如表 6-6 所示。

表 6-6 Path 常用方法

方法名称	方法描述
void lineTo（float x，float y）	从最后点到指定点画线
void moveTo（float x，float y）	移动到指定点
void reset( )	复位

6.2.2 图形绘制过程

Android 框架 API 提供了一组 2D 描画 API，使用这些 API 能够在一个画布（Canvas）上渲染自己的定制图形，也能够修改那些既存的 View 对象，来定制它们的外观和视觉效果。android. graphics. drawable 包中能够找到用于绘制二维图形的共同的类。在绘制 2D 图形时，通常要使用以下两种方法中的一种：

（1）把图形或动画绘制到布局中的一个 View 对象中。在这种方式中，图形的绘制是由系统通常的绘制 View 层次数据的过程来处理的，只需简单定义要绘制到 View 对象内的图形即可。

（2）把图形直接绘制在一个画布对象上（Canvas 对象）。这种方法，要亲自调用相应类的 onDraw( )方法（把图形传递给 Canvas 对象），或者调用 Canvas 对象的一个方法（如 drawPicture( )）。在这个过程中，还可以控制任何动画。

当想要把不需要动态变化和没有游戏性能要求的一个简单的图形绘制到 View 对象时，方法（1）是最好的选择。当应用程序需要经常重新绘制自己的时候，使用方法（2）把图形绘制到 Canvas 中，是一个比较好的选择。

1. 用 Canvas 对象来绘制图形（Draw with a Canvas）

当要编写专业的绘图或控制图形动画的应用程序时，应该使用 Canvas 对象来进行绘制操作。Canvas 用一个虚拟的平面来工作，以便把图形绘制在实际的表面上。Canvas 对象持有所有的用 draw 开头的方法调用。对 Canvas 对象的操作，实际上是执行一个底层的位图绘制处理，这个位图被放置到窗口中。

在 onDraw( )回调方法的绘制事件中，会提供一个 Canvas 对象，并且只需要把要绘制的内容交给 Canvas 对象就可以了。在处理 SurfaceView 对象时，还可以从 SurfaceHolder. lockCanvas( )方法来获取一个 Canvas 对象。但是，如果需要创建一个新的 Canvas 对象，那么就必须在实际执行绘制处理的 Canvas 对象上定义 Bitmap 对象。对于 Canvas 对象来说，这个 Bitmap 对象始终是必须的，应该像以下示例这样建立一个新的 Canvas 对象：

```
Bitmap b = Bitmap.createBitmap(100, 100, Bitmap.Config.ARGB_8888);
Canvas c = new Canvas(b);
```

现在就可以在被定义的 Bitmap 对象上绘图了。在 Canvas 对象上绘制图形之后，能够用 Canvas. drawBitmap（Bitmap，…）的一个方法，把该 Bitmap 对象绘制到另一个 Canvas 对象中。通过 View. onDraw( )方法或 SufaceHolder. lockCanvas( )方法提供的 Canvas 对象来完成最终的图形绘制处理是被推荐的。

Canvas 类有自己的一组绘图方法，如 drawBitmap（ ）、drawRect（ ）、drawText（ ）等。还可以使用其他的 draw( )方法类。例如，可能想要把某些 Drawable 对象放到 Canvas 对象上。Drawable 类就有带有 Canvas 对象作为参数的 draw( )方法。

2. 在 View 对象上绘图

如果应用程序不需要大量的图形处理或很高的帧速率（如一个棋类游戏、Snake 游戏或另外的慢动画类应用程序），那么就应该考虑创建一个定制的 View 组件，并且用该组件的 View. onDraw( )方法的 Canvas 参数来进行图形绘制。这么做最大的方便是，Android 框架会提供一个预定义的 Canvas 对象，该对象用来放置绘制图形的调用。

从继承 View 类（或其子类）开始，并定义 onDraw( )回调方法。系统会调用该方法来完成 View 对象自己的绘制请求。这也是通过 Canvas 对象来执行所有的图形绘制调用的地方，这个 Canvas 对象是由 onDraw( )回调方法传入的。

Android 框架只在必要的时候才会调用 onDraw( )方法，每次请求应用程序准备完成图形绘制任务时，必须通过调用 invalidate( )方法让该 View 对象失效。这表明可以在该 View 对象上进行图形绘制处理了，然后 Android 系统会调用该 View 对象的 onDraw( )方法（尽管不保证该回调方法会立即被调用）。

在定制的 View 组件的 onDraw( )方法内部，使用给定的 Canvas 对象来完成所有的图形绘制处理（如 Canvas. draw( )方法或把该 Canvas 对象作为参数传递给其他类的 draw( )方法）。一旦 onDraw( )方法被执行完成，Android 框架就会使用这个 Canvas 对象来绘制一个有系统处理的 Bitmap 对象。

【例 6-3】 下面用 Canvas 对象来绘制五星图形。

（1）新建名为 Star 的工程，在 res/values/attr. xml 文件中添加如下代码：

```
1  <?xml version="1.0" encoding="utf-8"?>
2  <resources>
3  <declare-styleable name="Stars">
4  <attr name="color" format="color" />
5  <attr name="radius" format="float" />
6  </declare-styleable>
7  </resources>
```

（2）在 res/layout/activity_main. xml 文件中添加如下代码：

```
1   <RelativeLayout xmlns:android="http://schemas.android.com/apk/res/android"
2        xmlns:tools="http://schemas.android.com/tools"
3        xmlns:test="http://schemas.android.com/apk/res/com.cs.stars"
4        android:layout_width="match_parent"
5        android:layout_height="match_parent"
6        android:paddingBottom="@dimen/activity_vertical_margin"
7        android:paddingLeft="@dimen/activity_horizontal_margin"
8        android:paddingRight="@dimen/activity_horizontal_margin"
9        android:paddingTop="@dimen/activity_vertical_margin"
10       tools:context=".MainActivity" >
11        <com.cs.stars.Stars
12       android:id="@+id/stars1"
13       android:layout_marginTop="20dip"
14       android:layout_width="wrap_content"
15       android:layout_height="wrap_content"
16       test:color="@color/red"
17       test:radius="100"
18  />
19  <com.cs.stars.Stars
20       android:id="@+id/stars2"
21       android:layout_width="wrap_content"
22       android:layout_height="wrap_content"
23       android:layout_toRightOf="@id/stars1"
24       test:color="@color/red"
25       test:radius="40"
26  />
27  <com.cs.stars.Stars
28       android:id="@+id/stars3"
29       android:layout_width="wrap_content"
30       android:layout_height="wrap_content"
31       android:layout_toRightOf="@id/stars2"
32       android:layout_below="@id/stars2"
33       test:color="@color/red"
34       test:radius="40"
```

```
35    />
36    <com.cs.stars.Stars
37          android:id="@+id/stars4"
38          android:layout_width="wrap_content"
39          android:layout_height="wrap_content"
40          android:layout_marginTop="2dip"
41          android:layout_toRightOf="@id/stars2"
42          android:layout_below="@id/stars3"
43          test:color="@color/red"
44          test:radius="40"
45    />
46    <com.cs.stars.Stars
47          android:id="@+id/stars5"
48          android:layout_width="wrap_content"
49          android:layout_height="wrap_content"
50          android:layout_toRightOf="@id/stars1"
51          android:layout_below="@id/stars4"
52          test:color="@color/red"
53          test:radius="40"
54    />
55    </RelativeLayout>
```

（3）在 Stars/Stars/src/com/stars 文件夹下新建 Stars.java 并添加如下代码：

```
1   package com.cs.stars;
2   import android.content.Context;
3   import android.content.res.TypedArray;
4   import android.graphics.Bitmap;
5   import android.graphics.Canvas;
6   import android.graphics.Paint;
7   import android.graphics.Path;
8   import android.util.AttributeSet;
9   import android.view.View;
10  import android.view.ViewGroup.LayoutParams;
11  public class Stars extends View {
12      private float radius = 40;
13      private int color = 0xFF0000 ;
14      private final static float DEGREE = 36; //五角星角度

15      public Stars(Context context) {
16      super(context);
17      }

18  public Stars(Context context, AttributeSet attrs) {
19      super(context, attrs);
20      // TODO Auto-generated constructor stub
21      //获取自定义属性
22      try {
```

```
23            TypedArray a = context.obtainStyledAttributes(attrs, R.styleable.Stars);
24            this.color = a.getColor(R.styleable.Stars_color, color);
25            this.radius = a.getFloat(R.styleable.Stars_radius, radius);
26            a.recycle();
27            } catch (Exception e) {
28                // TODO: handle exception
29            }
30        }

31        /** 角度转弧度
32        * @param degree
33        * @return
34        */
35 private float degree2Radian(float degree){
36        return (float) (Math.PI* degree / 180);
37 }

38 @Override
39 public LayoutParams getLayoutParams() {
40        // TODO Auto-generated method stub
41        LayoutParams params = super.getLayoutParams();
42        try {
43            params.width = (int) (radius * Math.cos(degree2Radian(DEGREE) / 2) * 2);
44            params.height = (int) (radius + radius * Math.cos(degree2Radian(DEGREE)));
45        } catch (Exception e) {
46            // TODO: handle exception
47        }
48        return params;
49 }

50 @Override
51 protected void onLayout(boolean changed, int left, int top, int right, int bottom) {
52        // TODO Auto-generated method stub
53        super.onLayout(changed, left, top, right, bottom);
54 }

55 @SuppressWarnings("DrawAllocation")
56 @Override
57 protected void onDraw(Canvas canvas) {
58        // TODO Auto-generated method stub
59        super.onDraw(canvas);
60        Paint paint = new Paint();
61        paint.setColor(this.color);
62        paint.setAntiAlias(true);
63        Path path = new Path();
64        float radian = degree2Radian(DEGREE);
```

```
65        float radius_in = (float) (radius * Math.sin(radian / 2) / Math.cos(radian)); //中间五边形
的半径

66        path.moveTo((float) (radius * Math.cos(radian / 2)), 0);
67        path.lineTo((float) (radius * Math.cos (radian / 2) + radius_in * Math.sin (radian)),
(float) (radius - radius * Math.sin(radian / 2)));
68        path.lineTo((float) (radius * Math.cos (radian / 2) * 2), (float) (radius - radius *
Math.sin(radian / 2)));
69        path.lineTo((float) (radius * Math.cos(radian / 2) + radius_in * Math.cos( radian /2)),
(float) (radius + radius_in * Math.sin( radian /2)));
70        path.lineTo((float) (radius * Math.cos(radian / 2) + radius * Math.sin(radian)), (float)
(radius + radius * Math.cos(radian)));
71        path.lineTo((float) (radius * Math.cos(radian / 2)), (float) (radius + radius_in));
72        path.lineTo((float) (radius * Math.cos(radian / 2) - radius * Math.sin(radian)), (float)
(radius + radius * Math.cos(radian)));
73        path.lineTo((float) (radius * Math.cos (radian / 2) - radius_in * Math.cos( radian /2)),
(float) (radius + radius_in * Math.sin(radian / 2)));
74        path.lineTo(0, (float) (radius - radius * Math.sin(radian /2)));
75        path.lineTo((float) (radius * Math.cos(radian / 2) - radius_in * Math.sin(radian)) , (float)
(radius - radius * Math.sin(radian /2)));

76        path.close();
77        canvas.drawPath(path, paint);
78        canvas.restore();
79        Bitmap bitmap = Bitmap.createBitmap(10, 10, Bitmap.Config.ARGB_8888);
80        canvas.drawBitmap(bitmap, 10, 10, paint);
81        }
82  }
```

（4）运行结果如图 6-4 所示。

图 6-4　绘制几何图形示例

6.2.3 图形特效

Android 中提供了 Bitmap 类来获取图像文件信息，进行图像的平移、旋转及缩放等操作，并可以指定格式保存图像文件。

1. 图像绘制

在绘制图像之前，需要从项目目录下的 res \ drawable 中获取所需的图片资源。可以通过资源索引来获得该图像的 Bitmap 对象。具体方法如下（在项目目录下的 res \ drawable 中放置一张名为 fuwa. png 的图片）：

```
mBitmap = ((BitmapDrawable) getResources().getDrawable(R.drawable.fuwa)).getBitmap();
```

其中，getResources( )方法的作用是取得资源对象；getDrawable( )方法的作用是取得资源中的 Drawable 对象，参数为资源索引 ID；getBitmap( )方法的作用是得到 Bitmap 对象。

获得图像资源后，可以使用 drawBitmap( )方法将图像显示到屏幕的（x，y）坐标位置上，具体方法如下：

```
Canvas.drawBitmap(mBitmap, x, y, null);
```

此外，要获得图像的信息，可以通过 mBitmap. getHight( )方法获得该图像的高度，通过 mBitmap. getWidth( )方法获得该图像的宽度。

2. 图像的平移

由图像的绘制方法，我们知道使用 Canvas. drawBitmap（mBitmap，x，y，null）方法可以将图像绘制到屏幕的（x，y）坐标位置上。所以，要实现图像的平移，只需要改变图像绘制到屏幕上的（x，y）坐标位置即可。

3. 图像的旋转

在 Android 中，可以使用 Matrix 对象来进行图像旋转，Matrix 对象是一个 3×3 的矩阵，专门用于图像变换匹配。Matrix 对象没有结构体，必须被初始化，可以通过 reset( )或 set( )方法来实现，如下所示：

```
mMatrix.reset();
```

初始化之后就可以通过 setRotate( )方法来设置想要的旋转角度，如下所示：

```
mMatrix.setRotate();
```

旋转角度设置完毕后，可以使用 creatBitmap( )方法创建一个经过旋转处理的 Bitmap 对象，方法如下所示：

```
mBitmapRotate = Bitmap.creatBitmap(mBitmap, 0, 0, mBitmapWidth, mBitmapHight, mMatrix, true);
```

最后，将该 Bitmap 对象绘制到屏幕上，便实现了图像旋转的操作。

4. 图像的缩放

在 Android 中，同样可以使用 Matrix 对象来实现图像的缩放。使用 Matrix 对象的 postScale( )方法来设置图像缩放的倍数，如下所示：

```
mMatrix.postScale();
```

缩放倍数设置完毕后，同样需要使用 creatBitmap( )方法创建一个经过缩放处理的 Bitmap

对象。最后，将该 Bitmap 对象绘制到屏幕上，便实现了图像缩放的操作。

【例 6-4】 图像特效例程。

```
1       package com.example.image;
2       import android.os.Bundle;
3       import android.view.ViewGroup.LayoutParams;
4       import android.widget.ImageView;
5       import android.widget.ImageView.ScaleType;
6       import android.widget.LinearLayout;
7       import android.app.Activity;
8       import android.graphics.Bitmap;
9       import android.graphics.BitmapFactory;
10      import android.graphics.Matrix;
11      import android.graphics.drawable.BitmapDrawable;

12      public class MainActivity extends Activity {
13          @Override
14          protected void onCreate(Bundle savedInstanceState) {
15          super.onCreate(savedInstanceState);
16          setTitle("Android实现图片缩放与旋转。");
17          LinearLayout linLayout = new LinearLayout(this);

18          //加载需要操作的图片,这里是一张图片
19      Bitmap bitmapOrg = BitmapFactory.decodeResource(getResources(),R.drawable.te-st);

20          //获取这个图片的宽和高
21          int width = bitmapOrg.getWidth();
22          int height = bitmapOrg.getHeight();

23          //定义预转换成的图片的宽度和高度
24          int newWidth = 200;
25          int newHeight = 200;

26          //计算缩放率,新尺寸除原始尺寸
27          float scaleWidth = ((float) newWidth) / width;
28          float scaleHeight = ((float) newHeight) / height;

29          // 创建操作图片用的 matrix 对象
30          Matrix matrix = new Matrix();

31          // 缩放图片动作
32          matrix.postScale(scaleWidth, scaleHeight);

33          //旋转图片动作
34          matrix.postRotate(45);

35          // 创建新的图片
```

```
36            Bitmap resizedBitmap  = Bitmap.createBitmap(bitmapOrg, 0, 0,width, height, matrix,
true);

37            //将上面创建的 Bitmap 转换成 Drawable 对象,使得其可以使用在 ImageView, ImageButton 中
38            BitmapDrawable bmd = new BitmapDrawable(resizedBitmap);

39            //创建一个 ImageView
40            ImageView imageView = new ImageView(this);
41            // 设置 ImageView 的图片为上面转换的图片
42            imageView.setImageDrawable(bmd);

43            //将图片居中显示
44            imageView.setScaleType(ScaleType.CENTER);

45            //将 ImageView 添加到布局模板中
46            linLayout.addView(imageView,
47            new LinearLayout.LayoutParams(
48            LayoutParams.FILL_PARENT, LayoutParams.FILL_PARENT
49            )
50            );
51            // 设置为本 Activity 的模板
52            setContentView(linLayout);
53            }
54        }
```

运行结果如图6-5所示。

图6-5　图形缩放旋转示例

6.3　Web 视图

网络应用程序是一种使用网页浏览器在互联网或企业内部网上操作的应用软件。网络应

用程序采用网页语言（例如 HTML、JavaScript、Java 等编程语言）进行编写，需要通过浏览器来运行。网络应用程序流行的原因之一是因为可以直接在各种计算机平台上运行，不需要事先安装或定期升级程序。常见的网页应用程序有网络商店、网络拍卖、网络论坛、博客和网络游戏等。

Android 提供两种方式进行网页访问，一种是使用 Android SDK 开发并安装在一个 APK 客户端的应用程序访问网页，另一种方法采用 Web 浏览器访问网页。图 6-6 为 Android 提供的两种访问网页的方式。如果采用编写网络应用程序方式，可以利用 API 包中的 WebKit 模块。WebView 作为应用程序的 UI 接口，为用户提供一系列的网页浏览和用户交互功能。选择基于 Web 浏览器的方式，允许用户指定显示属性，适合手机的屏幕配置。

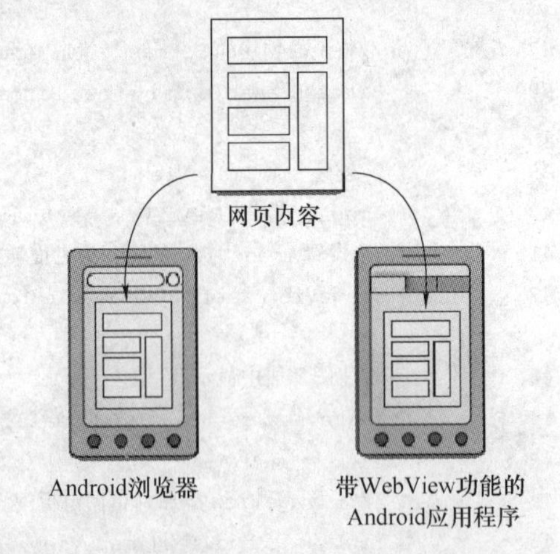

图 6-6　Android 提供两种方式进行网页访问

6.3.1　浏览器引擎 WebKit

Android 手机中内置了一款高性能 WebKit 内核浏览器，WebKit 是一个开源的浏览器引擎，它们都是自由软件，在 GPL 条约下授权，同时支持 BSD 系统的开发。所以 WebKit 也是自由软件，同时开放源代码。WebKit 的优势在于高效稳定，兼容性好，且源码结构清晰，易于维护。WebKit 内核拥有非常好的网页解析机制，WebKit 内核在手机上的应用也十分广泛，例如谷歌的 Android 手机、苹果的 iPhone，诺基亚的 Series 60 等所使用的 Browser 内核引擎，都是基于 WebKit。Android 平台的 Webkit 模块由 Java 层和 WebKit 库两个部分组成，Java 层负责与 Android 应用程序进行通信，而 WebKit 类库负责实际的网页排版处理。

6.3.2　Web 视图对象

在 WebKit 的 API 包中，最重要、最常用的类是 Android. WebKit. WebView。WebView 类是 WebKit 模块 Java 层的视图类，所有需要使用 Web 浏览功能的 Android 应用程序都要创建该视图对象，用于显示和处理请求的网络资源。目前，WebKit 模块支持 HTTP、HTTPS、FTP 以及 JavaScript 请求。WebKit 作为应用程序的 UI 接口，为用户提供了一系列的网页浏览、用户交互接口，客户程序通过这些接口访问 WebKit 核心代码。

1. 通过 webView. getSettings 设置 WebView 的一些属性、状态。

```
setAllowFileAccess:是否能访问文件数据；
setBuiltInZoomControls:设置是否支持缩放；
setCacheMode:设置缓冲的模式；
setJavaScriptEnabled:设置是否支持 JavaScript。
```

2. 通过 WebViewClient 来自定义网页浏览程序。webView. setWebChromeClient 为专门辅助 WebView 处理各种通知、请求等事件的类。

```
doUpdateVisitedHistory:更新历史记录;
onFormResubmission:应用程序重新请求网页数据;
onLoadResource:加载指定地址提供的资源;
onPageFinished:网页加载完毕;
onPageStarted:网页开始加载;
onReceivedError:报告错误信息;
onScaleChanged:WebView 发生改变;
shouldOverrideUrlLoading:控制新的连接在当前 WebView 中打开。
```

3. WebChromeClient 专门用来辅助 WebView 处理 JavaScript 的对话框、图标、网站标题、加载进度栏等控件。

```
onCloseWindow:关闭 WebView;
onCreateWindow:创建 WebView;
onJsAlert:处理 JavaScript 中的 Alert 对话框;
onJsConfirm:处理 JavaScript 中的 Confirm 对话框;
onJsPrompt:处理 JavaScript 中的 Prompt 对话框;
onProgressChanged:加载进度条改变;
onReceivedIcon:网页图标更改;
onReceivedTitle:网页标题更改;
onRequestFocus:WebView 显示焦距。
```

下面给出一个使用 WebView 的实例。

6.3.3 Web 视图实例

【例 6-5】 应用 WebView 对象浏览网页。

创建一个 Webview 的一般步骤如下：

（1）在布局文件中声明 WebView。

（2）在 Activity 中实例化 WebView。

（3）调用 WebView 的 loadUrl( )方法，设置 WebView 要显示的网页。

（4）为了让 WebView 能够响应超链接功能，调用 setWebViewClient( )方法，设置 WebView 视图。

（5）用 WebView 单击链接看了很多页以后为了让 WebView 支持回退功能，需要覆盖 Activity 类的 onKeyDown( )方法，如果不做任何处理，单击系统回退键，整个浏览器会调用 finish( )而结束自身，而不是回退到上一页面。

（6）需要在 AndroidManifest. xml 文件中添加权限，否则出现 Web page not available 错误。

```
<uses-permission android:name="android.permission.INTERNET"/>
```

下面通过一个实例来说明其用法：

（1）创建工程 WebView，并向其 MainActivity. java 中添加如下代码：

```
1    package com.example.webview;

2    import android.app.Activity;
3    import android.os.Bundle;
4    import android.view.KeyEvent;
```

```
5       import android.webkit.WebView;
6       import android.webkit.WebViewClient;

7       public class MainActivity extends Activity {
8             /** Called when the activity is first created. */
9             private WebView webview;
10            @Override
11            public void onCreate(Bundle savedInstanceState) {
12            super.onCreate(savedInstanceState);
13            setContentView(R.layout.activity_main);

14            webview = (WebView)findViewById(R.id.webview);
15            //设置 WebView 属性,能够执行 JavaScript 脚本
16            webview.getSettings().setJavaScriptEnabled(true);
17            //加载 URL 内容
18            webview.loadUrl("http://neu.edu.cn");
19            //设置 Web 视图客户端
20            webview.setWebViewClient(new MyWebViewClient());
21      }
22      //设置回退
23      public boolean onKeyDown(int keyCode,KeyEvent event){
24            if((keyCode == KeyEvent.KEYCODE_BACK)&&webview.canGoBack()){
25                webview.goBack();
26                return true;
27            }
28            return super.onKeyDown(keyCode,event);
29      }
30      //Web 视图客户端
31      public class MyWebViewClient extends WebViewClient{
32            public boolean shouldOverviewUrlLoading(WebView view,String url){
33                view.loadUrl(url);
34                return true;
35                }
36            }
37      }
```

（2）在 activity. main. xml 中添加如下代码

```
1     <?xml version="1.0" encoding="utf-8"?>
2     <LinearLayout xmlns:android="http://schemas.android.com/apk/res/android"
3          android:orientation="vertical"
4          android:layout_width="fill_parent"
5          android:layout_height="fill_parent"
6     >
7     <WebView
8          android:id="@+id/webview"
9          android:layout_width="fill_parent"
10         android:layout_height="fill_parent"
```

```
11  />
12  </LinearLayout>
```

（3）在 AndroidManifest. xml 文件中添加如下代码获取网络权限：

```
<uses-permission android:name="android.permission.INTERNET"/>
```

（4）运行结果如图 6-7 所示。

图 6-7　用 WebView 显示网页

6.4　HTTP 和 URL 网络资源获取

获得网络资源主要的方式是 HTTP 和 URL 请求两种。URL 对象全称是符号描述符，是指向互联网“资源”的指针，资源可以是简单的文件目录也可以是更复杂的对象的引用，URL 可以由协议名、主机、端口和资源组成。HTTP 通信技术是网络通信中最常用的技术之一，客户端向服务器发出 HTTP 请求，服务器接收到客户端的请求后，处理客户端的请求，处理完成后再通过 HTTP 应答回去给客户端。这里的客户端一般是浏览器。本章中客户端是 Android 手机，服务器一般是 HTTP 服务器，HTTP 定义了与服务器交互的不同方法，最基本的方法有四种，分别是 GET、POST、PUT、DELETE。对应着对这个资源的查、改、增、删操作，GET 一般用于获取/查询资源信息，而 POST 一般用于更新资源信息。

Android 平台提供了三种接口用于网络的访问，分别为：

（1）. java. net. * :标准 JAVA 接口。

（2）. org. apache：Apache 接口。

（3）android. net. * :Android 网络接口。

其中使用最多的就是前两种接口，以下就介绍使用 HTTP 和 URL 的基本用法。

1. HttpURLConnection 接口

HttpURLConnection 是 Java 的标准类，继承自 URLConnection 类，两者都是抽象类所以无法直接实例化。获取 HttpURLConnection 对象，主要通过 URL 类的 openConnection 来实现。创建一个 HttpURLConnection 链接的标准步骤如下代码所示：

```
URL url = new URL("http://www.baidu.com");
HttpURLConnection urlConnection = (HttpURLConnection)url.openConnection();
```

这样只是创建了 HttpRULConnection 的实例，并没有真正的进行联网的操作。因此，可以在链接之前对其进行一些参数的设置，比如：

```
urlConnection.setRequestMethod("POST");  //设置请求方法为 POST,默认为 GET
urlConnection.setUseCasches(false); //请求是否使用缓存,默认为 true
```

当链接完成后应该关闭这个链接，如下：

```
urlConnectin.disconnect();
```

以下是使用 HttpURLConnection 访问网络资源的标准流程：

```
1      //http 地址
2      String httpUrl = "http://www,baidu.com";
3      StringBuffer buffer = new StringBuffer();
4      String line = null;
5      BufferedReader reader = null;
6      HttpURLConnection urlConn = null;
7      try{
8          //创建 URL 对象
9          URL url = new URL(httpUrl);
10         //通过 URL 对象创建一个 HttpURLConnection 对象
11         urlConn = (HttpURLConnection) url.openConnection();
12         //得到读取内容的输入流
13         InputStream in = new  InputStreamReader(urlConn.getInputStream());
14         reader = new BufferedReader (in);
15         while((line = reader.readLine())! =null)     //逐行读取文件,将每行数据存入 line 中,当
   文件读取完毕后,line 为 null
16             {
17                 buffer.append(line);   //逐行将 line 添加到 StringBuffer 对象中
18             }
19             }catch (Exception e)
20             e.printStackTrace();
21             }finally{
22             if(reader! =null&&urlConn! =null){
23                 try {
24                     reader.close();   //关闭 BufferedReader 对象
25                     urlConn.disconnect();   //关闭 HttpURLConnection 对象
26                     } catch (Exception e1) {
27                     e1.printStackTrace();
28                 }
29             }
```

```
30        }
```

以上是标准的GET请求的流程，参数的传递是直接放在URL后面；如果要使用POST，则参数要放在HTTP请求数据中，则需要先调用setRequestMethod方法设置请求类型为POST，然后将需要传递的参数内容通过writeBytes方法写入数据流中。

2. HttpClient接口

标准java. net. *可以完成一些基本的网络操作，但对于更复杂的操作，则需要用到A-pache的HttpClient接口。首先需要了解一些重要的类：

（1）ClientConnectionManager：客户端连接管理接口，它提供了一系列的方法用于连接对象的管理。

（2）DefaultHttpClient：一个默认的HTTP客户端，可以使用它创建一个HTTP连接。

（3）HttpGet，HttpPost：对应GET请求和POST请求。

（4）HttpResponse：是一个HTTP连接的响应。当执行一个HTTP连接后，就会返回一个HttpResponse，它封装了相应的所有信息。

下面代码将演示如何使用HttpClient来执行GET请求，示例如下：

```
1   String httpUrl = "http://www.baidu.com?par=abcd";
2   //创建HttpGet连接对象
3   HttpGet httpRequest = new HttpGet(httpUrl);
4         try{
5             //取得HttpClient对象
6             HttpClient httpClient = new DefaultHttpClient();
7             //请求HttpClient,获得HttpResponse
8             HttpResponse httpResponse = httpClient.excute(httpRequest);
9             //判断请求是否成功
10            if(httpResponse.getStatusLine().getStatusCode()==HttpStatus.SC_OK){
11                //取得响应的内容
12                String strResult = EntituUtils.toString(httpResponse.getEntity());
13                }else{
14                        System.out.println("请求出错!");
15                }
16            }catch (Exception e){
17        e.printStackTrace();
18 }
```

同样，如果要执行POST请求，则需要先构建一个HttpPost对象，至于参数的传递，可以使用List来保存要传递的参数对象NameValuePair，然后用BasicNameValuePair类来构造一个需要被传递的参数，最后就是调用add方法将这个参数保存到List中，在执行POST请求前将请求参数传递给HttpPost对象，示例如下：

```
1   String httpUrl = "http://www.baidu.com?";
2   //创建HttpPost连接对象
3   HttpPost httpRequest = new HttpPostt(httpUrl);
4   //所要传递的所有参数
5   List<NameValuePair> params = new ArrayList<NameValuePair>();
6   //添加参数
```

```
7  params.add(new BaseNameValuePair("par","abcdefg"));
8      try{
9          //设置字符集
10         HttpEntity httpEntity = new UrlEncodedFormEntity(params,"utf-8");

11         //将请求参数赋给 HttpPost 对象
12         httpReques.setEntity(httpEntity);
13         //取得 HttpClient 对象
14         HttpClient httpClient = new DefaultHttpClient();
15         //获得 HttpResponse
16         HttpResponse httpResponse = httpClient.excute(httpRequest);
17         //判断请求是否成功
18         if(httpResponse.getStatusLine().getStatusCode = =HttpStatus.SC_OK){
19             //取得响应的内容
20             String strResult = EntituUtils.toString(httpResponse.getEntity());
21         }else{
22             System.out.println("请求出错!");
23         }
24     }catch (Exception e){
25         e.printStackTrace();
26 }
```

同样的，如果要抓取网络上的一张图片，使用 HttpClient 接口，应该怎么做呢？步骤与上面的演示基本一致，下面只贴出关键的步骤：

方法一：获得 HttpEntity 对象后，通过 getContent( )方法获得一个输入流。

```
1  InputStream in = entity.getContent();
2  bitmap = BitmapFactory.decodeStream(in);
3  in.close();
```

方法二：直接获得字节数组。

```
1  byte [] bytes = EntityUtils.toByteArray(entity);
2  bitmap = BitmapFactory.decodeByteArray(bytes, 0, bytes.length);
```

【例 6-6】 具体实例如下（读取百度主页的脚本文件）。

新建工程 HttpTest，并在 mainactivity.java 文件中添加如下代码：

```
1  package com.example.httptest;

2  import java.io.BufferedInputStream;
3  import java.io.InputStream;
4  import java.net.URL;
5  import java.net.URLConnection;
6  import org.apache.http.util.ByteArrayBuffer;
7  import org.apache.http.util.EncodingUtils;
8  import com.example.httptest.R;
9  import android.app.Activity;
10 import android.os.Bundle;
11 import android.widget.TextView;
```

```
12 public class MainActivity extends Activity {
13       /* *  Called when the activity is first created. * /
14       @ Override
15       public void onCreate(Bundle savedInstanceState) {
16           super.onCreate(savedInstanceState);
17           setContentView(R.layout.activity_main);
18           TextView tv = new TextView(this);
19           String myString = null;
20           try {
21               // 定义获取文件内容的 URL
22               URL myURL = new URL(
23               "HTTP://www.baidu.com/hello.txt&quot");
24               // 打开 URL 链接
25               URLConnection ucon = myURL.openConnection();
26               // 使用 InputStream,从 URLConnection 读取数据
27               InputStream is = ucon.getInputStream();
28               BufferedInputStream bis = new BufferedInputStream(is);
29               // 用 ByteArrayBuffer 缓存
30               ByteArrayBuffer baf = new ByteArrayBuffer(50);
31               int current = 0;
32               while ((current = bis.read()) ! = -1) {
33                   baf.append((byte) current);
34               }
35               // 将缓存的内容转化为 String,用 UTF-8 编码
36               myString = EncodingUtils.getString(baf.toByteArray(), "UTF-8");
37           } catch (Exception e) {
38               myString = e.getMessage();
39           }
40           // 设置屏幕显示
41           tv.setText(myString);
42           this.setContentView(tv);
43       }
44 }
```

运行结果如图 6-8 所示。

部分代码注释：

（1）实例 URL 类：myURL，表示要获取内容的网址：

```
URL myURL = new URL(HTTP://www.baidu.com/hello.txt);
```

（2）实例 URLConnection 类，表示一个打开的网络连接 ucon：

```
URLConnection ucon = myURL.openConnection();
```

（3）用字节流的形式表示从网络上读到的数据：

```
InputStream is = ucon.getInputStream();
```

为避免频繁读取字节流，提高读取效率，用 BufferedInputStream 缓存读到的字节流：

```
InputStream is = ucon. getInputStream();
BufferedInputStream bis = new BufferedInputStream(is);
```

（4）用 read 方法读入网络数据：

```
ByteArrayBuffer baf = new ByteArrayBuffer(50);
int current = 0;
while((current = bis. read())! =-1){
      baf. append((byte)current);
}
```

（5）由于读到的数据只是字节流，无法直接显示到屏幕上，所以需要在显示之前将字节流转换为可读取的字符串：

```
myString = EncodingUtils. getString(baf. toByteArray(), &quot;UTF-8&quot;);
```

（如果读取的 . txt 等文件是 UTF-8 格式的，就需要对数据进行专门的转换。）

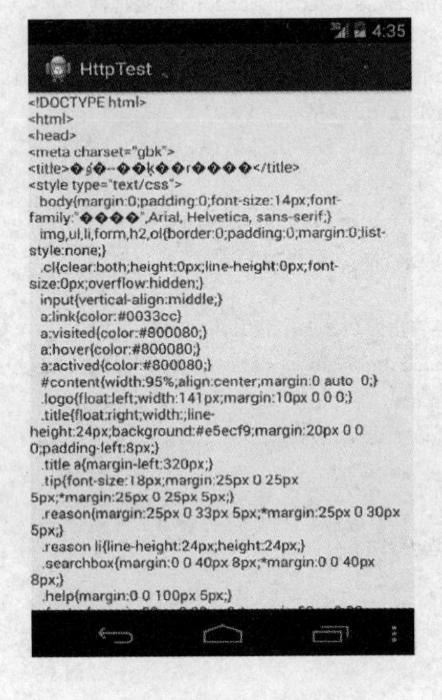

图 6-8　显示 HTTP 协议报文

本 章 小 结

本章介绍了基于 Android 平台的多媒体、图形及网络等应用程序的开发。多媒体包含了音频、视频等媒体，对于 Android 平台下的图形绘制与特效，介绍了图像的平移、旋转、缩放、保存指定格式图像文件等操作，以及如何编写专业的绘图或控制图形动画的应用程序。在 Android 网络应用程序开发方面，介绍了 Android 手机中内置的 WebKit 内核浏览器，及如何使用 HTTP 和 URL 获得网络资源等内容。

习 题

6-1 设计一个具有选歌功能的音频播放器。

6-2 设计一个具有播放、停止、暂停功能的视频播放器。

6-3 设计一个手绘图的画板。

6-4 设计一个图形编辑器，使其具有图像的显示、旋转、缩放功能。

6-5 编写一个可以打开任意输入的网址并显示其页面的用户界面。

第 7 章　Android NDK

本章介绍基于 Android 系统的嵌入式硬件系统的组成、嵌入式微处理器的特点，由于 Android 的应用层的类都是以 Java 写的，在执行过程中，如果 Java 类需要与 C 组件沟通，VM 就会通过 JNI 载入 C 组件，然后让 Java 的函数顺利地调用 C 组件的函数。本章首先通过 Android NDK 自带的 hello-jni 示例程序来展示 NDK 程序的编译和运行，然后介绍 Android NDK 系统的搭建。

7.1　Android NDK 简介

Android NDK 是在 SDK 前面又加上了“原生”二字，即 Native Development Kit，因此又被 Google 称为“NDK”。Android NDK 是 Android SDK 的伴随工具，可以允许用户采用诸如 C/C ++原生编程语言开发 Android 应用程序。一般情况下，Android 应用程序开发主要使用 Java 语言，编译后产生的托管代码在 Dalvik 虚拟机上运行，Android 平台的第三方应用程序均是依靠基于 Java 的 Dalvik 特制虚拟机进行开发的。但在一些需要较高执行效率的地方，程序员希望能够使用非托管代码，以提高 Android 应用程序的核心部分的运行速度。不仅如此，程序员还希望使用传统的 C 或 C ++ 语言编写程序，并在程序封包文件（. apks）中直接嵌入原生库文件。原生 SDK 的公布可以让开发者更加直接地接触 Android 系统资源，并极大地提高 Android 应用程序开发的灵活性。

Google 表示在使用 Android NDK 的过程中程序开发人员应该清楚地认识到 Android NDK 的不足，使用原生 SDK 编程相比 Dalvik 虚拟机也有一些劣势，首先使用 C 或 C ++程序，并直接嵌入原生库文件中，会使程序更加复杂，程序的兼容性难以保障。其次由于无法访问 Framework API 等嵌入模块，增加了程序的调试难度。最后在程序设计中，需要考虑哪些核心代码部分适合 C 或 C ++语言编写，从而使非托管代码的运行效率最高。因此开发者需要自行斟酌使用。

Android NDK 需要安装全部 Android SDK 1. 5 或以上版本，到目前为止 Android NDK 已经修订了 9 个版本。Android NDK 集成了交叉编译器，支持 ARM、x86 及 MIPS 处理器指令集、JNI 接口和一些稳定的库文件。Android NDK 创建的原生库只能用于运行在特定的最低版本的 Android 平台的设备上。所需的最小平台版本取决于目标设备的 CPU 架构。这意味如果 ARM 的设备，需要运行 Android 1. 5 或更高版本的 NDK 生成的原生库嵌入的应用程序。而如果使用 x86 或 MIPS 架构的设备，需要运行 Android 2. 3 或更高版本的应用程序。Android NDK 具体包括以下几部分：

- ARM、x86 和 MIPS 交叉编译器
- 构建系统
- Java 原生接口头文件
- C 库

- Math 库
- POSIX 线程
- 最小的 C ++ 库
- ZLib 压缩库
- 动态链接库
- Android 日志库
- Android 像素缓冲区库
- Android 原生应用 APIs
- OpenGL ES 3D 图形库
- Open ES 原生音频库
- OpenMAX AL 最小支持

7.2 构建 NDK 系统

7.2.1 Android NDK 开发环境构建

Android NDK 的构建系统是基于 GNU Make 的，提供了头文件、库和交叉编译器工具链，可以在 Microsoft Windows、Apple Mac OS X 和 Linux 三种操作系统平台上运行。在安装前要确保已经是最新版的 Android SDK 和升级应用程序环境。NDK 兼容旧的平台版本，但没有旧版本的 SDK 工具。在安装 NDK 时，首先根据开发系统和软件要求比如 CPU 架构的不同，选择 NDK 包。其次下载解压缩所选择的工具包，本书仅介绍 Microsoft Windows 系统的 NDK 系统构建，请按照下列步骤操作：

第一步，在安装完成 SDK 工具后，下载并安装 Cygwin。Android NDK 最初设计在类 UNIX 系统上工作，NDK 的一些组件是 shell 脚本，这些脚本不能直接在 Windows 系统下进行交叉编译，因此需要在 Windows 系统下安装 Cygwin 才能进行完整的操作。Android NDK 要求 GNU Make 的版本高于或等于 3. 18，因此安装 Cygwin 1. 7 版本才能运行。访问 http://cygwin. com/install. html 网站上下载 Cygwin 安装程序 setup. exe。Cygwin 不是单个的应用程序，而是包含多个应用程序的巨大的软件集合。Cygwin 安装程序会为用户提供一个可用包列表，在搜索框中输入关键字 Make 对包列表进行过滤，展开 Devel 目录，选择 GNU Make 包如图 7-1 所示，否则，Cygwin 将无法编译C/C ++ 代码程序。

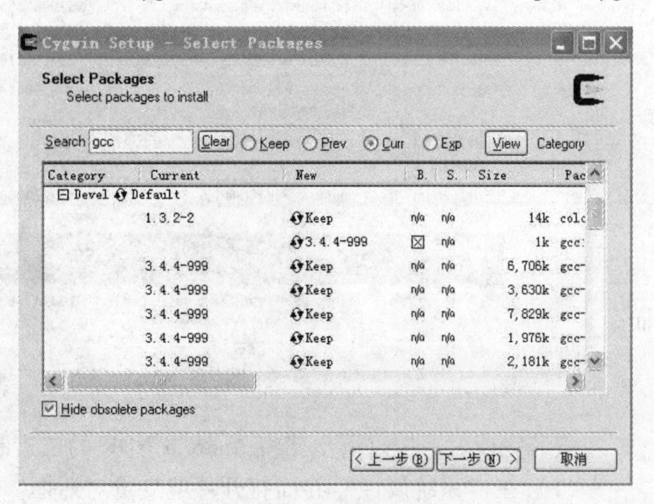

图 7-1 Cygwin 安装选项

安装完成后，要把 Cygwin 二进制路径添加到系统可执行搜索路径中。操作如下：

（1）在“系统属性”界面打开环境变量对话框。

（2）在系统变量部分单击 New 按钮定义一个新的环境变量，将变量名设置成 CYGWIN_HOME，将变量值设置成前面记下的 Cygwin 安装目录。

（3）单击 OK 按键保存新环境变量。

（4）在环境变量对话框中的系统变量列表中，双击 PATH 变量，并将;% CYGWIN_HOME% \ bin 追加到变量值后面。

完成了上述安装步骤后，Cygwin 工具成为系统可执行搜索路径的一部分。为了验证安装是否成功，打开一个命令提示窗口，在命令提示符下执行 make -version，如果安装成功，则会显示 GNU Make 的版本号，如图 7-2 所示。

```
Administrator@20110927-1150 /cygdrive/c
$ make -version
GNU Make 3.81
Copyright (C) 2006  Free Software Foundation, Inc.
This is free software; see the source for copying conditions.
There is NO warranty; not even for MERCHANTABILITY or FITNESS FOR A
PARTICULAR PURPOSE.

This program built for i686-pc-cygwin

Administrator@20110927-1150 /cygdrive/c
$
```

图 7-2　验证 Cygwin 安装结果

第二步，下载最新版 Android NDK 开发包，目前本书编写时，最新版本是 R9，请根据需要到 Google 的官方网站 http://developer.android.com/tools/sdk/ndk/index.html 下载 Android NDK，下载页面如图 7-3 所示。将下载的 ZIP 文件解压缩到用户的 Android 开发目录中，记住 Android NDK 文件的路径目录名以备后面设置环境变量时使用。

Platform	Package	Size (Bytes)	MD5 Checksum
Windows 32-bit	android-ndk-r9d-windows-x86.zip	491440074	b16516b611841a075685a10c59d6d7a2
Windows 64-bit	android-ndk-r9d-windows-x86_64.zip	520997454	8cd244fc799d0e6e59d65a59a8692588
Mac OS X 32-bit	android-ndk-r9d-darwin-x86.tar.bz2	393866116	ee6544bd8093c79ea08c2e3a6ffe3573
Mac OS X 64-bit	android-ndk-r9d-darwin-x86_64.tar.bz2	400339614	c914164b1231c574dbe40debef7048be
Linux 32-bit (x86)	android-ndk-r9d-linux-x86.tar.bz2	405218267	6c1d7d99f55f0c17ecbcf81ba0eb201f
Linux 64-bit (x86)	android-ndk-r9d-linux-x86_64.tar.bz2	412879983	c7c775ab3342965408d20fd18e71aa45
Additional Download	Package	Size (Bytes)	MD5 Checksum
STL debug info	android-ndk-r9d-cxx-stl-libs-with-debug-info.zip	104947363	906c8d88e0f02295c3bfe6b8e98a1a35

图 7-3　Android NDK 下载页面

安装完成后，按以下步骤将 Android NDK 的二进制路径加到系统可执行搜索路径中：

（1）在“系统属性”界面打开环境变量对话框。

（2）在系统变量部分单击 New 按钮定义一个新的环境变量，将变量名设置成 ANDROID_

NDK_HOME，将变量值设置成前面记下的 Android NDK 安装目录。

（3）单击 OK 按键保存新环境变量。

（4）在环境变量对话框中的系统变量列表中，双击 PATH 变量，并将;% ANDROID_NDK_HOME% 追加到变量值后面。

现在可以很容易地访问 Android NDK。为了验证安装是否成功。打开命令提示窗口，在命令提示符下执行 ndk-build，如果安装成功，就会看到 NDK 给出的关于项目目录的提示。

在安装过程中，所有的 Android NDK 组件都被安装在目标目录下，下面介绍一些重要文件和子目录。

- ndk-build：该 shell 脚本是 Android NDK 构建系统的起始点。
- ndk-gdb：该 shell 脚本允许用 GNU 调试器调试原生组件。
- ndk-stack：该 shell 脚本可以帮助分析原生组件崩溃时的堆栈追踪。
- build：该目录包含可 Android NDK 构建系统的所有模块。
- platforms：该目录包含了支持不同 Android 目标版本的头文件和库文件。Android NDK 构建系统会根据具体的 Android 版本自动引用这些文档。
- samples：该目录包含了一些示例应用程序，这些程序可以体现 Android NDK 的性能。示例项目对于学习使用 Android NDK 的特性很有帮助。
- sources：该目录包含了可供开发人员导入到现有的 Android NDK 项目的一些共享模块。
- toolchains：该目录包含目前 Android NDK 支持的不同目标机体系结构的交叉编译器。Android NDK 目前支持 ARM、x86 和 MIPS 体系结构。Android NDK 构建系统根据选定的体系结构使用不同的交叉编译器。

Android NDK 最重要的组件是它的构建系统，它包含了所有的其他组件。

第三步是配置 Cygwin 的 NDK 开发环境。在默认情况下，Cygwin 安装在 C 盘的根目录下，修改在用户目录下的 . bash_profile 文件，在该文件的末尾处添加如下代码：

```
ANDROID_NDK_ROOT = /cygdrive/d/android/android-nkd-r9d
Export ANDROID_NDK_ROOT
```

上述代码说明 Android NDK 安装的目录，即 D 盘的 android/android-nkd-r9d 目录下。

最后测试开发环境是否可以正常工作，首先双击 Cygwin 图标，出现 Cygwin 后运行 ndk-build 出现如图 7-4 所示的运行结果，说明 Android NDK 的开发环境已经可以正常工作了。

```
Administrator@20110927-1150 /cygdrive/d/android/android-ndk-r9d
$ ndk-build
Android NDK: Could not find application project directory !
Android NDK: Please define the NDK_PROJECT_PATH variable to point to it.
/cygdrive/d/android/android-ndk-r9d/build/core/build-local.mk:148: *** Android NDK: Aborting    .  Stop.

Administrator@20110927-1150 /cygdrive/d/android/android-ndk-r9d
$
```

图 7-4　ndk-build 对原生组件进行构建

7.2.2 解析 hello-jni 例程

Android NDK 最重要的组件是它的构建系统，它包含了所有的其他组件。想要更好地了解系统地构建，我们先通过 NDK 中自带的 hello-jni 来展示 NDK 程序的编译和运行过程。

【例 7-1】 hello-jni 例程解析

首先使用 import 向 Eclipse 中导入工程，本机目录为 F:\android-ndk-r9d\samples\hello-jni。

注意：在导入工程中会同时出现实际项目和名为 test 的项目，这里我们只选择实际项目，并且为了不改变原始项目我们在导入时应勾选 Copy projects into workplace 选项。即如图 7-5 所示。

图 7-5 选择工程目录

导入工程之后我们会在项目目录下发现两个比较重要的文件：一个是 src/com.example.hellojni/HelloJni.java 文件，一个是 jni/hello-jni.c 文件。其中 C 文件内容如下：

```
1   #include <string.h>
2   #include <jni.h>

3   /*  This is a trivial JNI example where we use a native method
4    *  to return a new VM String. See the corresponding Java source
5    *  file located at:
6    *
7    *     apps/samples/hello-jni/project/src/com/example/hellojni/HelloJni.java
8    * /
9   jstring
10  Java_com_example_hellojni_HelloJni_stringFromJNI(JNIEnv*  env,jobjectthiz )
11  {
12  #if defined(__arm__)
13        #if defined(__ARM_ARCH_7A__)
14            #if defined(__ARM_NEON__)
15                #define ABI "armeabi-v7a/NEON"
16            #else
17                #define ABI "armeabi-v7a"
18            #endif
19        #else
20            #define ABI "armeabi"
21        #endif
22        #elifdefined(__i386__)
23        #define ABI "x86"
24        #elifdefined(__mips__)
```

```
25          #define ABI "mips"
26          #else
27              #define ABI "unknown"
28              #endif
29   return (* env)->NewStringUTF(env, "Hello from JNI !   Compiled with ABI " ABI ".");
30   }
```

这里主要的是定义了输出字符串" Hello from JNI！ Compiled with ABI " ABI " . " 的方法，而 HelloJni. java 文件则通过调用上述 hello. c 的动态链接库来获取输出的字符串，并显示到模拟器上。而 HelloJni. java 文件内容如下：

```
1    package com.example.hellojni;

2    import android.app.Activity;
3    import android.widget.TextView;
4    import android.os.Bundle;

5    public class HelloJni extends Activity{
6          /** Called when the activity is first created. */
7          @Override
8          public void onCreate(Bundle savedInstanceState){
9              super.onCreate(savedInstanceState);

10             /* Create a TextView and set its content.
11              * the text is retrieved by calling a native
12              * function.
13              */
14             TextView tv = new TextView(this);
15             tv.setText(stringFromJNI() );
16             setContentView(tv);
17         }

18        /* A native method that is implemented by the
19         * 'hello-jni' native library, which is packaged
20         * with this application.
21         */
22        public native String  stringFromJNI();

23            /* This is another native method declaration that is *not*
24             * implemented by 'hello-jni'. This is simply to show that
25             * you can declare as many native methods in your Java code
26             * as you want, their implementation is searched in the
27             * currently loaded native libraries only the first time
28             * you call them.
29             *
30             * Trying to call this function will result in a
31             * java.lang.UnsatisfiedLinkError exception !
32             */
```

```
33          public native String  unimplementedStringFromJNI();

34          /*  this is used to load the 'hello-jni' library on application
35          *  startup. The library has already been unpacked into
36          *  /data/data/com.example.hellojni/lib/libhello-jni.so at
37          *  installation time by the package manager.
38          */
39          static {
40              System.loadLibrary("hello-jni");
41          }
42      }
```

接着启动 Cygwin 命令行输入 cd f:/android-ndk-r9d/samples/hello-jni 进入到 NDK 主目录，然后编译 C 代码，其命令如下：../../ndk-build，在 Cygwin 中显示如图 7-6 所示。

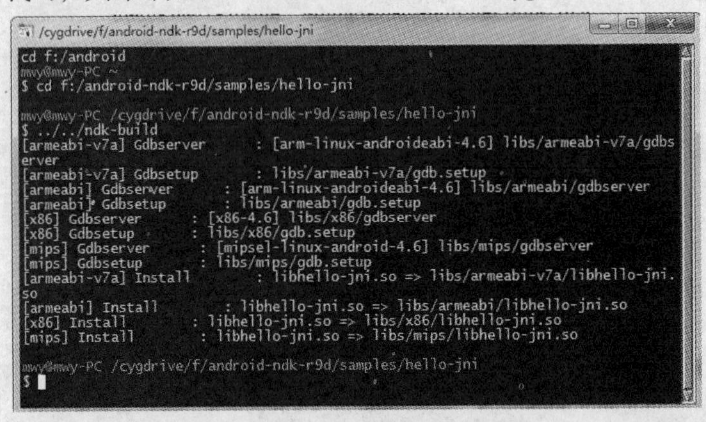

图 7-6　编译结果

此时表明已生成所需的动态链接库，接下来要对其进行加载。在 Eclipse 的 Project Explorer 视图中右击 HelloJni，依次选择 Android tools、Add Native Support 在弹出的对话框中显示信息如图 7-7 所示。由于该项目已经包含了一个原生项目，所以库名可以保持不变，单击 Finish 即可完成加载。

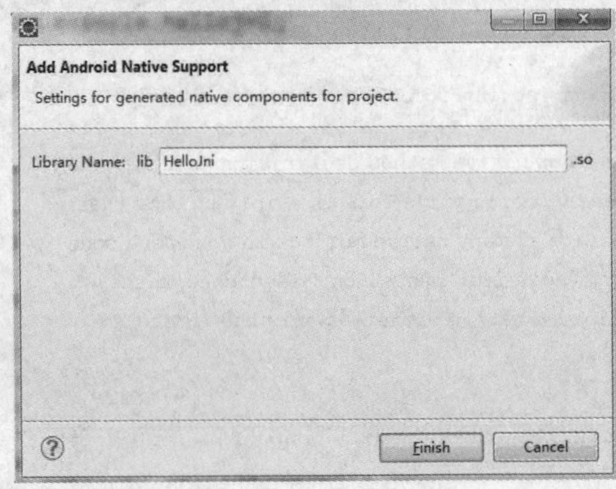

图 7-7　添加原生支持

如果是第一次向 Java-only 项目中添加原生支持，可以在该对话框中指定首选的共享库名，再将构建文件自动生成为进程的一部分时会使用该首选项共享库名称。

至此已经完成了项目的编写，可以在 Android 模拟器上运行该项目，其运行结果如图 7-8 所示。

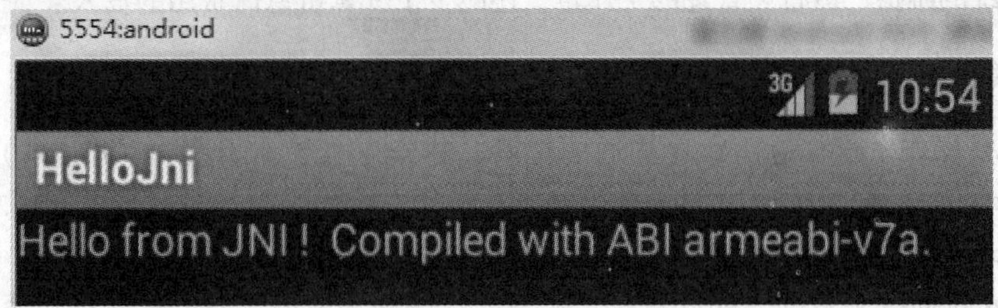

图 7-8　运行结果

7.3　NDK 开发过程详解

上面已经通过一个简单的“Hello World”介绍了 NDK 的大致使用方法。下面将以利用 Android NDK 来开发 OpenGL 的 3D 项目为例来展示 Android NDK 构建系统所提供的不同功能，例如：

- 建立一个共享库
- 建立多种共享库
- 建立静态库
- 利用共享库通用模块
- 在多种 NDK 项目间共享模块
- 使用预建库
- 建立独立的可执行文件
- 其他构建系统变量和宏
- 定义新变量和条件操作

在进行 NDK 开发时，一般需要同时建立 Android 工程和 C/C ++ 工程，然后使用 NDK 编译 C/C ++ 工程，形成可以被调用的共享库，最后共享库文件会被复制到 Android 工程中，并被打包到 apk 文件中。其开发过程可分为以下几步：

（1）设计 JNI 接口。

（2）用 C/C ++ 实现本地的方法。

（3）编译文件的实现。

（4）生成动态链接库。

7.3.1　中间件的概念

在介绍 Android 系统中间件之前我们需要了解什么是中间件。中间件（Middleware）是处于操作系统和应用程序之间的软件，它包括一组服务，以便于运行在一台或多台机器上的多个软件通过网络进行交互。该架构通常用于支持分布式应用程序并简化其复杂度，它包括 Web 服务器、事务监控器和消息队列软件。中间件能够屏蔽操作系统和网络协议的差异，为应用程序提供多种通信机制；并提供相应的平台以满足不同领域的需要。因此，中间件为应用程序提供了一个相对稳定的高层应用环境。同时，中间件也使程序开发人员面对一个简单而统一的开发环境，减少程序设计的复杂性，不必再为程序在不同系统软件上的移植而重复工作，从而大大减少了技术上的负担。中间件带给应用系统的，不只是开发的简便、开发周期的缩短，也减少了系统的维护、运行和管理的工作量，还减少了计算机总体费用的投入。

7.3.2 Android 系统的中间件

中间件是操作系统与应用程序沟通的桥梁，共分为两层：函数层（Library）和虚拟机（Virtual Machine）。应用程序用 Java 语言开发，操作系统代码则是 C 代码，它们之间的通信需要用 JNI 来实现。JNI（Java Native Interface）中文为“Java 本地接口”，从 Java 1.1 开始，JNI 标准就成为 Java 平台的一部分，它允许 Java 代码和其他语言写的代码进行交互。JNI 是本地编程接口，它使得在 Java 虚拟机（VM）内部运行的 Java 代码能够与其他编程语言编写的应用程序和库进行交互操作。JNI 一开始是为了本地已编译语言，尤其是 C 和 C ++ 而设计的，但是它并不妨碍使用其他语言，只要支持调用时约定的语言类型就可以了。

JNI 一般有以下一些应用场景：

（1）在一些情况下因为处理运算量非常大，为了获取高性能，直接使用 Java 是不能胜任的，如一些图形的处理。

（2）调用一些硬件的驱动或者一些软件的驱动，比如调用一些外部系统接口的驱动，如读卡器的驱动和 OCI 驱动。

（3）需要使用大内存，远远超过 VM 所能分配的内存，如进程内 Cache。

（4）调用 C 或者操作系统提供的服务，如 Java 调用搜索服务，搜索是由 C/C ++实现的。

中间件的开发首先要实现 JNI，为上层 Java 应用程序提供函数接口，然后要实现本地 C/C ++代码，调用内核代码提供的接口。JNI 的源文件为普通的 cpp 文件，编译后为 so 文件。JNI 标准 so 文件和应用程序一般通过 NDK 这个工具一起打包在 apk 文件里。

由于 Android 的应用层的类都是以 Java 写的，在执行过程中，如果 Java 类需要与 C 组件沟通时，VM 就会通过 JNI 载入 C 组件，然后让 Java 的函数顺利地调用 C 组件的函数。Android 中，主要的 JNI 代码放在以下的路径中 frameworks/base/core/jni/，这个路径中的内容被编译成库 libandroid_runtime.so，这是个普通的动态库，被放置在目标系统的/system/lib 目录下。此外，Android 还有其他的 JNI 库。JNI 中的各个文件，实际上就是普通的 C ++ 源文件；在 Android 中实现的 JNI 库，需要链接动态库 libnativehelper.so 文件。

对于开发者自己实现的 JNI 库 *.so 文件与 Java 应用一起打包到 apk 文件中。应用程序加载动态库后就可以调用本地函数。

通常 JNI 的使用自上至下有四层：本地库、JNI 库、声明本地接口的 Java 类和 Java 调用者，JNI 在 Android 层次结构中的作用如图 7-9 所示。实现 JNI 需要在 C ++ 代码中实

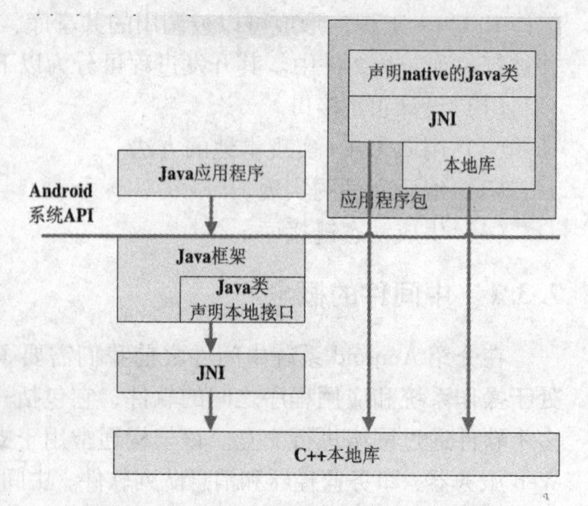

图 7-9 JNI 在 Android 层次结构中的作用

现 JNI 的各种方法，并注册到系统中，另外还要在 Java 源代码中声明。

【例 7-2】 实现 JNI 接口的设计实例。

下面通过一个实例来说明 JNI 接口的设计方法。

在此之前需要先将 NDK 与 Cygwin 关联起来，即打开 < Cygwin 安装目录 > \home\Administrator\. bash_profile 文件（本文 F: \cygwin\home\mwy\. bash_profile，在文件中加入如下代码：

```
1   NDK_ROOT = /cygdrive/f/ android-ndk-r9d/
2   export NDK_ROOT
3   PATH + = : $ NDK_ROOT
4   export PATH
```

注意：第一次进入 < Cygwin 安装目录 > \home\Administrator 目录时目录下并没有 . bash_profile 文件，打开 Cygwin 后就会自动生成 Administrator 目录以及 . bash_profile 文件。另外必须强调的是 . bash_profile 文件不能用记事本或者写字板软件打开进行编译，否则重启 Cygwin 后会报错，原因是由于写字板或记事本编辑后 . bash_profile 文件的格式被转为 DOC（默认该文件为 UNIX 格式），回车符多了一个'\r',所以 Cygwin 会报错。这里选用 UE 对其进行编辑。

接口设计过程如下：

(1)在 NDK 安装目录下新建文件夹 apps,在 apps 下新建目录 san-angeles(这个名字可以自己任意取),在 Eclipse 中新建工程 SanAngeles。建立工程之后需要创建并编写 JNI 类代码，用来声明要调用的本地方法,关键字 native。在这里只需要声明不需要具体实现。此时在 Eclipse 中新建名为 JNITest 的类。其代码如下：

```
1   package com.example.SanAngeles;
2   public class JNITest {
3   private static native void nativeInit();
4   private static native void nativeResize(int w, int h);
5   private static native void nativeRender();
6   private static native void nativeDone();
7   }
```

(2) 编译 Java 文件，并生成相应的头文件

将新建的 JNItest. java 文件复制到工程目录的 bin 目录下并用命令行编译。首先进入 NDK 主目录，接着输入 javac JNITest. java，即可生成 JNItest. class 文件。将此文件复制到 F: \android-ndk-r9d\apps\san-angeles\project\bin\classes\com\example\SanAngeles 中替换原有文件；从命令行进入 bin/classes 目录，输入 javah - jnicom_example_angles_JNITest 生成 com_example_angles_JNITest. h 的 C 头文件，最后在此工程目录下（即 project 目录下）创建 jni 文件夹，并将生成的头文件复制到此处。这个头文件主要是来处理 C/C ++ 和 Java 中一些定义的差别。而其中的 JNIEXPORT 和 JNICALL 都是 JNI 的关键字，表示函数式要被 JNI 调用的，其内容如下：

```
1   #include <jni.h>
2   /* Header for class com_example_angles_JNITest */
```

```
3   #ifndef _Included_com_example_angles_JNITest
4   #define _Included_com_example_angles_JNITest
5   #ifdef __cplusplus
6       extern "C" {
7   #endif
8   /*
9   * Class:     com_example_angles_JNITest
10  * Method:    nativeInit
11  * Signature: ()V
12  * /
13  JNIEXPORT void JNICALL Java_com_example_angles_JNITest_nativeInit  (JNIEnv * , jclass);

14  /*
15    * Class:     com_example_angles_JNITest
16    * Method:    nativeResize
17    * Signature: (II)V
18    * /
19    JNIEXPORT void JNICALL Java_com_example_angles_JNITest_nativeResize  (JNIEnv * , jclass,
jint, jint);

20    /*
21     * Class:     com_example_angles_JNITest
22    * Method:    nativeRender
23     * Signature: ()V
24    * /
25    JNIEXPORT void JNICALL Java_com_example_angles_JNITest_nativeRender  (JNIEnv * , jclass);

26    /*
27     * Class:     com_example_angles_JNITest
28     * Method:    nativeDone
29     * Signature: ()V
30    * /
31    JNIEXPORT void JNICALL Java_com_example_angles_JNITest_nativeDone  (JNIEnv * , jclass);

32    #ifdef __cplusplus
33    }
34    #endif
35    #endif
```

7.3.3 使用 C/C ++ 实现本地方法

通过上节的介绍可以得知 JNI 的接口已经实现，Android 程序便可通过 JNI 来调用本地的 C/C ++ 程序，接下来重点介绍该代码的编写规范。

在 jni 目录下新建 app-android.c 文件并编写其代码，这里重点介绍这几个原生方法的实现：

```
1    void   Java_com_example_angles_JNITest_nativeInit(JNIEnv*   env){
2        ImportGLInit();
3        appInit();
4        gAppAlive=1;
5        sDemoStopped=0;
6        sTimeOffsetInit=0;
7    }
8    void  Java_com_example_angles_JNITest_nativeResize(JNIEnv*   env,jobjectthiz,jint w,jint h){
9        sWindowWidth=w;
10       sWindowHeight=h;
11       __android_log_print(ANDROID_LOG_INFO,"SanAngeles","resize w=%d h=%d",w,h);
12   }
13   /* Call to finalize the graphics state* /
14   void   Java_com_example_angles_JNITest_nativeDone(JNIEnv*   env){
15       appDeinit();
16       importGLDeinit();
17       }
18   /* This is called to indicate to the render loop that it should
19   * stop as soon as possible.
20   * /
21   void   Java_com_example_angles_DemoGLSurfaceView_nativePause(JNIEnv*   env){
22       sDemoStopped=! sDemoStopped;
23       if(sDemoStopped){
24           /* we paused the animation,so store the current
25           * time in sTimeStopped for future nativeRender calls* /
26           sTimeStopped=_getTime();
27       }else{
28           /* we resumed the animation,so adjust the time offset
29           * to take care of the pause interval.* /
30           sTimeOffset-=_getTime()-sTimeStopped;
31       }
32   }
33   /* Call to render the next GL frame* /
34   void   Java_com_example_angles_JNITest_nativeRender(JNIEnv*   env){
35       longcurTime;
36       /* NOTE:if sDemoStopped is TRUE, then we re-render the same frame
37       *       on each iteration.
38       * /
39       if(sDemoStopped){
40           curTime=sTimeStopped+sTimeOffset;
41       }else{
42           curTime=_getTime()+sTimeOffset;
43           if(sTimeOffsetInit==0){
```

```
44          sTimeOffsetInit =1;
45          sTimeOffset =-curTime;
46          curTime =0;
47      }
48  }

49  appRender(curTime,sWindowWidth,sWindowHeight);
50  }
```

注意：Java_com_example_angles_JNITest_nativeInit 这个命名必须遵循“函数名 = Java_开头 + 包名和类名 + 接口名”这一规律。这个名字要是错了就无法被调用到。

在上述方法中有两个常见的参数，其中一个参数 JNIEnv 是指向可用 JNI 函数表的接口指针；另一个参数 jobject 是 Java 引用。下面将对这两个参数做详细介绍。

（1）JNIEnv 接口指针，指向一个函数表，函数表中的每一个入口指向一个 JNI 函数。JNIEnv 是一个与线程相关的变量，不同线程的 JNIEnv 彼此独立。在 C 与 C ++ 中使用 JNIEnv * env 分别如下：

```
C:(* env)- > 方法名(env,参数列表)
C ++ :env- > 方法名(参数列表)
```

上面这二者的区别是，在 C 中必须先对 env 间接寻址（得到的内容仍然是一个指针），在调用方法时要将 env 传入作为第一个参数。C ++ 则直接利用 env 指针调用其成员。图 7-10 演示了 JNIEnv 这个指针：

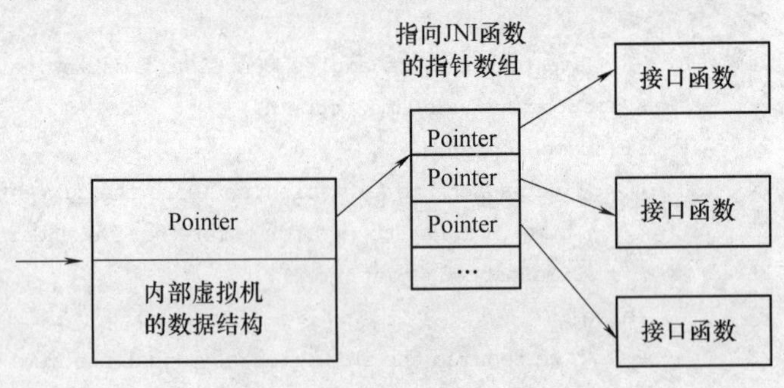

图 7-10　JNIEnv 接口指针

（2）在原生代码中的第二个参数 jobject 代表的是一个原生实例的方法，它与类实例相关，它们只能在类实例中调用，通过第二个参数获取实例引用。原生代码不仅可以通过实例方法实现也可以通过静态方法实现，与实例方法不同的是第二参数为 jclass 值类型，而且它没有与实例绑定，可以在静态上下文直接引用。

接下来便是通过 C 语言来实现 OpenGL 编程，具体可参见工程中其余的代码。

7.3.4　依赖关系建立

Android NDK 构建系统依赖另外两个文件：Android. mk 和 Application. mk，这两个文件应该作为 NDK 项目的一部分由开发人员提供，下面重点介绍。

1. Android. mk 文件

Android. mk 是向 Android NDK 系统项目中 GUN Makefile 文件，用于描述实现嵌入的必备信息，因此也是每个 NDK 项目的必备组件。该文件在 jni 子目录中，针对上述的工程，需添加的 Android. mk 的文件内容如下。

```
1  LOCAL_PATH: =$(call my-dir)
```

```
2 include $(CLEAR_VARS)
3 LOCAL_MODULE:=hello-jni
4 LOCAL_SRC_FILES:=hello-jni.c
5 include $(BUILD_SHARED_LIBRARY)
```

include $ (BUILD_SHARED_LIBRARY) 因为这是 GNU Makefile 文件，所以句法和 Makefile 是一样的。每行都包含一个单独的指令，上述代码中以“#”开头作为注释行，GNU Make 工具不处理它们，根据命名规范，变量名要大写。

第 1 行，Android 系统用 LOCAL_PATH 来表示需要编译的 C/C++代码所在目录，my-dir 是 Android NDK 编译系统提供的宏功能，通过调用该宏功能返回当前目录的位置。

第 2 行，Android 系统将 CLEAR_VARS 变量设置为脚本文件 clear-vars.mk 的位置，$ (CLEAR_VARS) 表示清除 LOCAL_PATH 以外所有 LOCAL_开始的变量，例如 LOCAL_MODULE 与 LOCAL_SRC_FILES 等变量。因为所有的脚本都将在同一个 GNU Make 的执行上下文中，而且所有变量都是全局变量，因此必须在每次使用前清空所有以前用过的变量。

第 3 行，变量 LOCAL_MODILE 用来声明模块名称，模块名称必须唯一，而且中间不能存在空格。因为模块名称也被用于给 NDK 编译系统生成的共享库文件命名，所以系统自动在模块名称前添加 lib 前缀，然后生成.so 文件。本例中，模块名称为 hello-jni，生成的共享库文件名称为 libhello-jni.so。需要注意的是，如果开发人员使用具有 lib 前缀的模块名称时，NDK 系统将不再添加前缀。

第 4 行，变量 LOCAL_SRC_FILES 表示编译模块时使用的 C/C++文件列表。本例中只有一个 C 文件，因此文件名称为 hello-jni.c。当源文件由多个文件组成时，可以用空格分开多个源文件名。默认情况下，.c 文件为 C 语言源文件，.cpp 文件是 C++语言源文件。

第 5 行，代码 include $ (BUILD_SHARED_LIBRARY) 表示 NDK 系统构建的共享库，Android NDK 编译系统将 BUILD_SHARED_LIBRARY 变量设置成 build-shared-library.mk 文件的保存位置。

2. Application.mk 文件

在进行 NDK 开发时，在应用程序目录中一定要有 Application.mk 文件，用来声明 Android 工程需要调用的共享库或静态库模块。Application.mk 文件通常被放置在 $ PROJECT/jni/Application.mk 下，$ PROJECT 指的是项目。另一种方法是将其放在顶层的子目录（$ NDK/apps 目录）下，例如：$ NDK/apps/<myapp>/Application.mk，<myapp>是一个简称，用于描述 NDK 编译系统的应用程序（这个名字不会生成共享库或者最终的包）。针对上述工程需添加的 Application.mk 文件内容如下：

```
APP_MODULES:=JNITest
APP_PROJECT_PATH:=$(call my-dir)/project
```

Application.mk 其实是一个小型 GNU Makefile 片段，它必须定义一些这样的变量：

• APP_PROJECT_PATH：这个变量是强制性的，并且会给出应用程序工程的根目录的一个绝对路径。这是用来复制或者安装一个没有任何版本限制的 JNI 库，从而给 APK 生成工具一个详细的路径。

注意：此变量对于第一种方法是可选的，但对于第二种方法却是必须的。

• APP_MODULES：这个变量是可选的，如果没有定义，NDK 将由在 Android. mk 中声明的默认的模块编译，并且包含所有的子文件（makefile 文件）。如果 APP_MODULES 定义了，它不许是一个空格分隔的模块列表，这个模块名字被定义在 Android. mk 文件中的 LOCAL_MODULE 中。注意 NDK 会自动计算模块的依赖。

注意：NDK 在 R4 开始改变了这个变量的行为，在此之前，在 Application. mk 中，该变量是强制的；必须明确列出所有需要的模块。

• APP_OPTIM：这个变量是可选的，用来定义“release”或“debug”。在编译应用程序模块的时候，可以用来改变优先级。“release”模式是默认的，并且会生成高度优化的二进制代码。“debug”模式生成的是未优化的二进制代码，但可以检测出很多的 BUG，可以用于调试。需要注意的是如果应用程序是可调试的（即如果清单文件中设置了 android：debuggable 的属性是“true”）。默认的是“debug”而不是“release”。则可以通过设置 APP_OPTIM 为“release”来将其覆盖。

注意：可以在“release”和“debug”模式下一起调试，但是“release”模式编译后将会提供更少的 BUG 信息。在清除 BUG 的过程中，有一些变量被优化了，或者根本就无法被检测出来，代码的重新排序会让这些代码变得更加难以阅读，并且让这些轨迹更加不可靠。

• APP_CFLAGS：当编译模块中有任何 C 文件或者 C ++ 文件的时候，C 编译器的信号就会被发出。这里可以在应用中需要这些模块时，进行编译的调整，这样就不需要直接更改 Android. mk 文件本身。

• APP_CPPFLAGS：C ++ 代码的编译选项。在 android-ndk-1. 5_r1 版本中，此变量只适用于 C ++，但是现在可以同时适用于 C 和 C ++。

• APP_ABI：默认情况下，Android NDK 编译系统会产生“armeabi”ABI 二进制文件。可以用该变量改变为其他 ABI 的二进制文件，例如：

支持 IA32 指令集：

APP_ABI：= x86

支持基于 armv7 FPU 指令集的设备：

APP_ABI：= armeabi-v7a

同时支持三种：

APP_ABI：= armeabi armeabi-v7a x86

从 NDK-r7 版本后，同时支持三种还可以这样写：

APP_ABI：= all

• APP_STL：在默认情况下，NDK 通过 Android 自带的最小化的 C ++ 运行库（system/lib/libstdc ++. so）来提供标准 C ++ 头文件。然而，NDK 提供了可供选择的 C ++ 实现，可以通过此变量来选择使用哪个链接到程序。做法可按如下形式：

```
APP_STL：= stlport_static-- > 静态 STLport 库
APP_STL：= stlport_shared-- > 动态 STLport 库
APP_STL：= system-- > 默认的 C ++ 运行时库
```

• APP_GNUSTL_FORCE_CPP_FEATURES：在先前的 NDK 版本中，当使用 GNU libstdc ++ runtime 运行库（通过设置 APP_STL 变量为 gnustl_static 或 gnustl_shared）都会强制支持

异常和 RTTI，在有些极少情况下可能会出现问题，同时会使生成的机器码包含不必须的内容。这种问题在 NDK r7b 中得到了解决，但是这也意味着如果真的需要支持异常和 RTTI 的话，则必须显式声明。要么通过 APP_CPPFLAGS，要么通过 LOCAL_CPPFLAGS 或 LOCAL_CPP_FEATURES。本变量就是为了解决此问题的。有两个选项供选择，也可同时都选择。

exceptions：强制所有模块支持异常。

rtti：强制所有模块支持 RTTI。

比如：

APP_GNUSTL_FORCE_CPP_FEATURES：= exceptions rtti。

- APP_BUILD_SCRIPT：在默认情况下，NDK 会在 jni 目录下查找 Android. mk 文件并使用它，如果想修改它，那么在此变量中可以指定一个自己的脚本来执行，路径还是以工程顶层目录为相对路径。

7. 3. 5 NDK 程序的链接与运行

【例 7-3】 通过 NDK 手法显示一个 OpenGL 视图。

经过上述的准备工作，现在便可以得到需要的 . so 文件并在 Android 中调用。首先打开 Cygwin 输入“cd $ NDK_ROOT”进入 NDK 安装目录，输入“cd apps/mwy”进入到配置文件目录，然后输入“ndk-build”使其生成 . so 文件，具体如图 7-11 所示。

```
mwy@mwy-PC /cygdrive/f/android-ndk-r9d/apps/san-angeles/project
$ ../../ndk-build
bash: ../../ndk-build: No such file or directory

mwy@mwy-PC /cygdrive/f/android-ndk-r9d/apps/san-angeles/project
$ ndk-build
[armeabi] Cygwin         : Generating dependency file converter script
[armeabi] Compile thumb  : JNITest <= importgl.c
[armeabi] Compile thumb  : JNITest <= demo.c
[armeabi] Compile thumb  : JNITest <= app-android.c
[armeabi] SharedLibrary  : libJNITest.so
[armeabi] Install        : libJNITest.so => libs/armeabi/libJNITest.so

mwy@mwy-PC /cygdrive/f/android-ndk-r9d/apps/san-angeles/project
$
```

图 7-11 编译结果

其运行成功后将生成的 libs 文件夹复制到工程中（与 src 同级），接下来便可以调用本地原生方法加载 C/C ++ 中的数据，其部分代码如下：

```
1    public class DemoActivity extends Activity{
2        @ Override
3        protected void onCreate(Bundle savedInstanceState){
4            super. onCreate(savedInstanceState);
5            mGLView = new DemoGLSurfaceView(this);
6            setContentView(mGLView);
7        }
8        @ Override
9        protected void onPause(){
```

```
10          super.onPause();
11          mGLView.onPause();
12      }
13      @Override
14      protected void onResume(){
15          super.onResume();
16          mGLView.onResume();
17      }
18      private GLSurfaceViewmGLView;
19      //装在动态链接库
20      static{
21          System.loadLibrary("JNITest");
22      }
23  }

24  class DemoGLSurfaceView extends GLSurfaceView{
25      public DemoGLSurfaceView(Context context){
26          super(context);
27          mRenderer=new DemoRenderer();
28          setRenderer(mRenderer);
29      }
30      public booleanonTouchEvent(final MotionEvent event){
31          if(event.getAction()==MotionEvent.ACTION_DOWN) {
32              nativePause ();
33          }
34          return true;
35      }
36      DemoRenderermRenderer;
37      private static native void nativePause ();
38  }

39  class DemoRenderer implements GLSurfaceView.Renderer {
40      public void onSurfaceCreated (GL10 gl, EGLConfigconfig) {
41          nativeInit ();
42      }
43      public void onSurfaceChanged (GL10 gl, int w, int h) {
44          nativeResize (w, h);
45      }
46      public void onDrawFrame (GL10 gl) {
47          nativeRender ();
48      }
49      //声明原生方法
50      private static native void nativeInit ();
51      private static native void nativeResize (int w, int h);
52      private static native void nativeRender ();
53      private static native void nativeDone ();
54  }
```

代码 25 ~ 38 行用一个 GLSurfaceView 类来显示一个 OpenGL 视图，而在 GLSurfaceView 类中实现了一个 Render 接口。代码 39 ~ 48 行则是 Render 接口的具体实现。代码第 51 ~ 54 行是对原生方法的声明，该声明中含有关键字 native 以通知 Java 编译器，它用另一种语言提供该方法的具体实现。尽管现在虚拟机知道该方法被原生实现，但是仍然不知道到哪儿去找该方法的实现。因此需要加载共享库，此时需要用到 java. lang. System 类的两个静态方法 load 和 loadLibrary 来加载共享库，比如代码中的第 21 行。此处尽管 Android NDK 生成的共享库名为 libJNITest. so，但是 loadLibrary 方法只采用 JNITest 这个库名，再按照所使用的具体操作系统的要求加上必要的前缀和后缀。库名与 Android. mk 文件中使用的 LOCAL_MODULE 构建系统变量定义的模块名相同。loadLibrary 的参数也不包含共享库的位置。Java 库路径，也就是系统属性 Java. library. path 保存 loadLibrary 方法在共享库搜索的目录列表，Android 上的 Java 库路径包含/vendor/lib 和/system/lib。其运行结果如图 7-12 所示。

图 7-12　运行结果

本章小结

本章主要介绍了 Android NDK 的相关知识，包括 NDK 的安装、开发环境的配置以及开发流程等内容，使用 NDK 实现一些对代码性能要求较高的模块并将这些模块嵌入到 Android 应用程序中会大大地提高程序效率。本章解析了利用 NDK 开发 OpenGL 实现 3D 特效的例程，介绍了 Android NDK 开发中如何实现依赖关系建立、NDK 程序的链接与运行等。

习　题

7-1　简述 NDK 系统的搭建过程。

7-2　NDK 开发过程一般分为哪几步？

7-3　Android. mk 和 Application. mk 文件有什么作用？它们各自必须包含哪些内容？

7-4　设计一个音乐播放器，其主体是利用 NDK 实现的。

实 践 篇

第 8 章 Android 通信应用

本章通过讨论 Socket 通信、蓝牙和 WiFi 对 Android 的通信 API 进行探索。Android 为 Socket 通信提供了两种基于网络传输协议 TCP/UDP 的通信方式供用户选择，满足了不同用户的需求。Android 提供了一些 API 用于管理及监视用户的蓝牙设备，以控制及发现邻近的蓝牙设备，在应用程序中使用蓝牙作为近距离的点对点传输层。Android 还提供了一个完整的 WiFi 数据包。使用这些 API 可以扫描热点，创建并修改 WiFi 配置。

8.1 Socket 通信

Android 与服务器的通信方式主要有两种，一种是 HTTP 通信，另一种是 Socket 通信。两者的最大差异在于，HTTP 连接使用的是“请求—响应方式”，即在请求时建立连接通道，当客户端向服务器发送请求后，服务器端才能向客户端返回数据。而 Socket 通信则是在双方建立起连接后就可以直接进行数据的传输，在连接时可实现信息的主动推送，而不需要每次由客户端向服务器发送请求。

那么，什么是 Socket? Socket 又称套接字，在程序内部提供了与外界通信的端口，即端口通信。通过建立 Socket 连接，可为通信双方的数据传输提供通道。Socket 通信的主要特点是数据丢失率低，使用简单且易于移植。

8.1.1 Socket 简介

Socket 通常也称作“套接字”，应用程序通常通过“套接字”向网络发出请求或者应答网络请求。以 J2SDK-1.3 为例，Socket 和 ServerSocket 类库位于 java.net 包中。ServerSocket 用于服务器端，Socket 是建立网络连接时使用的。在连接成功时，应用程序两端都会产生一个 Socket 实例，操作这个实例，完成所需的会话。对于一个网络连接来说，套接字是平等的，并没有差别，不因为在服务器端或在客户端而产生不同级别。不管是 Socket 还是 ServerSocket 它们的工作都是通过 SocketImp1 类及其子类完成的。

Socket 通信在双方建立起连接后就可以直接进行数据的传输，在连接时可实现信息的主动推送，而不需要每次由客户端向服务器发送请求。Socket 是一种抽象层，应用程序通过它来发送和接收数据，使用 Socket 可以将应用程序添加到网络中，与处于同一网络中的其他应用程序进行通信。简单来说，Socket 提供了程序内部与外界通信的端口并为通信双方提供了数据传输通道。

Socket 是一种低级、原始的通信方式，要编写服务器端代码和客户端代码，自己开端口，自己制定通信协议、验证数据安全和合法性，而且通常还应该是多线程的，开发起来比较繁琐。但是它也有其优点：灵活，不受编程语言、设备、平台和操作系统的限制，通信速度快而且高效。在介绍其具体的通信过程之前，需要先了解它的通信模型及重要的 API。

8.1.2 Socket 通信模型及重要的 API

在介绍 Socket 通信模型之前，需要先了解网络的传输协议。

1. 网络传输协议分为两种：TCP 和 UDP

（1）传输控制协议（Transmission Control Protocol，TCP）是基于连接的协议，也就是说，在正式收发数据前，必须和对方建立可靠的连接。TCP 能为应用程序提供可靠的通信连接，使一台计算机发出的字节流无差错地发往网络上的其他计算机，对可靠性要求高的数据通信系统往往使用 TCP 传输数据。

（2）用户数据报协议（User Data Protocol，UDP）是与 TCP 相对应的协议。它是面向非连接的协议，它不与对方建立连接，而是直接就把数据包发送过去。UDP 适用于一次只传送少量数据、对可靠性要求不高的应用环境。

2. Socket API

在 Socket 通信中能用到的重要的 Socket API：java. net. Socket 继承于 java. lang. Object，有八个构造器用于负责类中成员变量（域）的初始化。其方法并不多，下面介绍使用最频繁的三个方法。

（1）Accept 方法用于产生“阻塞”，直到接收到一个连接，并且返回一个客户端的 Socket 对象实例。“阻塞”是一个术语，它使程序运行暂时“停留”在这个地方，直到一个会话产生，然后程序继续；通常“阻塞”是由循环产生的。

（2）getInputStream 方法获得网络连接输入，同时返回一个 InputStream 对象实例。

（3）getOutputStream 方法连接的另一端将得到输出，同时返回一个 OutputStream 对象实例。注意：其中 getInputStream 和 getOutputStream 方法均可能会产生一个 IOException，它必须被捕获，因为它们返回的流对象，通常都会被另一个流对象使用。

8.1.3 ServerSocket 类

进行 Socket 通信必然用到 ServerSocket 类，在这里首先介绍一下 ServerSocket 类的各个构造方法以及成员方法的用法。

1. 构造 ServerSocket

ServerSocket 的构造方法有以下几种重载形式：

```
• ServerSocket()
• ServerSocket(int port)
• ServerSocket(int port, int backlog)
• ServerSocket(int port, int backlog, InetAddress bindAddr)
```

在以上构造方法中，参数 port 指定服务器要绑定的端口（服务器要监听的端口），参数 backlog 指定客户连接请求队列的长度，参数 bindAddr 指定服务器要绑定的 IP 地址。

（1）绑定端口

除了第一个不带参数的构造方法以外，其他构造方法都会使服务器与特定端口绑定，该端口由参数 port 指定。例如，以下代码创建了一个与 80 端口绑定的服务器。

```
ServerSocket ServerSocket = new ServerSocket(80);
```

如果运行时无法绑定到 80 端口，以上代码会抛出 IOException，更确切地说，是抛出 BindException，它是 IOException 的子类。BindException 一般是由以下原因造成的：

- 端口已经被其他服务器进程占用。
- 在某些操作系统中，如果没有以超级用户的身份来运行服务器程序，那么操作系统不允许服务器绑定到 1～1023 之间的端口。

如果把参数 port 设为 0，表示由操作系统来为服务器分配一个任意可用的端口。由操作系统分配的端口也称为匿名端口。对于多数服务器，会使用明确的端口，而不会使用匿名端口，因为客户程序需要事先知道服务器的端口，才能方便地访问服务器。

（2）设定客户连接请求队列的长度

ServerSocket 构造方法的 backlog 参数用来显式设置连接请求队列的长度，它将覆盖操作系统限定的队列的最大长度。值得注意的是，在以下几种情况中，仍然会采用操作系统限定的队列的最大长度。

- backlog 参数的值大于操作系统限定的队列的最大长度。
- backlog 参数的值小于或等于 0。
- 在 ServerSocket 构造方法中没有设置 backlog 参数。

（3）设定绑定的 IP 地址

如果主机只有一个 IP 地址，那么默认情况下，服务器程序就与该 IP 地址绑定。ServerSocket 的第四个构造方法 ServerSocket（int port，int backlog，InetAddress bindAddr）有一个 bindAddr 参数，它显式指定服务器要绑定的 IP 地址，该构造方法适用于具有多个 IP 地址的主机。假定一个主机有两个网卡，一个网卡用于连接到 Internet，IP 地址为 222. 67. 5. 94，还有一个网卡用于连接到本地局域网，IP 地址为 192. 168. 3. 4。如果服务器仅仅被本地局域网中的客户访问，那么可以按如下方式创建 ServerSocket：

```
ServerSocket ServerSocket = new ServerSocket(8000,10,InetAddress.getByName("192.168.3.4"));
```

（4）默认构造方法的作用

ServerSocket 有一个不带参数的默认构造方法。通过该方法创建的 ServerSocket 不与任何端口绑定，接下来还需要通过 bind( )方法与特定端口绑定。这个默认构造方法的用途是，允许服务器在绑定到特定端口之前，先设置 ServerSocket 的一些选项。因为一旦服务器与特定端口绑定，有些选项就不能再改变了。在以下代码中，先把 ServerSocket 的 SO_REUSEADDR 选项设为 true，然后再把它与 8000 端口绑定：

```
ServerSocket ServerSocket = new ServerSocket();
ServerSocket.setReuseAddress(true);//设置 ServerSocket 的选项
serversocket.bind(new InetSocketAddress(8000));//与 8000 端口绑定
```

如果把以上程序代码改为：

```
ServerSocketServerSocket = new ServerSocket(8000);
ServerSocket.setReuseAddress(true);//设置 ServerSocket 的选项
```

那么 ServerSocket. setReuseAddress（true）方法就不起任何作用了，因为 SO_REUSEAD-

DR 选项必须在服务器绑定端口之前设置才有效。

2. 获取 ServerSocket 的信息

ServerSocket 的以下两个 get 方法可分别获得服务器绑定的 IP 地址，以及绑定的端口：

```
• public InetAddressgetInetAddress()
• public intgetLocalPort()
```

前面已经讲到，在构造 ServerSocket 时，如果把端口设为 0，那么将由操作系统为服务器分配一个端口（称为匿名端口），程序只要调用 getLocalPort( )方法就能获知这个端口号。多数服务器会监听固定的端口，这样才便于客户程序访问服务器。匿名端口一般适用于服务器与客户之间的临时通信，通信结束，就断开连接，并且 ServerSocket 占用的临时端口也被释放。

3. ServerSocket 选项

ServerSocket 有以下三个选项。

- SO_TIMEOUT：表示等待客户连接的超时时间。
- SO_REUSEADDR：表示是否允许重用服务器所绑定的地址。
- SO_RCVBUF：表示接收数据的缓冲区的大小。

（1）SO_TIMEOUT 选项

```
• 设置该选项:public void setSoTimeout(int timeout)throws SocketException
• 读取该选项:public int getSoTimeout()throws IOException
```

SO_TIMEOUT 表示 ServerSocket 的 accept( )方法等待客户连接的超时时间，以毫秒为单位。如果 SO_TIMEOUT 的值为 0，表示永远不会超时，这是 SO_TIMEOUT 的默认值。当服务器执行 ServerSocket 的 accept( )方法时，如果连接请求队列为空，服务器就会一直等待，直到接收到了客户连接才从 accept( )方法返回。如果设定了超时时间，那么当服务器等待的时间超过了超时时间，就会抛出 SocketTimeoutException，它是 InterruptedException 的子类。

（2）SO_REUSEADDR 选项

```
• 设置该选项:public void setResuseAddress(boolean on)throws SocketException
• 读取该选项:public boolean getResuseAddress()throws SocketException
```

这个选项与 Socket 的 SO_REUSEADDR 选项相同，用于决定如果网络上仍然有数据向旧的 ServerSocket 传输数据，是否允许新的 ServerSocket 绑定到与旧的 ServerSocket 同样的端口上。SO_REUSEADDR 选项的默认值与操作系统有关，在某些操作系统中，允许重用端口，而在某些操作系统中不允许重用端口。当 ServerSocket 关闭时，如果网络上还有发送到这个 ServerSocket 的数据，这个 ServerSocket 不会立刻释放本地端口，而是会等待一段时间，确保接收到了网络上发送过来的延迟数据，然后再释放端口。

（3）SO_RCVBUF 选项

```
• 设置该选项:public void setReceiveBufferSize(int size)throws SocketException
• 读取该选项:public int getReceiveBufferSize()throws SocketException
```

SO_RCVBUF 表示服务器端的用于接收数据的缓冲区的大小，以字节为单位。一般说来，传输大的连续的数据块（基于 HTTP 或 FTP 的数据传输）可以使用较大的缓冲区，这样可以减少传输数据的次数，从而提高传输数据的效率。而对于交互式的通信（Telnet 和网络游戏），则应该采用小的缓冲区，确保能及时把小批量的数据发送给对方。

8.1.4　Socket 连接过程

根据连接启动的方式以及本地套接字要连接的目标，套接字之间的连接过程可以分为三个步骤：服务器监听、客户端请求和连接确认。

服务器监听：是服务器端套接字并不定位具体的客户端套接字，而是处于等待连接的状态，实时监控网络状态。

客户端请求：是指由客户端的套接字提出连接请求，要连接的目标是服务器端的套接字。为此，客户端的套接字必须首先描述它要连接的服务器的套接字，指出服务器端套接字的地址和端口号，然后就向服务器端套接字提出连接请求。

连接确认：是指当服务器端套接字监听到或者说接收到客户端套接字的连接请求，它就响应客户端套接字的请求，建立一个新的线程，把服务器端套接字的描述发给客户端，一旦客户端确认了此描述，连接就建立好了。而服务器端套接字继续处于监听状态，继续接收其他客户端套接字的连接请求。

8.1.5　Android 中的 Socket 通信

Android 中的 Socket 通信主要是基于上述的两种网络通信协议，其基本通信模型主要分为两类，即 TCP 和 UDP 模型。下面将简单介绍这两种类型的通信方式。

1. 基于 TCP 的 Socket

服务器端首先声明一个 ServerSocket 对象并且指定端口号，然后调用 ServerSocket 的 accept( ) 方法接收客户端的数据。accept( ) 方法在没有数据进行接收时处于堵塞状态。

```
Socket socket = ServerSocket.accept()
```

一旦接收到数据，通过 inputstream 读取接收的数据。客户端创建一个 Socket 对象，指定服务器端的 IP 地址和端口号。

```
Socket socket = new Socket("172.168.10.108",8080);
```

通过 inputstream 读取数据，获取服务器发出的数据。

```
OutputStream outputstream = socket.getOutputStream();
```

最后将要发送的数据写入到 outputstream 即可进行 TCP 的 Socket 数据传输。

2. 基于 UDP 的数据传输

服务器端首先创建一个 DatagramSocket 对象，并且指定监听的端口。接下来创建一个空的 DatagramSocket 对象用于接收数据。

```
bytedata[] = new byte[1024;]DatagramSocket packet = new DatagramSocket(data,data.length)
```

使用 DatagramSocket 的 receive( ) 方法接收客户端发送的数据，receive( ) 与 ServerSocket 的 accepet( ) 类似，在没有数据进行接收时处于堵塞状态。客户端也创建一个 DatagramSocket 对象，并且指定监听的端口。接下来创建一个 InetAddress 对象，这个对象类似于一个网络的发送地址。

```
InetAddress serveraddress = InetAddress.getByName("172.168.1.120")
```

定义要发送的一个字符串，创建一个 DatagramPacket 对象，并指定要将这个数据报发送到网络的哪个地址以及端口号，最后使用 DatagramSocket 对象的 send( ) 发送数据。

```
String str = "hello";bytedata[] = str.getByte();
DatagramPacket packet = newDatagramPacket(data, data.length, serveraddress, 4567); socket.send
(packet);
```

3. 以 TCP 通信模型为例，介绍其基本实现方式

- TCP 客户端实现：

```
1    //创建一个 Socket 对象,指定服务器端的 IP 地址和端口号
2    Socket socket = new Socket("192.168.1.104",4567);
3    //使用 InputStream 读取硬盘上的文件
4    InputStream inputStream = new FileInputStream("f://file/words.txt");
5    //从 Socket 当中得到 OutputStream
6    OutputStream outputStream = socket.getOutputStream();
7    bytebuffer[] = newbyte[4* 1024];inttemp = 0;
8    //将 InputStream 当中的数据取出,并写入到 OutputStream 当中
9    while((temp = inputStream.read(buffer))! =-1)
10   {
11       outputStream.write(buffer,0,temp);}
12   outputStream.flush();
```

- TCP 服务器端实现：

```
1    //声明一个 ServerSocket 对象
2    ServerSocket ServerSocket = null;try{
3    //创建一个 ServerSocket 对象,并让这个 Socket 在 4567 端口监听
4    ServerSocket = newServerSocket(4567);
5    //调用 ServerSocket 的 accept()方法,接收客户端所发送的请求,//如果客户端没有发送数据,那么该线程就停
滞不继续
6    Socket socket = ServerSocket.accept();
7    //从 Socket 当中得到 InputStream 对象
8    InputStream inputStream = socket.getInputStream();
9    bytebuffer[] = newbyte[1024* 4];inttemp = 0;
10   //从 InputStream 当中读取客户端所发送的数据
11   while((temp = inputStream.read(buffer))! = -1){
12       System.out.println(newString(buffer,0,temp));}
13   }catch(IOExceptione){
14       //TODOAuto-generatedcatchblocke.printStackTrace();}
15       ServerSocket.close();
16   }
```

【例 8-1】 实现服务器与客户端的 Socket 通信功能。

(1) 服务器端代码

建立 SocketServer 工程并向 Mysever.java 文件中添加如下代码：

```
1    import java.io.BufferedReader;
2    import java.io.BufferedWriter;
3    import java.io.IOException;
4    import java.io.InputStreamReader;
5    import java.io.OutputStreamWriter;
6    import java.io.PrintWriter;
```

```
7     import java.net.ServerSocket;
8     import java.net.Socket;
9     import java.util.ArrayList;
10    import java.util.List;
11    import java.util.concurrent.ExecutorService;
12    import java.util.concurrent.Executors;

13    public class MyServer{
14        private static final int PORT = 9999;
15        private List<Socket> mList = new ArrayList<Socket>();
16        private ServerSocket server = null;
17        private ExecutorService mExecutorService = null; //thread pool

18        public static void main(String[] args){
19        new MyServer();
20    }
21    public MyServer(){
22        try{
23            server = new ServerSocket(PORT);
24            mExecutorService = Executors.newCachedThreadPool(); //create a thread pool
25            System.out.println("服务器已启动...");
26            Socket client = null;
27            while(true){
28                client = server.accept();
29                //把客户端放入客户端集合中
30                mList.add(client);
31                 mExecutorService.execute(new Service(client)); //start a new thread to handle
the connection
32            }
33        }catch(Exception e){
34            e.printStackTrace();
35        }
36    }
37    class Service implements Runnable{
38        private Socket socket;
39        private BufferedReader in = null;
40        private String msg = "";

41        public Service(Socket socket){
42            this.socket = socket;
43            try{
44            in = new BufferedReader(new InputStreamReader(socket.getInputStream()));
45              //客户端只要一连到服务器,便向客户端发送下面的信息。
46              msg = "服务器地址:" + this.socket.getInetAddress() + "come toal:"
                   + mList.size() + "(服务器发送)";
47              this.sendmsg();
48            }catch(IOException e){
```

```
49                e.printStackTrace();
50            }
51        }
52        @Override
53        public void run(){
54            try{
55                while(true){
56                    if((msg = in.readLine())! = null){
57                        //当客户端发送的信息为 exit 时,关闭连接
58                        if(msg.equals("exit")){
59                            System.out.println("ssssssss");
60                            mList.remove(socket);
61                            in.close();
62                            msg = "user:" + socket.getInetAddress() + "exit total:" + mList.size();
63                            socket.close();
64                            this.sendmsg();
65                            break;
66                            //接收客户端发过来的信息 msg,然后发送给客户端
67                        }else{
68                            msg = socket.getInetAddress() + ":" + msg + "(服务器发送)";
69                            this.sendmsg();
70                        }
71                    }
72                }
73            }catch(Exception e){
74                e.printStackTrace();
75            }
76        }
77        /* *
78        * 循环遍历客户端集合,给每个客户端都发送信息
79        * /
80        public void sendmsg(){
81            System.out.println(msg);
82            int num = mList.size();
83            for(int index = 0;index < num;index ++){
84                Socket mSocket = mList.get(index);
85                PrintWriter pout = null;
86                try{
87                    pout = new PrintWriter(new BufferedWriter(new OutputStreamWriter(mSock-
et.getOutputStream())),true);
88                    pout.println(msg);
89                }catch(IOException e){
90                    e.printStackTrace();
91                }
92            }
93        }
```

```
94        }
95    }
```

（2）客户端代码

1）新建名为 SocketClient 的 Android 项目，在 res/layout/activity_main. xml 文件中分别声明一个 TextView，EditText 和 Button，其代码如下：

```
1     <LinearLayoutxmlns:android = "http://schemas.android.com/apk/res/android"
2         android:orientation = "vertical"
3         android:layout_width = "fill_parent"
4         android:layout_height = "fill_parent"
5     >
6     <TextView
7         android:id = "@ + id/TextView"
8         android:layout_width = "wrap_content"
9         android:layout_height = "wrap_content"
10        android:text = "@ string/hello_world"/>
11    <Button
12        android:id = "@ + id/Button02"
13        android:layout_width = "wrap_content"
14        android:layout_height = "wrap_content"
15        android:text = "@ string/send" >
16    </Button>
17    <EditText
18        android:id = "@ + id/EditText01"
19        android:layout_width = "wrap_content"
20        android:layout_height = "wrap_content"/>
21    </LinearLayout>
```

2）在 MainActivity. java 中添加如下代码：

```
1     package com.example.socketclient;

2     import java.io.BufferedReader;
3     import java.io.BufferedWriter;
4     import java.io.IOException;
5     import java.io.InputStreamReader;
6     import java.io.OutputStreamWriter;
7     import java.io.PrintWriter;
8     import java.net.Socket;
9     import android.app.Activity;
10    import android.app.AlertDialog;
11    import android.content.DialogInterface;
12    import android.os.Bundle;
13    import android.os.Handler;
14    import android.os.Message;
15    import android.view.View;
16    import android.widget.Button;
17    import android.widget.EditText;
```

```
18    import android.widget.TextView;

19    public class MainActivity extends Activity implements Runnable{
20        private TextView tv_msg = null;
21        private EditText ed_msg = null;
22        private Button btn_send = null;
23        //private Button btn_login = null;
24        private static final String HOST = "10.0.2.2";
25        private static final int PORT = 9999;
26        private Socket socket = null;
27        private BufferedReader in = null;
28        private PrintWriter out = null;
29        private String content = "";
30        //接收线程发送过来信息,并用 TextView 显示
31        public Handler mHandler = new Handler(){
32            public void handleMessage(Message msg){
33                super.handleMessage(msg);
34                tv_msg.setText(content);
35            }
36        };
37        @Override
38        public void onCreate(Bundle savedInstanceState){
39            super.onCreate(savedInstanceState);
40            setContentView(R.layout.activity_main);

41            tv_msg = (TextView)findViewById(R.id.TextView);
42            ed_msg = (EditText)findViewById(R.id.EditText01);
43            btn_send = (Button)findViewById(R.id.Button02);

44            try{
45                socket = new Socket(HOST, PORT);
46                in = new BufferedReader(new InputStreamReader(socket.getInputStream()));
47                 out = new PrintWriter (new BufferedWriter (new OutputStreamWriter (sock-
et.getOutputStream())), true);
48            }catch(IOException ex){
49                ex.printStackTrace();
50                ShowDialog("login exception" + ex.getMessage());
51            }
52            btn_send.setOnClickListener(new Button.OnClickListener(){
53            @Override
54                public void onClick(View v){
55                    //TODO Auto-generated method stub
56                    String msg = ed_msg.getText().toString();
57                    if(socket.isConnected()){
58                        if(! socket.isOutputShutdown()){
59                            out.println(msg);
60                        }
```

```
61                  }
62              }
63          });
64          //启动线程,接收服务器发送过来的数据
65          new Thread(MainActivity.this).start();
66      }
67      /**
68      * 如果连接出现异常,弹出 AlertDialog!
69      */
70      public void ShowDialog(String msg){
71          new AlertDialog.Builder(this).setTitle("notification").setMessage(msg)
            .setPositiveButton("ok", new DialogInterface.OnClickListener(){
72          @Override
73          public void onClick(DialogInterface dialog, int which){

74          }
75      }).show();
76  }
77  /**
78  * 读取服务器发来的信息,并通过 Handler 发给 UI 线程
79  */
80      public void run(){
81          try{
82              while(true){
83                  if(!socket.isClosed()){
84                      if(socket.isConnected()){
85                          if(!socket.isInputShutdown()){
86                              if((content=in.readLine())!=null){
87                                  content+="\n";
88                              mHandler.sendMessage(mHandler.obtainMessage());
89                              }else{
90                              }
91                          }
92                      }
93                  }
94              }
95          }catch(Exception e){
96              e.printStackTrace();
97          }
98      }
99  }
```

3）在 AndroidManifest.xml 文件中加入对网络的访问权限，代码如下：

```
<uses-permission android:name="android.permission.INTERNET"></uses-permission>
```

（3）单击 SocketServer 选择运行 Java Application。

（4）单击 SocketClient 选择运行 Android Application。

（5）运行结果

首次运行 SocketServer 时在 Eclipse/console 中 Android 模拟器中显示如图 8-1 所示。服务器地址设定为 127.0.0.1。在文本框内输入 Socket test 并单击发送按钮后在 Eclipse/console 中显示如下内容，/127.0.0.1：Socket test（服务器发送），在 Android 模拟器中显示如图 8-2 所示。

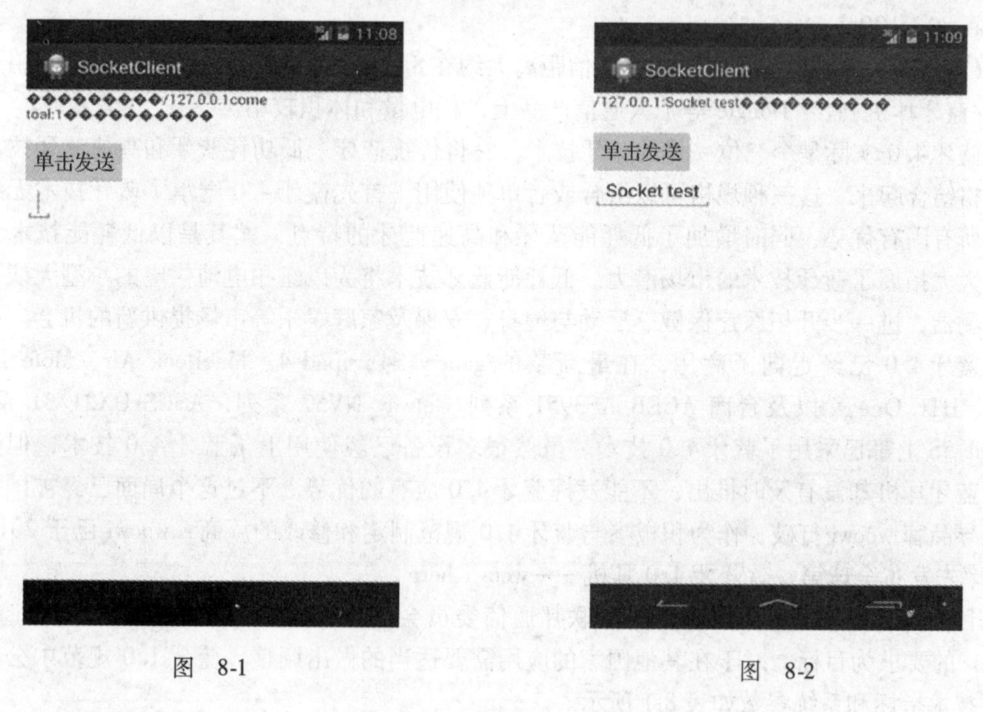

图 8-1　　图 8-2

8.2 蓝牙通信

蓝牙是一种支持设备短距离通信的无线电技术。蓝牙起源于 1994 年，最早由瑞典的手机业巨头爱立信开始研发，目的是以无线通信的方式取代手机与各种周边设备的有线连接。1998 年 2 月，爱立信、诺基亚、IBM、东芝和英特尔五大跨国公司组成研究小组，目标是建立一个全球性的小范围无线通信技术，现在蓝牙技术是一种无线数据与语音通信的开放性全球规范，它以低成本的近距离无线连接为基础，为固定与移动设备通信环境建立一个特别的连接。利用蓝牙技术，能够有效地简化移动通信终端设备之间的通信，也能够成功地简化设备与因特网之间的通信，从而使数据传输变得更加迅速高效，为无限通信拓宽了道路。蓝牙采用分散式网络结构以及快跳频和短包技术，支持点对点以及点对多点通信，工作在全球通信的 2.4GHz ISM（即工业、科学、医学）频段，其数据速率为 1Mbit/s，采用时分双工传输方案实现全双工传输。

8.2.1 蓝牙简介

蓝牙技术联盟（Bluetooth Special Interest Group，Bluetooth SIG）是一家由电信、计算机、汽车制造、工业自动化和网络行业的领先厂商组成的蓝牙技术协会。该小组致力于推动蓝牙无线技术的发展，为短距离连接移动设备制定低成本的无线规范，并将其推向市场。截止

2010 年 7 月，蓝牙共有六个版本 V1.1/1.2/2.0/2.1/3.0/4.0，以通信距离来看在不同版之上本可再分为 Class A 和 Class B。

（1）Class A 是用在大功率/远距离的蓝牙产品上，但因成本高和耗电量大，不适合用于个人通信产品（手机/蓝牙耳机/蓝牙 Dongle 等），故多用在部分商业特殊用途上，通信距离大约在 80 ~ 100m。

（2）Class B 是最流行的制式，通信距离大约在 8 ~ 30m，视产品的设计而定，多用于手机内/蓝牙耳机/蓝牙 Dongle 等个人通信产品上，耗电量和体积较小。

蓝牙 4.0 实际是个三位一体的蓝牙技术，它将传统蓝牙、低功耗蓝牙和高速蓝牙技术三种规格结合起来，这三种规格可以组合或者单独使用。首先蓝牙 4.0 继承了蓝牙技术无线连接的所有固有优势，同时增加了低耗能蓝牙和高速蓝牙的特点，尤其是以低耗能技术为核心，大大拓展了蓝牙技术的市场潜力。低耗能蓝牙技术将为以纽扣电池供电的小型无线产品及感测器，进一步开拓医疗保健、运动与健身、安保及家庭娱乐等市场提供新的机会。

蓝牙 4.0 已经走向了商用，在最新款的 galaxy S4，ipad 4、MacBook Air、Moto Droid Razr、HTC One X 以及台商 ACER AS3951 系列/Getway NV57 系列，ASUS UX21/31 系列，iPhone 5S 上都已采用了蓝牙 4.0 技术。虽然很多设备已经使用上了蓝牙 4.0 技术，但是相应的蓝牙耳机却没有及时推出，不能发挥蓝牙 4.0 应有的优势。不过这个局面已经被国内蓝牙领导品牌 woowi 打破，作为积极参与蓝牙 4.0 规范制定和修改的厂商，woowi 已于 2012 年 6 月率先发布全球第一款蓝牙 4.0 耳机——woowi hero。

目前，蓝牙技术主要以满足美国联邦通信委员会（Federal Communications Commission，FCC）的要求为目标。对于在其他国家的应用需要适当的做出调整。蓝牙 1.0 规范中公布的主要技术指标和系统参数如表 8-1 所示。

表 8-1　蓝牙技术指标和系统参数

工作频段	ISM 频段：2.402 ~ 2.480GHz
双工方式	全双工，TDD 时分双工
业务类型	支持电路交换和分组交换业务
数据速率	1Mbit/s
非同步信道速率	非对称连接 721Kbit/s、57.6Kbit/s、对称连接：432.6Kbit/s
同步信道速率	64Kbit/s
功率	没过 FFC 要求小于 0dbm（1mW），其他国家可扩展为 100mW
跳频频率数	79 个频点/MHz
跳频速率	1600 跳/秒
工作模式	PARK/HOLD/SNIFF
数据连接方式	面向连接业务 SCO，无连接业务 ACL
纠错方式	1/3FEC、2/3FEC、ARQ
鉴权	采用反应逻辑算数
信道加密	采用 0 位、40 位、60 位加密字符
语音编码方式	连续可变斜率调制 CVSD
发射距离	一般可达到 10m，增加功率情况下可达到 100m

8.2.2 蓝牙系统的组成

蓝牙系统一般由天线单元、链路控制（固件）单元、链路管理（软件）单元和蓝牙软件（协议栈）单元四个功能单元组成。

蓝牙的天线单元是微带天线，具有体积小、重量轻的特点。

链路控制(固件) 单元中使用了三个 IC 分别作为连接控制器、基带处理器和射频传输/接收器。

链路管理（软件）单元中的软件模块携带了链路的数据设置、鉴权、链路硬件配置和其他一些协议。

链路管理能够发现其他远端的链路并通过链路管理协议与之通信。软件（协议栈）单元是一个独立的操作系统，不与任何操作系统捆绑，它必须符合已经制定好的蓝牙规范。

8.2.3 蓝牙技术的特点

从目前的应用来看，由于蓝牙体积小、功率低，其应用已不局限于计算机外设，几乎可以被集成到任何数字设备之中，特别是那些对数据传输速率要求不高的移动设备和便携设备。蓝牙技术的特点可归纳为如下几点：

（1）全球范围适用　蓝牙工作在 2.4GHz 的 ISM 频段，全球大多数国家 ISM 频段的范围是 2.4 ~ 2.4835GHz，使用该频段无需向各国的无线电资源管理部门申请许可证。

（2）同时可传输语音和数据　蓝牙采用电路交换和分组交换技术，支持异步数据信道、三路语音信道以及异步数据与同步语音同时传输的信道。每个语音信道数据速率为 64Kbit/s，语音信号编码采用脉冲编码调制（PCM）或连续可变斜率增量调制（CVSD）方法。当采用非对称信道传输数据时，速率最高为 721Kbit/s，反向为 57.6Kbit/s；当采用对称信道传输数据时，速率最高为 342.6Kbit/s。蓝牙有两种链路类型：异步无连接（Asynchronous Connection-Less，ACL）链路和同步面向连接（Synchronous Connection-Oriented，SCO）链路。

（3）可以建立临时性的对等连接（Ad-hoc Connection）　根据蓝牙设备在网络中的角色，可分为主设备（Master）和从设备（Slave）。主设备是组网连接主动发起连接请求的蓝牙设备，几个蓝牙设备连接成一个皮网（Piconet）时，其中只有一个主设备，其余的均为从设备。皮网是蓝牙最基本的一种网络形式，最简单的皮网是由一个主设备和一个从设备组成的点对点的通信连接。通过时分复用技术，一个蓝牙设备便可以同时与几个不同的皮网保持同步，具体来说，就是该设备按照一定的时间顺序参与不同的皮网，即某一时刻参与某一皮网，而下一时刻参与另一个皮网。

（4）具有很好的抗干扰能力　工作在 ISM 频段的无线电设备有很多种，如家用微波炉、无线局域网（Wireless Local Area Network，WLAN）和 HomeRF 等产品，为了很好地抵抗来自这些设备的干扰，蓝牙采用了跳频（Frequency Hopping）方式来扩展频谱（Spread Spectrum），将 2.402 ~ 2.48GHz 频段分成 79 个频点，相邻频点间隔 1MHz。蓝牙设备在某个频点发送数据之后，再跳到另一个频点发送，而频点的排列顺序则是伪随机的，每秒钟频率改变 1600 次，每个频率持续 625μs。

（5）蓝牙模块体积很小、便于集成　由于个人移动设备的体积较小，嵌入其内部的蓝牙模块体积就应该更小，如爱立信公司的蓝牙模块 ROK101008 的外形尺寸仅为 32.8mm × 16.8mm × 2.95mm。

（6）低功耗　蓝牙设备在通信连接（Connection）状态下，有四种工作模式——激活（Active）模式、呼吸（Sniff）模式、保持（Hold）模式和休眠（Park）模式。Active 模式是正常的工作状态，另外三种模式是为了节能所规定的低功耗模式。

（7）开放的接口标准　SIG 为了推广蓝牙技术，将蓝牙的技术标准全部公开，全世界范围内的任何单位和个人都可以进行蓝牙产品的开发，只要最终通过 SIG 的蓝牙产品兼容性测试，就可以推向市场。

（8）成本低　随着市场需求的扩大，各个供应商纷纷推出自己的蓝牙芯片和模块，蓝牙产品价格飞速下降。

8.2.4 Android 蓝牙驱动架构

Android 的蓝牙协议栈是使用 BlueZ 实现的，它是 Linux 平台上一套完整的蓝牙协议栈。对 GAP、SDP 和 RECOMM 等应用规范提供支持，并获得 SIG 认证。广泛应用于各 Linux 发行版本中，并被移植到众多移动平台上。由于 BlueZ 使用了 GPL（GNU 通用公共许可证）授权，为了避免使用未经授权的代码，Android 框架通过 D-BUSIPC（进程间通信机制）来与 BlueZ 的用户控件代码进行交互。BlueZ 提供了很多分散的应用，包括守护进程和一些工具，BlueZ 通过 D-BUS IPC 机制来提供应用层接口。从 Bluetooth 的规格可以看出，整个蓝牙的构架除了硬件的 RF、Baseband、Link Manager 以外，BlueZ 实现 L2CAP、RFCOMM、SDP、TCS 等软件部分，并以标准 Socket 形式封装了 HCI、L2CAP、RFCOMM 协议，使得应用调用更加方便。

BlueZ 是 Linux 官方的 Bluetooth 栈，由主机控制接口（Host Control Interface，HCI）层、Bluetooth 协议核心、逻辑链路控制和适配协议（Logical Link Control and Adaptation Protocol，L2CAP）、SCO 音频层、其他 Bluetooth 服务、用户空间后台进程以及配置工具组成。其 Bluetooth 架构如图 8-3 所示。Linux 程序运行状态分为内核态和用户态，蓝牙协议栈 BlueZ 可分为两部分：Kernel 层的内核代码和 Library 层的用户态程序及工具集，其中内核代码由 BlueZ 核心协议和驱动程序等模块组成。Bluetooth 协议实现在内核代码/net/bluetooth 中，包括 HCI、L2CAP、HID、RFCOMM、SCO、SDP、BNEP 等协议的实现。驱动程序放在/drivers/bluetooth 中，包括 Linux Kernel 对各种接口的 Bluetooth 设备的驱动，例如 USB 接口、串口等。用户态程序及工具集包括应用程序接口和 BlueZ 工具集，位于"externel/bluez"目录中。BlueZ 提供函数库以及应用程序接口，便于开发 Bluetooth 应用程序。BlueZ utils 是主要工具集，实现对 Bluetooth 设备的初始化和控制。

蓝牙系统的核心是 BlueZ，因此 JNI 和上层都围绕跟 BlueZ 的沟通进行。JNI 和 Android 应用层，跟 BlueZ 沟通的主要手段是 D-BUS，D-BUS 是一套应用广泛的进程间通信（IPC）机制，相对于 Socket 等底层 IPC，它是更加复杂的 IPC 机制，支持更系统化的服务名、函数名等，同时也能对众多服务进程和客户端进行管理，调度通信消息的传递。跟 Android 框架使用的 Binder 类似。

如图 8-4 所示为 Android 蓝牙协议栈，它主要由底层硬件模块、中间协议层和高端应用层组成。对于某些应用可能只需要使用其中的某一列或多列，而没有必要使用全部协议。

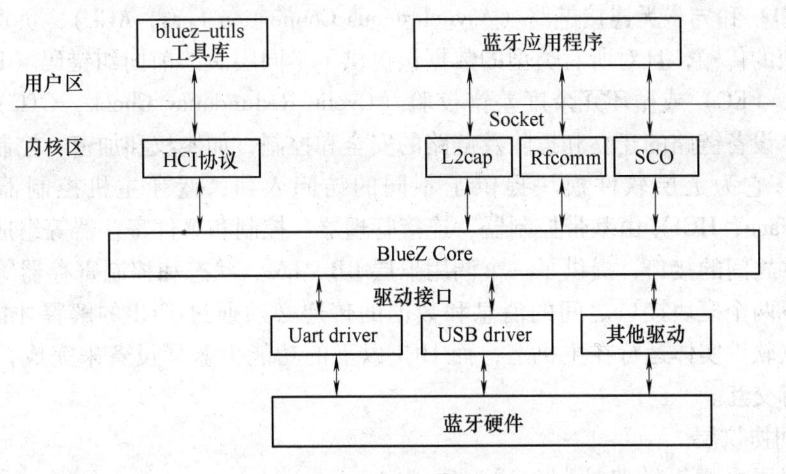

图 8-3 Bluetooth 架构

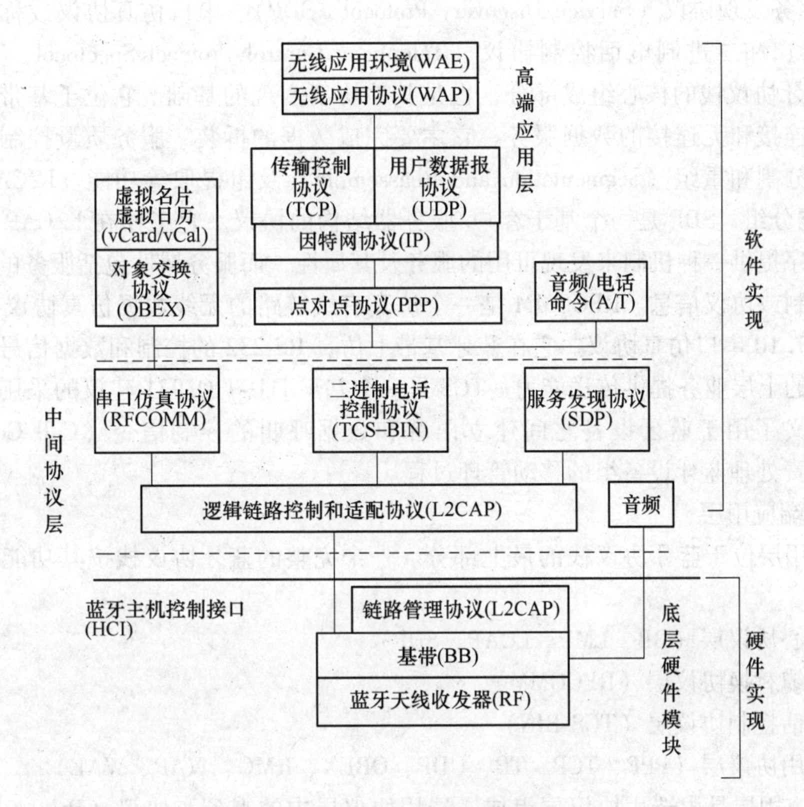

图 8-4 Android 蓝牙协议栈

(1) 底层硬件模块

底层模块是蓝牙技术的核心模块，所有嵌入蓝牙技术的设备都必须包括底层模块。它主要由链路管理协议层（Link Manager Protocol，LMP）、基带层（Base Band，BB）和射频（Rodio Fraquency，RF）组成。其功能是：无线连接层（RF）通过 2.4GHz 无需申请的 ISM 频段，实现数据流的过滤和传输；它主要定义了工作在此频段的蓝牙接收机应满足的需求；基带层（BB）提供了两种不同的物理链路，即同步面向连接链路（Synchronous Connection

Oriented, SCO）和异步无连接链路（Asynchronous Connection Less, ACL），负责跳频和蓝牙数据及信息帧的传输，且对所有类型的数据包提供了不同层次的前向纠错码（Frequency Error Correction, FEC）或循环冗余度差错校验（Cyclic Redundancy Check, CTC）；LMP 层负责两个或多个设备链路的建立和拆除及链路的安全和控制，如鉴权和加密、控制和协商基带包的大小等，它为上层软件模块提供了不同的访问入口；蓝牙主机控制器接口（Host Cntroller Interface, HCI）由基带控制器、连接管理器、控制和事件寄存器等组成。它是蓝牙协议中软硬件之间的接口，提供了一个调用下层 BB、LM、状态和控制寄存器等硬件的统一命令，上、下两个模块接口之间的消息和数据的传递必须通过 HCI 的解释才能进行。HCI 层以上的协议软件实体运行在主机上，而 HCI 以下的功能由蓝牙设备来完成，二者之间通过传输层进行交互。

（2）中间协议层

中间协议层由逻辑链路控制与适配协议（Logical Link Control and Adaptation Protocol, L2CAP）、服务发现协议（Service Discovery Protocol, SDP）、串口仿真协议或称线缆替换协议（RFCOM）和二进制电话控制协议（Telephony Control Protocol Spectocol, TCS）组成。L2CAP 是蓝牙协议栈的核心组成部分，也是其他协议实现的基础。它位于基带之上，向上层提供面向连接和无连接的数据服务。它主要完成数据的拆装、服务质量控制、协议的复用、分组的分割和重组（Segmentation and Reassembly）及组提取等功能。L2CAP 允许高达 64KB 的数据分组。SDP 是一个基于客户/服务器结构的协议。它工作在 L2CAP 层之上，为上层应用程序提供一种机制来发现可用的服务及其属性，而服务属性包括服务的类型及该服务所需的机制或协议信息。RFCOMM 是一个仿真有线链路的无线数据仿真协议，符合 ETSI 标准的 TS 07.10 串口仿真协议。它在蓝牙基带上仿真 RS-232 的控制和数据信号，为原先使用串行连接的上层业务提供传送能力。TCS 是一个基于 ITU-T Q.931 建议的采用面向比特的协议，它定义了用于蓝牙设备之间建立语音和数据呼叫的控制信令（Call Control Signalling），并负责处理蓝牙设备组的移动管理过程。

（3）高端应用层

高端应用层位于蓝牙协议栈的最上部分。一个完整的蓝牙协议栈按其功能又可划分为四层：

（1）核心协议层（BB、LMP、LCAP、SDP）

（2）线缆替换协议层（RFCOMM）

（3）电话控制协议层（TCS-BIN）

（4）选用协议层（PPP、TCP、TP、UDP、OBEX、IrMC、WAP、WAE）

而高端应用层是由选用协议层组成。选用协议层中的点到点协议（Point-to-Point Protocol, PPP）是由封装、链路控制协议、网络控制协议组成，定义了串行点到点链路应当如何传输因特网协议数据，它主要用于 LAN 接入、拨号网络及传真等应用规范；传输控制协议/网络层协议（TCP/IP）、对象交换协议（User Datagram Protocol, UDP）是三种已有的协议，它定义了因特网与网络相关的通信及其他类型计算机设备和外围设备之间的通信。蓝牙采用或共享这些已有的协议去实现与连接因特网的设备通信，这样，既可提高效率，又可在一定程度上保证蓝牙技术和其他通信技术的互操作性；OBEX（Object Exchange Protocol）是对象交换协议，它支持设备间的数据交换，采用客户/服务器模式提供与超文本传输协议（HT-

TP）相同的基本功能。该协议作为一个开放性标准还定义了可用于交换的电子商务卡、个人日程表、消息和便条等格式；WAP（Wireless Application Protocol）是无线应用协议，它的目的是要在数字蜂窝电话和其他小型无线设备上实现因特网业务。它支持移动电话浏览网页、收取电子邮件和其他基于因特网的协议；WAE（Wireless Application Environment）是无线应用环境，它提供用于 WAP 电话和个人数字助理（PDA）所需的各种应用软件。

8.2.5 蓝牙在 Android 下的应用

Android 平台包括蓝牙网络栈，它允许设备以无线方式与其他蓝牙设备进行数据交换。应用程序框架提供了通过 Android 蓝牙 API 访问蓝牙的功能。这些 API 允许应用程序以无线方式连接至其他蓝牙设备，可实现点对点和点对多点无线通信功能。

使用蓝牙的 API，Android 应用程序可以执行以下操作：

- 扫描其他蓝牙设备
- 查询本地蓝牙适配器配对的蓝牙设备
- 建立 RFCOMM 通道
- 通过服务发现并连接到其他设备
- 和其他设备传输数据
- 管理多个连接

蓝牙技术的目的就是在小型移动设备及其外围设备之间建立无线连接，使得它们之间的通信变得迅速、便捷。蓝牙设备的使用意味着在工作、生活中无需布设专用的线缆和连接器，通过蓝牙装置可以形成一个微网，网内的设备可以互相通信。

现代家庭在科学技术的帮助下，生活方便而高效。人们不仅将技术融入工作办公，还将技术应用扩展到家庭生活的其他方面。

通过使用蓝牙技术产品，人们可以免除办公室电缆缠绕的苦恼。鼠标、键盘、打印机、耳机和扬声器等均可以在 PC 环境中无线使用。此外，通过在移动设备和 PC 之间同步联系人和日历信息，用户可以随时随地存取最新的信息。蓝牙设备不仅可以使办公更加轻松，还能使家庭娱乐更加便利。例如用户可以在蓝牙使用范围内无线控制存储在 PC 或 Apple iPod 上的音频文件。蓝牙技术还可以用在适配器中，允许人们从相机、手机、计算机向电视发送照片以与朋友共享。

在这里以蓝牙在 Android 手机上的通信为例，介绍一下蓝牙之间是如何通信的。Android 蓝牙通信 API 是对蓝牙无线频率通信协议 RFCOMM 进行的封装。RFCOMM 支持在逻辑链路控制和适配协议层上进行 RS232 串行通信。RFCOMM 是一个面向连接，通过蓝牙模块进行的数据流传输方式，它也被称为串行端口规范（Serial Port Profile，SPP）。对蓝牙端口的监听类似于 TCP 端口：使用 Socket 和 ServerSocket 类。在服务器端，使用 BluetoothServerSocket 类来创建一个监听服务端口。当一个连接被 BluetoothServerSocket 所接受，它会返回一个新的 BluetoothSocket 来管理该连接。在客户端，使用一个单独的 BluetoothSocket 类去初始化一个外接连接和管理该连接。为了创建一个对准备好的新来的连接去进行监听 BluetoothServerSocket 类，使用 BluetoothAdapter. listenUsingRfcommWithServiceRecord( ) 方法。然后调用 accept( ) 方法去监听该连接的请求。在连接建立之前，该调用会被阻断，也就是说，它将返回一个 BluetoothSocket 类去管理该连接。每次获得该类之后，如果不再需要接受连接，最后调用在 BluetoothServerSocket 类下的 close( )方法。关闭 BluetoothServerSocket 类不会关闭这个已经返回的 BluetoothSocket 类。BluetoothSocket 类线程安全。特别的，close( )方法总会马上放弃外界操作并关闭服务器端口。

在 Android 中要对蓝牙设备进行操作需要用到 Android SDK 中与蓝牙相关的 API，在介绍这些 API 之前先介绍一个比较重要的概念—介质访问控制（Medium/Media Access Control，MAC）。MAC 地址是烧录在网卡（NetworkInterfaceCard，NIC）里的。MAC 地址，也叫硬件地址，是由 48Byte 长（6B），十六进制的数字组成。0 ~ 23 位叫做组织唯一标志符（organizationally unique）是识别 LAN（局域网）节点的标志。24 ~ 47 位是由厂家自己分配的。其中第 40 位是组播地址标志位。网卡的物理地址通常是由网卡生产厂家烧入网卡的 EPROM（一种闪存芯片，通常可以通过程序擦写），它存储的是传输数据时真正发出数据的电脑和接收数据的主机的地址。两个蓝牙设备之间的通信过程大致如下，首先是两个设备上都要有蓝牙设备（或者专业一点叫做蓝牙适配器）。其通信具体过程如图 8-5 所示。

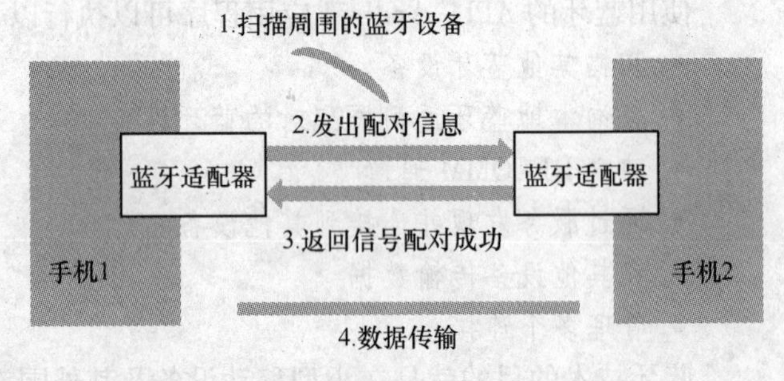

图 8-5　蓝牙通信过程示意图

也就是说首先手机 1 扫描周围的蓝牙设备，找到有可用设备时就会向其发出配对信息（密钥），手机 2 在接到请求后输入相应密钥即配对完成，接下来就可以进行数据传输了。

1. Android 蓝牙通信中常用的方法

Android 平台提供蓝牙 API 来实现蓝牙设备之间的通信，蓝牙设备之间的通信主要包括了四个步骤：设置蓝牙设备、寻找局域网内可用或者匹配的设备、连接设备和设备之间的数据传输。以下是建立蓝牙连接所需要的一些基本类：

BluetoothAdapter 类：代表了一个本地的蓝牙适配器。它是所有蓝牙交互的入口点。利用它可以发现其他蓝牙设备，查询绑定了的设备，使用已知的 MAC 地址实例化一个蓝牙设备和建立一个 BluetoothServerSocket（作为服务器端）监听来自其他设备的连接。

BluetoothDevice 类：代表了一个远端的蓝牙设备，使用它请求远端蓝牙设备连接或者获取远端蓝牙设备的名称、地址、种类和绑定状态（其信息封装在 BluetoothSocket 中）。

BluetoothSocket 类：代表了一个蓝牙套接字的接口（类似于 TCP 中的套接字），它是应用程序通过输入、输出流与其他蓝牙设备通信的连接点。

BlueboothServerSocket 类：代表打开服务连接来监听可能到来的连接请求（属于 Server 端），为了连接两个蓝牙设备必须有一个设备作为服务器打开一个服务套接字。当远端设备发起连接请求时，并且请求送达时，BlueboothServerSocket 类将会返回一个 BluetoothSocket。

BluetoothClass 类：描述了一个蓝牙设备的一般特点和能力。它的只读属性集定义了设备的主、次设备类和一些相关服务。然而，它并没有准确地描述所有该设备所支持的蓝牙文件和服务，而是作为对设备种类来说的一个小小暗示。

要操作蓝牙，先要在 AndroidManifest. xml 里加入权限：

```
<uses-permissionandroid:name="android.permission.BLUETOOTH_ADMIN"/> <uses-permissionandroid:name="android.permission.BLUETOOTH"/>
```

Android 所有关于蓝牙开发的类都在 android. bluetooth 包下，只有 8 个类。常用的四个类如下所示：

（1）BluetoothAdapter

蓝牙适配器，直到建立 BluetoothSocket 连接之前，都要不断操作它。BluetoothAdapter 里的方法很多，常用的有以下几个：

cancelDiscovery()：取消发现，也就是说当正在搜索设备的时候调用这个方法将不再继续搜索。

disable()：关闭蓝牙。

enable()：打开蓝牙，这个方法打开蓝牙不会弹出提示，更多的时候需要问下用户是否打开，以下两行代码同样是打开蓝牙，但会提示用户。

```
Intent enabler=new Intent(BluetoothAdapter.ACTION_REQUEST_ENABLE);
startActivityForResult(enabler,reCode);//同 startActivity(enabler);
```

getAddress()；//获取本地蓝牙地址

getDefaultAdapter()；　//获取默认 BluetoothAdapter

实际上，也只有这一种方法获取 BluetoothAdapter；getName()获取本地蓝牙名称；getRemoteDevice（String address）根据蓝牙地址获取远程蓝牙设备；getState()获取本地蓝牙适配器当前状态（调试时使用）；isDiscovering()判断当前是否正在查找设备，是则返回 true；isEnabled()判断蓝牙是否打开，已打开返回 true，否则返回 false；listenUsingRfcommWithServiceRecord（String name，UUID uuid）根据名称，UUID 创建并返回 BluetoothServerSocket，这是创建 BluetoothSocket 服务器端的第一步；startDiscovery()开始搜索，这是搜索的第一步。

（2）BluetoothDevice

BluetoothDevice 类被用于描述一个远程蓝牙设备，它拥有设备的名字、地址、连接状态等属性。它是对蓝牙设备硬件地址的一个轻量级包装，可从 BluetoothAdapter 的 getRemoteDevice 等方法中获取当前蓝牙设备列表。BluetoothDevice 类中重要的方法是 createRfcommSocketToServiceRecord（UUID uuid），根据 UUID 创建一个此蓝牙设备的 RFCOMM 蓝牙安全输出连接。然后可以利用该连接进行安全的写操作。这个类其他的方法，如 getAddress()、getName()等，同 BluetoothAdapter 一样。

（3）BluetoothServerSocket

如果去除了 Bluetooth 相信大家一定再熟悉不过了，既然是 Socket，方法就应该都差不多，这个类一共只有三个方法：

1）public BluetoothSocket accept（int timeout）：阻塞直到超时时间内的连接建立。在一个成功建立的连接上返回一个已连接的 BluetoothSocket 类。每当该调用返回时，它可以再次调用去接收以后新来的连接。close()方法可以用来放弃从另一线程来的调用。

2）public BluetoothSocket accept()：阻塞直到一个连接已经建立。在一个成功建立的连接上返回一个已连接的 BluetoothSocket 类。每当该调用返回时，它可以再次调用去接收以后新来的连接。close( )方法可以用来放弃从另一线程来的调用。

3）public void close( )：马上关闭端口，并释放所有相关的资源。在其他线程的该端口中引起阻塞，从而使系统马上抛出一个 IO 异常。关闭 BluetoothServerSocket 不会关闭接收自 accept( )的任意 BluetoothSocket。

两个重载的 accept( )和 accept（int timeout）的区别在于后面的方法指定了过时时间，需要注意的是，执行这两个方法的时候，直到接收到了客户端的请求（或是过期之后），都会阻塞线程，应该放在新线程里运行。还有一点需要注意的是，这两个方法都返回一个 BluetoothSocket，最后的连接也是服务器端与客户端的两个 BluetoothSocket 的连接。

（4）BluetoothSocket 是客户端，跟 BluetoothServerSocket 相对应

一共五种方法，不出意外，都会用到 close( )关闭、connect( )连接、getInputStream( )获取输入流、getOutputStream( )获取输出流和 getRemoteDevice( )获取远程设备，这里指的是获取 bluetoothSocket 指定连接的那个远程蓝牙设备。

2. Android 蓝牙通信各步骤的实现

（1）启动蓝牙功能

首先通过调用静态方法 getDefaultAdapter( )获取蓝牙适配器 BluetoothAdapter，以后就可以使用该对象了。如果返回为空，则无法继续执行了。例如：

```
1    BluetoothAdaptermBluetoothAdapter = BluetoothAdapter.getDefaultAdapter();
2    if(mBluetoothAdapter = =null){//Device does not support Bluetooth};
```

其次，调用 isEnabled( )来查询当前蓝牙设备的状态，如果返回 false，则表示蓝牙设备没有开启，接下来需要封装一个 ACTION_ REQUEST_ ENABLE 请求到 Intent 里面，调用 startActivityForResult( )方法使能蓝牙设备，例如：

```
1    if(! mBluetoothAdapter.isEnabled()){
2    Intent enableBtIntent = new Intent(BluetoothAdapter.ACTION_REQUEST_ENABLE);
3    startActivityForResult(enableBtIntent,REQUEST_ENABLE_BT);
4    }
```

（2）查找设备

使用 BluetoothAdapter 类里的方法，可以查找远端设备（大概 10m 以内）或者查询在手机上已经匹配（或者说绑定）的其他手机了。当然需要确定对方蓝牙设备已经开启或者已经开启了“被发现使能”功能（对方设备是可以被发现的是能够发起连接的前提条件）。如果该设备是可以被发现的，会反馈回来一些对方的设备信息，比如名字、MAC 地址等，利用这些信息，设备就可以选择去向对方初始化一个连接。

如果是第一次与该设备连接，那么一个配对的请求就会自动地显示给用户。当设备配对好之后，它的一些基本信息（主要是名字和 MAC）被保存下来并可以使用蓝牙的 API 来读取。使用已知的 MAC 地址就可以对远端的蓝牙设备发起连接请求。匹配好的设备和连接上的设备的不同点：匹配好只是说明对方设备发现了你的存在，并拥有一个共同的识别码，并且可以连接。连接上表示当前设备共享一个 RFCOMM 信道并且两者之间可以交换数据。也就是说蓝牙设备在建立 RFCOMM 信道之前，必须是已经配对好的。

（3）查询匹配好的设备

在建立连接之前必须先查询配对好的蓝牙设备集（周围的蓝牙设备可能不止一个），以

便选取哪一个设备进行通信，例如可以查询所有配对的蓝牙设备，并使用一个数组适配器将其打印显示出来。

```
1  Set<BluetoothDevice>pairedDevices=mBluetoothAdapter.getBondedDevices();
2  //If there are paired devices
3  if(pairedDevices.size()>0){
4      //Loop through paired devices
5      for(BluetoothDevice device:pairedDevices){
6          //Add the name and address to an array adapter
7          ListViewmArrayAdapter.add(device.getName()+"\n"+device.getAddress());
8      }
9  }
```

建立一个蓝牙连接只需要 MAC 地址就足够了。

（4）扫描设备

扫描设备，只需要简单地调用 startDiscovery( ) 方法，扫描的过程大概持续 12 秒，应用程序为了使用 ACTION_FOUND 动作需要注册一个 BroadcastReceiver 来接收设备扫描到的信息。对于每一个设备，系统都会广播 ACTION_FOUND 动作。例如：

```
1  //Create a BroadcastReceiver for ACTION_FOUND
2  private final BroadcastReceivermReceiver=new BroadcastReceiver(){
3      public void onReceive(Context context, Intent intent){
4          String action=intent.getAction();
5          //When discovery finds a device
6          if(BluetoothDevice.ACTION_FOUND.equals(action)){
7              //Get the BluetoothDevice object from the Intent
8              BluetoothDevice device
=intent.getParcelableExtra(BluetoothDevice.EXTRA_DEVICE);
9              //Add the name and address to an array adapter
10             ListViewmArrayAdapter.add(device.getName()+"\n"+device.getAddress());
11         }
12     }
13 };
14 //Register the BroadcastReceiver
15 IntentFilter filter=new IntentFilter(BluetoothDevice.ACTION_FOUND);
16 registerReceiver(mReceiver,filter);//Don't forget to unregister during onDestroy
```

注意：扫描的过程是一个很耗费资源的过程，一旦找到需要的设备之后，在发起连接请求之前，确保程序调用 cancelDiscovery( ) 方法停止扫描。显然，如果已经连接上一个设备，启动扫描会减少通信带宽。

（5）使能被发现（Enabling discoverability）

如果想使设备能够被其他设备发现，将 ACTION_REQUEST_DISCOVERABLE 动作封装在 Intent 中并调用 startActivityForResult（Intent，int）方法就可以了。它将在不使应用程序退出的情况下使设备能够被发现。默认情况下的使能时间是 120 秒，当然可以通过添加 EXTRA_DISCOVERABLE_DURATION 字段来改变使能时间（最大不超过 300 秒，这是出于对设备上的信息安全考虑）。例如：

```
1   Intent discoverableIntent = new Intent(BluetoothAdapter.ACTION_REQUEST_DISCOVERABLE);
2   discoverableIntent.putExtra(BluetoothAdapter.EXTRA_DISCOVERABLE_DURATION,300);
3   startActivity(discoverableIntent);
```

运行该段代码之后，系统会弹出一个对话框来提示启动设备使能被发现（此过程中如果蓝牙功能没有开启，系统会开启），并且如果准备对该远端设备发现一个连接，不需要开启使能设备被发现功能，因为该功能只是在应用程序作为服务器端的时候才需要。

（6）连接设备

在应用程序中，想建立两个蓝牙设备之间的连接，必须实现客户端和服务器端的代码（因为任何一个设备都可以作为服务端或者客户端）。一个开启服务来监听，一个发起连接请求（使用服务器端设备的 MAC 地址）。当它们都拥有一个蓝牙套接字在同一 RFECOMM 信道上时，可以认为它们之间已经连接上了。服务器端和客户端通过不同的方式使用蓝牙套接字。当一个连接监听到的时候，服务器端获取到蓝牙套接字。当客户可打开一个 FR-COMM 信道给服务器端时，客户端获取到蓝牙套接字。

注意：在此过程中，如果两个蓝牙设备还没有配对好，Android 系统会通过一个通知或者对话框的形式来通知用户。RFCOMM 连接请求会在用户选择之前阻塞。

（7）服务器端的连接

当要连接两台设备时，一个必须作为服务端（通过持有一个打开的 BluetoothServerSocket），目的是监听外来连接请求，当监听到以后提供一个连接上的 BluetoothSocket 给客户端，当客户端从 BluetoothServerSocket 得到 BluetoothSocket 以后就可以销毁 BluetoothServerSocket，除非还想监听更多的连接请求。

建立服务套接字和监听连接的基本步骤：首先通过调用 listenUsingRfcommWithServiceRecord（String，UUID）方法来获取 BluetoothServerSocket 对象，参数 String 代表了该服务的名称，UUID 代表了和客户端连接的一个标识（128 位格式的字符串 ID，相当于 PIN 码），UUID 必须双方匹配才可以建立连接。其次调用 accept( )方法来监听可能到来的连接请求，当监听到以后，返回一个连接上的蓝牙套接字 BluetoothSocket。最后，在监听到一个连接以后，需要调用 close( )方法来关闭监听程序（一般蓝牙设备之间是点对点的传输）。注意：accept( )方法不应该放在主 Acitvity 里面，因为它是一种阻塞调用（在没有监听到连接请求之前程序就一直停在那里）。解决方法是新建一个线程来管理。例如：

```
1    private class AcceptThread extends Thread{
2        private final BluetoothServerSocket mmServerSocket;
3        public AcceptThread(){
4            //Use a temporary object that is later assigned to mmServerSocket,
5            //because mmServerSocket is final
6            BluetoothServerSocket tmp = null;
7            try{
8                //MY_UUID is the app's UUID string, also used by theclient code
9                tmp = mAdapter.listenUsingRfcommWithServiceRecord(NAME, MY_UUID);
10           }catch(IOException e){}
11           mmServerSocket = tmp;
12       }
```

```
13        public void run(){
14              BluetoothSocket socket = null;
15              //Keep listening until exception occurs or a socket is returned
16              while(true){
17                      try{
18                              socket = mmServerSocket.accept();
19                      }catch(IOException e){
20                              break;
21                      }
22                      //If a connection was accepted
23                      if(socket! =null){
24                              //Do work to manage the connection(in a separate thread)
25                              manageConnectedSocket(socket);
26                              mmServerSocket.close();
27                              break;
28                      }
29              }
30        }
31        /* * Will cancel the listening socket, and cause the thread to finish* /
32        public void cancel(){
33              try{
34                      mmServerSocket.close();
35              }catch(IOException e){}
36        }
37    }
```

（8）客户端的连接

为了初始化一个与远端设备的连接，需要先获取代表该设备的一个 BluetoothDevice 对象。通过 BluetoothDevice 对象来获取 BluetoothSocket 并初始化连接，具体步骤如下：

使用 BluetoothDevice 对象里的 createRfcommSocketToServiceRecord（UUID）方法来获取 BluetoothSocket。UUID 就是匹配码。然后，调用 connect( )方法。如果远端设备接收了该连接，它们将在通信过程中共享 RFCOMM 信道，并且 connect( )方法返回。例如：

```
1     private class ConnectThread extends Thread{
2         private final BluetoothSocket mmSocket;
3         private final BluetoothDevice mmDevice;
4         public ConnectThread(BluetoothDevice device){
5               //Use a temporary object that is later assigned to mmSocket,
6               //because mmSocket is final
7               BluetoothSockettmp = null;
8               mmDevice = device;
9               //Get a BluetoothSocket to connect with the given BluetoothDevice
10              try{
11                      //MY_UUID is the app's UUID string, also used by the server code
12                      tmp = device.createRfcommSocketToServiceRecord(MY_UUID);
13              }catch(IOException e){}
```

```
14              mmSocket = tmp;
15          }
16          public void run(){
17              //Cancel discovery because it will slow down the connection
18              mAdapter.cancelDiscovery();
19              try{
20                  //Connect the device through the socket. This will block
21                  //until it succeeds or throws an exception
22                  mmSocket.connect();
23              }catch(IOExceptionconnectException){
24              //Unable to connect; close the socket and get out
25              try{
26                  mmSocket.close();
27              }catch(IOExceptioncloseException){}
28              return;
29          }
30          //Do work to manage the connection(in a separate thread)
31          manageConnectedSocket(mmSocket);
32      }
```

注意：conncet()方法也是阻塞调用，一般建立一个独立的线程来调用该方法。在设备discover过程中不应该发起连接connect()，这样会明显减慢速度以至于连接失败。且数据传输完成只有调用close()方法来关闭连接，才可以节省系统内部资源。

（9）管理连接（主要涉及数据的传输）

当设备连接上以后，每个设备都拥有各自的BluetoothSocket。现在就可以实现设备之间数据的共享了。

1）首先通过调用getInputStream()和getOutputStream()方法来获取输入、输出流。然后通过调用read（byte [ ]）和write（byte [ ]）方法来读取或者写数据。

2）实现细节：因为读取和写操作都是阻塞调用，需要建立一个专用线程来管理。例如：

```
1   private class ConnectedThread extends Thread{
2     private final BluetoothSocket mmSocket;
3     private final InputStream mmInStream;
4     private final OutputStream mmOutStream;
5     public ConnectedThread(BluetoothSocket socket){
6         mmSocket = socket;
7         InputStream tmpIn = null;
8         OutputStream tmpOut = null;
9         //Get the input and output streams, using temp objects because
10        //member streams are final
11        try{
12            tmpIn = socket.getInputStream();
13            tmpOut = socket.getOutputStream();
14        }catch(IOException e){}
15        mmInStream = tmpIn;
16        mmOutStream = tmpOut;
```

```
17      }
18      public void run(){
19          byte[]buffer = new byte[1024];
20          //buffer store for the stream int bytes;
21          //bytes returned from read()
22          //Keep listening to the InputStream until an exception occurs
23          while(true){
24               try{
25                        //Read from the InputStream
26                        bytes = mmInStream.read(buffer);
27                        //Send the obtained bytes to the UI Activity
28                        mHandler.obtainMessage(MESSAGE_READ, bytes, -1, buffer).sendToTarget();
29               }catch(IOException e){
30                        break;
31               }
32          }
33      }
34      /* Call this from the main Activity to send data to the remote device* /
35      public void write(byte[]bytes){
36          try{
37               mmOutStream.write(bytes);
38          }catch(IOException e){}
39      }
40      /* Call this from the main Activity to shutdown the connection* /
41      public void cancel(){
42          try{
43               mmSocket.close();
44          }catch(IOException e){}
45      }
46  }
```

【例 8-2】 实现蓝牙扫描操作功能

根据上述介绍就可以进行一个最简单的蓝牙操作——扫描已配对的蓝牙设备。主要有以下几步：

（1）在 AndroidManifest 文件中声明对蓝牙使用的权限。

（2）获得 BluetoothAdapter 对象。

（3）判断当前设备中是否拥有蓝牙设备。

（4）判断当前设备中的蓝牙设备是否已经打开。

（5）得到所有已经配对的蓝牙设备对象。

首先在 Eclipse 中新建工程 SearchBluetoothdDevice，其 Androidmanifest 文件如下：

```
1     <?xml version="1.0"encoding="utf-8"?>
2     <manifest xmlns:android="http://schemas.android.com/apk/res/android"
3        package="com.example.searchbluetoothdevice"
4        android:versionCode="1"
5        android:versionName="1.0">
```

```
6         <uses-sdk
7             android:minSdkVersion="8"
8             android:targetSdkVersion="14"/>

9         <application
10            android:allowBackup="true"
11            android:icon="@drawable/ic_launcher"
12            android:label="@string/app_name"
13            android:theme="@style/AppTheme">
14            <activity
15                android:name="com.example.searchbluetoothdevice.MainActivity"
16                android:label="@string/app_name">
17                <intent-filter>
18                        <action android:name="android.intent.action.MAIN"/>
19                        <category android:name="android.intent.category.LAUNCHER"/>
20                </intent-filter>
21            </activity>
22        </application>
23    <uses-permissionandroid:name="android.permission.BLUETOOTH"/>
24    </manifest>
```

向其中加入了对蓝牙设备使用的权限。

其布局文件如下：

```
1      <RelativeLayoutxmlns:android="http://schemas.android.com/apk/res/android"
2      xmlns:tools="http://schemas.android.com/tools"
3      android:layout_width="match_parent"
4      android:layout_height="match_parent"
5      android:paddingBottom="@dimen/activity_vertical_margin"
6      android:paddingLeft="@dimen/activity_horizontal_margin"
7      android:paddingRight="@dimen/activity_horizontal_margin"
8      android:paddingTop="@dimen/activity_vertical_margin"
9      tools:context=".MainActivity">

10     <TextView
11         android:id="@+id/textViewId"
12         android:layout_width="wrap_content"
13         android:layout_height="wrap_content"
14         android:text="@string/hello_world"/>
15         <Button
16             android:id="@+id/scanButtonId"
17             android:layout_width="fill_parent"
18             android:layout_height="wrap_content"
19             android:layout_below="@id/textViewId"
20             android:text="扫描蓝牙设备"
21     />
22     </RelativeLayout>
```

在这里定义了一个按钮，单击便可搜索附近的蓝牙设备。最主要的工作则是在 Mainactivity. java 中实现，下面详细进行介绍。

```
1    package com.example.searchbluetoothdevice;

2    import java.util.Iterator;
3    import java.util.Set;
4    import android.os.Bundle;
5    import android.app.Activity;
6    import android.bluetooth.BluetoothAdapter;
7    import android.bluetooth.BluetoothDevice;
8    import android.content.Intent;
9    import android.view.Menu;
10   import android.view.View;
11   import android.view.View.OnClickListener;
12   import android.widget.Button;

13   public class MainActivity extends Activity{

14       private Button buttonId = null;
15       @Override
16       public void onCreate(Bundle savedInstanceState){
17               super.onCreate(savedInstanceState);
18               setContentView(R.layout.activity_main);
19               buttonId = (Button)findViewById(R.id.scanButtonId);
20               buttonId.setOnClickListener(new buttonIdOnClickListener());
21       }
22       private class buttonIdOnClickListener implements OnClickListener{
23               public void onClick(View v){
24                       System.out.println("listener");
25                       BluetoothAdapter adapter = BluetoothAdapter.getDefaultAdapter();
26                       if(adapter! =null){
27                               System.out.println("本机有蓝牙设备!");
28                               if(! adapter.isEnabled()){
29                                 Intent intent =new Intent(BluetoothAdapter.ACTION_REQUEST_ENABLE);
30                                         startActivity(intent);
31                               }
32                               Set<BluetoothDevice> devices = adapter.getBondedDevices();
33                               if(devices.size()>0){
34                                 for(Iterator iterator =devices.iterator(); iterator.hasNext();){
35                                 BluetoothDevicebluetoothDevice = (BluetoothDevice)iterator.next();
36                                 System.out.println(bluetoothDevice.getAddress());
37                                         }
38                               }
39                       }
40                       else{
```

```
41                              System.out.println("本机没有蓝牙设备!");
42                  }
43              }
44      }

45      @Override
46      public boolean onCreateOptionsMenu(Menu menu) {
47              //Inflate the menu; this adds items to the action bar if it is present.
48              getMenuInflater().inflate(R.menu.main, menu);
49              return true;
50      }
51 }
```

在这里首先通过 getDefaultAdapter()获得系统默认的蓝牙适配器，当然也可以自己指定适配器，得到 BluetoothAdapter 对象并判断这个对象是否为空，如果是空的话则说明这台机器根本没有蓝牙设备，不为空则调用 isEnable()方法，判断当前蓝牙设备是否可用，然后创建一个 Intent 对象来启动蓝牙适配器，接着用 adapter.getBondedDevices()方法来得到已配对的蓝牙设备，最后利用一个迭代把远程的 BluetoothDevice 取出来。

当前蓝牙未开启时单击按钮则会如图 8-6 所示，单击允许，则如图 8-7 所示，在图 8-8 中可以看到蓝牙已打开的结果。

图 8-6　结果 1

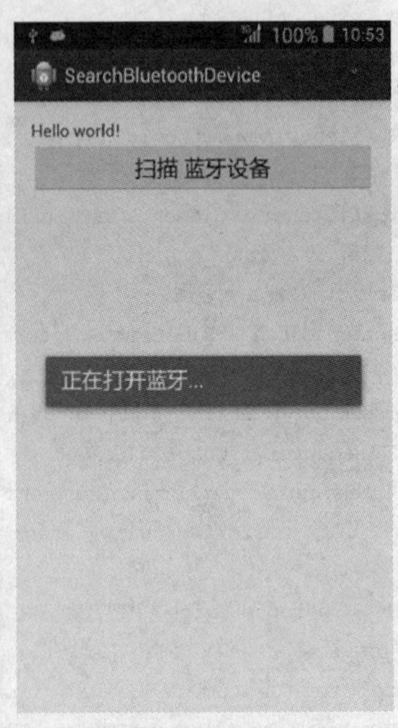

图 8-7　结果 2

图 8-8　结果 3

本机已与手机 OPPO R821T 配对，再单击按钮时在 Eclipse DDMS 中会显示如图 8-9 所示信息。

其中的 8C：0E：E3：6C：F3：E4 是与之配对的蓝牙设备的 MAC 地址。

Log			
Time	pid	tag	Message
06-09 10:55... I	2...	View	Touch up dispatch to android.widg...
06-09 10:55... I	2...	Syste...	listener
06-09 10:55... I	2...	Syste...	本机有蓝牙设备！
06-09 10:55... I	2...	Syste...	8C:0E:E3:6C:F3:E4

图 8-9　结果 4

8.3　WiFi 通信

WiFi 是一种无线联网技术，常见的是使用无线路由器。那么在这个无线路由器的信号覆盖的范围内都可以采用 WiFi 连接的方式进行联网。如果无线路由器连接了一个 ADSL 线路或其他的联网线路，则又被称为“热点”。无线网络在智能手机等掌上设备上应用越来越广泛。与早前应用于手机上的蓝牙技术不同，WiFi 具有更大的覆盖范围和更高的传输速率，因此 WiFi 手机成为了移动通信业界的时尚潮流。Android 当然也不会缺少这项功能，本节将对其进行详细介绍。

8.3.1　WiFi 包

android. net. WiFi 包为 Android 提供了对 WiFi 操作的相关方法。主要包括以下几个类和接口：

1. ScanResult

主要用来描述已经检测出的接入点，包括接入点的地址、接入点的名称、身份认证、频率、信号强度等信息。

2. WiFiConfiguration

WiFi 网络的配置，包括安全设置等。

3. WiFiInfo

WiFi 无线连接的描述，包括接入点、网络连接状态、隐藏的接入点、IP 地址、连接速度、MAC 地址、网络 ID、信号强度等信息。表 8-2 为 WiFiInfo 类的主要方法。

表 8-2　WiFiInfo 类的主要方法

String getBSSID( )	获取 BSSID
Static NetworkInfo. DetailedState getDetailedStateOf(SupplicantState suppState)	获取客户端的连通性
boolean getHiddenSSID( )	获得 SSID 是否被隐藏
int getIpAddress( )	获取 IP 地址
int getLinkSpeed( )	获得连接的速度
String getMacAddress( )	获得 Mac 地址
int getRssi( )	获得 802. 11n 网络的信号
String getSSID( )	获得 SSID
SupplicantState getSupplicanState( )	返回具体客户端状态的信息

4. WiFiManager

WiFiManager 是 WiFi 管理的主要类，此类提供了 WiFi 连接管理各方面的基本 API。通过调用 Context. getSystemServices（Context. WIFI_SERVICE）获得实例操作。其主要功能如下：

（1）配置网络列表：该列表可以查看和更新，各个条目的属性也可以被修改。

（2）查看网络状态：查看目前活跃的 WiFi 网络，可以查询网路状态的动态信息。

（3）扫描网络：扫描网络，查看接入点信息。

（4）定义广播：定义了几种 WiFi 状态改变的广播。

8.3.2 网卡状态

如果要对 WiFi 进行操作就必须了解 WiFi 网卡有几种状态，因为连接无线路由器时主要是通过 WiFi 网卡进行的，WiFi 的状态是通过一系列整型常量来表示的，如表 8-3 所示。

表 8-3 WiFi 网卡状态

int（1）	WIFI_STATE_DISABLED	WiFi 网卡不可用
int（0）	WIFI_STATE_DISABLING	WiFi 正在关闭
int（3）	WIFI_STATE_ENABLED	WiFi 网卡可用
int（2）	WIFI_STATE_ENABLING	WiFi 网卡正在打开
int（4）	WIFI_STATE_UNKNOWN	未知网卡状态

通常通过以下代码获取网卡状态：

```
WiFimanager = (WiFiManager) TextWiFi. this. getSystemService (Context. WIFI_SERVICE);
WiFimanager. getWiFiState ();
```

8.3.3 WiFi 网卡操作权限

对 WiFi 网卡的操作免不了需要得到某些控制权限，下面介绍跟 WiFi 操作相关的权限，如表 8-4 所示。

表 8-4 WiFi 网卡操作权限

String	ACCESS_NETWORK_STATE	访问网络的权限
String	ACCESS_WIFI_STATE	访问 WiFi 的权限
String	CHANGE_NETWORK_STATE	修改网络状态的权限
String	CHANGE_WIFI_MULTICAST_STATE	允许应用程序输入的 WiFi 组播模式
String	CHANGE_WIFI_STATE	改变 WiFi 连接状态的权限

8.3.4 更改 WiFi 状态

对 WiFi 网卡进行操作时要用到 WiFiManager，其中所用到的基本代码如下：

```
WiFimanager = (WiFiManager) TextWiFi. this. getSystem Service (Context. WIFI_SERVICE);
                                        //获取对象
WiFimanager. setWiFiEnabled (true)//打开网卡
WiFimanager. getWiFiState ();        //获取网卡当前状态
WiFimanager. setWiFiEnabled (false)//关闭网卡
```

【例 8-3】 实现对 WiFi 的操作

(1) 新建名为 WiFi 的工程，并在其布局文件 res/layout/activity_main. xml 中添加如下代码：

```
1        <? xml version = "1.0"encoding = "utf-8"? >
2        <ScrollView xmlns:android = "http://schemas.android.com/apk/res/android"
3       android:layout_width = "fill_parent"
4       android:layout_height = "fill_parent"
5        >
6        <LinearLayout
7             android:orientation = "vertical"
8             android:layout_width = "fill_parent"
9             android:layout_height = "fill_parent"
10       >
11       <Button
12            android:id = "@ +id/scan"
13            android:layout_width = "wrap_content"
14            android:layout_height = "wrap_content"
15            android:text = "扫描网络"
16       />
17       <Button
18            android:id = "@ +id/start"
19            android:layout_width = "wrap_content"
20            android:layout_height = "wrap_content"
21            android:text = "打开 WiFi"
22       />
23       <Button
24            android:id = "@ +id/stop"
25            android:layout_width = "wrap_content"
26            android:layout_height = "wrap_content"
27            android:text = "关闭 WiFi"
28       />
29       <Button
30            android:id = "@ +id/check"
31            android:layout_width = "wrap_content"
32            android:layout_height = "wrap_content"
33            android:text = "WiFi 状态"
34       />
35       <TextView
36            android:id = "@ +id/allNetWork"
37            android:layout_width = "fill_parent"
38            android:layout_height = "wrap_content"
39            android:text = "当前没有扫描到 WiFi 网络"
40       />
41       </LinearLayout>
42       </ScrollView>
```

（2）接下来把 WiFi 的相关操作都封装在了一个 WiFiAdmin 类中，以后开启或关闭等相关操作可以直接调用这个类的相关方法，在 src/com. example. WiFi 下新建名为 WiFiAdmin. Java 文件，并添加如下代码：

```
1    package com.example.WiFi;
2    import java.util.List;
3    import android.content.Context;
4    import android.net.WiFi.ScanResult;
5    import android.net.WiFi.WiFiConfiguration;
6    import android.net.WiFi.WiFiInfo;
7    import android.net.WiFi.WiFiManager;
8    import android.net.WiFi.WiFiManager.WiFiLock;
9    public class WiFiAdmin{
10       //定义一个 WiFiManager 对象
11       private WiFiManager mWiFiManager;
12       //定义一个 WiFiInfo 对象
13       private WiFiInfo mWiFiInfo;
14       //扫描出的网络连接列表
15       private List <ScanResult> mWiFiList;
16       //网络连接列表
17       private List <WiFiConfiguration> mWiFiConfigurations;
18       WiFiLock mWiFiLock;
19       public WiFiAdmin(Context context){
20           //取得 WiFiManager 对象
21           WiFiManager = (WiFiManager)context.getSystemService(Context.WIFI_SERVICE);
22           //取得 WiFiInfo 对象
23           mWiFiInfo = mWiFiManager.getConnectionInfo();
24       }
25       //打开 WiFi
26       public void openWiFi(){
27           if(! mWiFiManager.isWiFiEnabled()){
28               mWiFiManager.setWiFiEnabled(true);
29           }
30       }
31       //关闭 WiFi
32       public void closeWiFi(){
33           if(! mWiFiManager.isWiFiEnabled()){
34               mWiFiManager.setWiFiEnabled(false);
35           }
36       }
37       //检查当前 WiFi 状态
38       public int checkState(){
39       return mWiFiManager.getWiFiState();
40       }
41       //锁定 WiFiLock
42       public void acquireWiFiLock(){
```

```
43          mWiFiLock.acquire();
44          }
45          //解锁 WiFiLock
46            public void releaseWiFiLock(){
47                //判断是否锁定
48                if(mWiFiLock.isHeld()){
49                mWiFiLock.acquire();
50                }
51          }
52          //创建一个 WiFiLock
53          public void createWiFiLock(){
54                mWiFiLock=mWiFiManager.createWiFiLock("test");
55          }
56          //得到配置好的网络
57          public List<WiFiConfiguration>getConfiguration(){
58                return mWiFiConfigurations;
59          }
60          //指定配置好的网络进行连接
61          public void connetionConfiguration(int index){
62                if(index>mWiFiConfigurations.size()){
63                        return ;
64                }
65                //连接配置好指定 ID 的网络
66    mWiFiManager.enableNetwork(mWiFiConfigurations.get(index).networkId, true);
67          }
68          public void startScan(){
69                mWiFiManager.startScan();
70                //得到扫描结果
71                mWiFiList=mWiFiManager.getScanResults();
72                //得到配置好的网络连接
73                mWiFiConfigurations=mWiFiManager.getConfiguredNetworks();
74          }
75          //得到网络列表
76          public List<ScanResult>getWiFiList(){
77                return mWiFiList;
78          }
79          //查看扫描结果
80          public StringBuffer lookUpScan(){
81                StringBuffer sb=new StringBuffer();
82                for(int i=0;i<mWiFiList.size();i++){
83                        sb.append("Index_"+new Integer(i+1).toString()+":");
84                        //将 ScanResult 信息转换成一个字符串包
85                        //其中包括:BSSID、SSID、capabilities、frequency、level
86                        sb.append((mWiFiList.get(i)).toString()).append("\n");
87                }
88                return sb;
```

```
89          }
90          public String getMacAddress(){
91                return(mWiFiInfo==null)?"NULL":mWiFiInfo.getMacAddress();
92          }
93          public String getBSSID(){
94                return(mWiFiInfo==null)?"NULL":mWiFiInfo.getBSSID();
95          }
96          public int getIpAddress(){
97                return(mWiFiInfo==null)? 0:mWiFiInfo.getIpAddress();
98          }
99          //得到连接的 ID
100         public int getNetWordId(){
101               return(mWiFiInfo==null)? 0:mWiFiInfo.getNetworkId();
102         }
103         //得到 WiFiInfo 的所有信息
104         public String getWiFiInfo(){
105               return(mWiFiInfo==null)?"NULL":mWiFiInfo.toString();
106         }
107         //添加一个网络并连接
108         public void addNetWork(WiFiConfiguration configuration){
109               int wcgId=mWiFiManager.addNetwork(configuration);
110               mWiFiManager.enableNetwork(wcgId, true);
111         }
112         //断开指定 ID 的网络
113         public void disConnectionWiFi(int netId){
114               mWiFiManager.disableNetwork(netId);
115               mWiFiManager.disconnect();
116         }
117   }
```

（3）在 AndroidManifest.xml 文件中加入使用 WiFi 访问网络所需要的权限，代码如下：

```
1  <uses-permission android:name="android.permission.CHANGE_NETWORK_STATE"></uses-permission>
2  <uses-permission android:name="android.permission.CHANGE_WIFI_STATE"></uses-permission>
3  <uses-permission android:name="android.permission.ACCESS_NETWORK_STATE"></uses-permission>
4  <uses-permission android:name="android.permission.ACCESS_WIFI_STATE"></uses-permission>
```

（4）最后在 MainActivity.java 中添加如下代码。

```
1   package com.example.WiFi;
2   import java.util.List;
3   import android.app.Activity;
4   import android.net.WiFi.ScanResult;
5   import android.os.Bundle;
```

```
6   import android.view.View;
7   import android.view.View.OnClickListener;
8   import android.widget.Button;
9   import android.widget.TextView;
10  import android.widget.Toast;

11  public class WiFiActivity extends Activity{
12    /* * Called when the activity is first created. * /
13    private TextView allNetWork;
14    private Button scan;
15    private Button start;
16    private Button stop;
17    private Button check;
18    private WiFiAdmin mWiFiAdmin;
19    //扫描结果列表
20    private List<ScanResult>list;
21    private ScanResult mScanResult;
22    private StringBuffer sb=new StringBuffer();
23    @Override
24    public void onCreate(Bundle savedInstanceState){
25        super.onCreate(savedInstanceState);
26        setContentView(R.layout.main);
27        mWiFiAdmin=new WiFiAdmin(WiFiActivity.this);
28        init();
29    }
30    public void init(){
31        allNetWork=(TextView)findViewById(R.id.allNetWork);
32        scan=(Button)findViewById(R.id.scan);
33        start=(Button)findViewById(R.id.start);
34        stop=(Button)findViewById(R.id.stop);
35        check=(Button)findViewById(R.id.check);
36        scan.setOnClickListener(new MyListener());
37        start.setOnClickListener(new MyListener());
38        stop.setOnClickListener(new MyListener());
39        check.setOnClickListener(new MyListener());
40    }
41    private class MyListener implements OnClickListener{

42        @Override
43        public void onClick(View v){
44              //TODO Auto-generated method stub
45              switch(v.getId()){
46                    case R.id.scan://扫描网络
47                          getAllNetWorkList();
48                          break;
49                    case R.id.start://打开 WiFi
```

```
50                          mWiFiAdmin.openWiFi();
51                          Toast.makeText(WiFiActivity.this, "当前 WiFi 状态为:" + mWiFiAd-
min.checkState(), 1).show();
52                          break;
53                  case R.id.stop://关闭 WiFi
54                          mWiFiAdmin.closeWiFi();
55                          Toast.makeText(WiFiActivity.this, "当前 WiFi 状态为:" + mWiFiAd-
min.checkState(), 1).show();
56                          break;
57                  case R.id.check://WiFi 状态
58                          Toast.makeText(WiFiActivity.this, "当前 WiFi 状态为:" + mWiFiAd-
min.checkState(), 1).show();
59                          break;
60                  default:
61                          break;
62              }
63          }
64      }
65      public void getAllNetWorkList(){
66          //每次单击扫描之前清空上一次的扫描结果
67          if(sb! =null){
68              sb=new StringBuffer();
69          }
70          //开始扫描网络
71          mWiFiAdmin.startScan();
72          list=mWiFiAdmin.getWiFiList();
73          if(list! =null){
74              for(int i=0;i<list.size();i++){
75                  //得到扫描结果
76                  mScanResult=list.get(i);
77                  sb = sb.append(mScanResult.BSSID + "   ").append(mScanResult.SSID + "   ")
.append(mScanResult.capabilities + "   ").append(mScanResult.frequency + "   ").append(mScanRe-
sult.level+"\n\n");
78              }
79              allNetWork.setText("扫描到的 WiFi 网络:\n"+sb.toString());
80          }
81      }
82  }
```

（5）程序在真机上运行结果。

1）打开应用时如图 8-10 所示。

2）选择打开 WiFi 时如图 8-11 所示。

3）接着选择 WiFi 状态如图 8-12 所示。

4）选择扫描网络如图 8-13 所示。

5）选择关闭 WiFi 如图 8-14 所示。

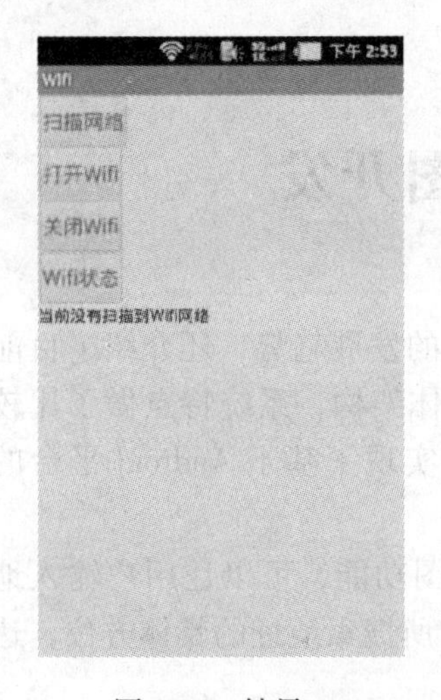

图 8-10 结果 1

图 8-11 结果 2

图 8-12 结果 3

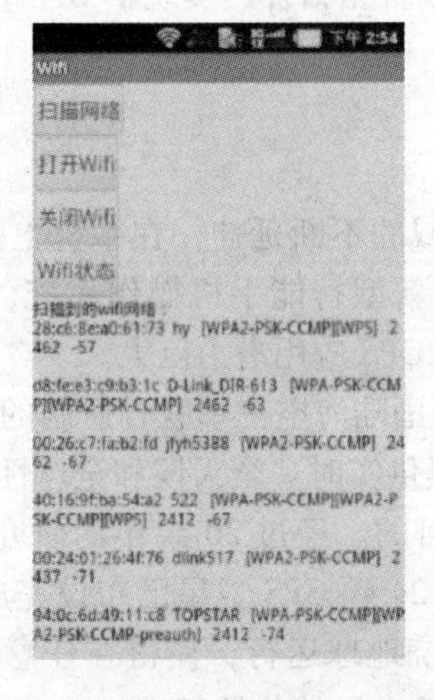

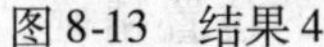
图 8-13 结果 4

图 8-14 结果 5

本章小节

本章主要介绍了 Android 平台下的几种通信方式，即 Socket 通信、蓝牙及 WiFi。其中涉及对它们通信方式的介绍、通信中所需的各种 API 及其使用方法。在 Socket 通信中主要介绍了它的通信模型以及通信各部分的实现并通过实例展示了其具体的通信过程。在蓝牙中主要介绍了蓝牙系统的基本构成，在 Android 下的各种 API 及通信方式。最后对 WiFi 的操作做了详细的介绍。

习 题

8-1 利用 Socket 设计一个可以发送和接收文字消息的聊天室程序。

8-2 设计一个蓝牙通信的应用程序。

8-3 编写一个利用 WiFi 传送图片的网络应用程序。

第 9 章　定位与 Google 地图开发

本章首先探讨了 GPS 技术及其应用现状，介绍了手机 GPS 的发展趋势，还介绍了目前 GPS 在民用中的地位；然后在 Android 平台的基础上，对它的整体架构，系统特点做了比较详细的分析；并且，结合 GoogleMaps 相关知识与应用，设计并实现了基于 Android 平台的 GPS 定位功能。

本系统实现的功能主要包括通过借助 GoogleMaps 的强大地图功能，可以让用户输入地址，显示出所要寻找的地方，并且可以放大缩小地图，看到用户所搜索地址的具体方位，达到 GPS 的定位功能。

本章将 GPS 定位技术与新兴的智能移动终端平台相结合，实现了 GPS 基本功能的拓展。

9.1　使用 GPS 定位

随着移动平台技术的飞速发展和 GPS 应用领域的不断延伸，在手机上拓展 GPS 功能已成为移动应用开发的一个热点。Android 作为一款新型智能手机操作系统，具有开放性好、软硬件功能扩展性强的特点，开发基于 Android 的 GPS 应用潜力巨大。

全球定位系统（GPS）是 20 世纪 70 年代由美国陆海空三军联合研制的新一代空间卫星导航定位系统。其主要目的是为陆、海、空三军提供实时、全天候和全球性的导航服务，并用于情报收集、核爆监测和应急通信等一些军事目的。经过 20 余年的研究实验，耗资 300 亿美元，到 1994 年 3 月，全球覆盖率高达 98% 的 24 颗 GPS 卫星已布设完成，24 颗 GPS 卫星在离地面 22000km 的高空上，以 12h 的周期环绕地球运行，使得在任意时刻，在地面的任意一点都可以同时观测到 4 颗以上的卫星。由于卫星的位置精确，在 GPS 观测中，可以得到卫星到接收机的距离，利用三维坐标中的距离公式和 3 颗卫星，就可以组成 3 个方程式，解出观测点的位置（x，y，z）。考虑到卫星的时钟与接收机时钟之间的误差，实际上有 4 个未知数，x、y、z 和时钟差，因而需要引入第 4 颗卫星，形成 4 个方程式求解，从而得到观测点的经纬度和高度。总体来说 GPS 定位的优点是准确、覆盖面广阔。本节将介绍如何通过 GPS 获得移动设备的位置信息，首先介绍 LocationManager 类，然后通过实例说明如何在程序代码中使用 Google 的位置服务。

所谓定位，就是获取当前设备的地理位置信息。Android 提供了地理定位服务的 API。该地理定位服务可以用来获取当前设备的地理位置。应用程序可以定时请求更新设备当前的地理位置信息。应用程序也可以借助一个 Intent 接收器来实现如下功能：以经纬度和半径划定的一个区域，当设备出入该区域时发出提醒信息。以下是 Android 中几个关于定位功能的比较重要的类。

LocationManager 类：本类提供访问定位服务的功能，也提供获取最佳定位提供者的功能。另外，临近警报功能（前面所说的那种功能）也可以借助该类来实现。LocationManager 系统服务是提供设备位置信息服务的核心组件，它提供了一系列方法来处理与位置相关的问题，包括查询上一个已知位置，注册或注销来自某个 LocationProvider 的周期性的位置更新，注册或注销接近某个坐标时对一个已定义的 Intent 的触发等。可以通过 getSystemService (Context. LOCATION_SERVICE）方法得到该类的实例。LocationManager 类常用的属性和方法如表 9-1 所示。

表 9-1　LocationManager 类常用的属性和方法

属性和方法	描　述
String GPS_PROVIDER	静态字符串常量，表明 LocationProvider 是 GPS
String NETWORK_PROVIDER	静态字符串常量，表明 LocationProvider 是网络
boolean addGpsStatusListener（GpsStatus. Listener listener）	添加一个 GPS 状态监听器
void addProximityAlert（double latitude，double longitude，float radius，long expiration，PendingIntent intent）	添加一个趋近警告
List < String > getAllProviders( )	获得所有的 LocationProvider 列表
String getBestProvider(Criteria criteria，booleanenabledOnly)	根据 Criteria 返回最适合的 LocationProvider
Location getLastKnownLocation（String provider）	根据 Provider 获得位置信息
LocationProvider getProvider（String name）	获得指定名称的 LocationProvider
List < String > getProvider（boolean enableOnly）	获得可利用的 LocationProvider 列表
void removeProximityAlert（PendingIntent intent）	删除趋近警告
void requestLocationUpdates（String provider，long minTime，float minDistance，PendingIntent intent）	通过给定的 Provider 名称，周期性地通知当前 Activity
void requestLocationUpdates（String provider，long minTime，float minDistance，LocationListener listener）	通过给定的 Provider 名称，并将其绑定指定的 LocationListener 监听器

LocationProvider 类：该类是定位提供者的抽象类。定位提供者具备周期性报告设备地理位置的功能。可以通过该类设置提供者的一些属性。通过 Criteria 类为 LocationProvider 设置条件，获得合适的 LocationProvider。LocationProvider 类常用的属性和方法如表 9-2 所示。

表 9-2　LocationProvider 类常用的属性和方法

属性和方法	描　述
int AVAILABLE	静态整型常量，标识是否可利用
int OUT_OF_SERVICE	静态整型常量，不在服务区
int TEMPORAILY_UNAVAILABLE	静态整型常量，临时不可利用
int getAccuarcy( )	获得精度
String getName( )	获得名称
int getPowerRequirement( )	获得电源需求
boolean hasMonetaryCost( )	付费的还是免费的
boolean requiresCell( )	是否需要访问基站网络
boolean requiresNetWork( )	是否需要 Intent 网络数据
boolean requiresSatelite( )	是否需要访问卫星
boolean supportsAltitude( )	是否能够提供高度信息
boolean supportsBearing( )	是否能够提供方向信息
boolean supportsSpeed( )	是否能够提供速度信息

LocationListener 类：为了实现自己的逻辑功能还需要对其设置监听器。该类定义了常见的 provider 状态变化和位置变化的方法，接下来只要在 LocationManager 中注册此监听器，就可以完成对各种状态的监听。

Criteria 类：该类使得应用能够通过在 LocationProvider 中设置的属性来选择合适的定位提供者。

要想实现定位除了用到 LocationProvider 和 LocationManager 之外还要用到 Location 类，它用于描述当前设备的地理位置信息，包括经纬度、方向、高度和速度等。可以通过 LocationManager. getLastKnownLocation（String provider）方法获得 Location 实例。

【例 9-1】 实例实现获得 GPS 信息。

（1）新建工程 getLocation，并向 MainActivity. java 中添加以下代码：

```
1    import android.location.Location;
2    import android.location.LocationListener;
3    import android.location.LocationManager;
4    import android.os.Bundle;
5    import android.app.Activity;
6    import android.content.Context;
7    import android.view.Menu;
8    import android.view.View;
9    import android.view.View.OnClickListener;
10   import android.widget.Button;
11   import android.widget.TextView;
12   public class MainActivity extends Activity implements OnClickListener{
13       Button getLocationButton;//声明获取地理位置按钮
14       Button stopGetLocation; //声明停止获取地址位置按钮
15       TextView showLocationTextView;//显示地理位置信息
16       private LocationManager locationManager;//位置管理器
17       private Location location;//位置信息,包含经纬度
18       //位置更新监听器
19       private LocationListener myLocationListener = new LocationListener(){
20           @Override
21           public void onStatusChanged(String provider, int status, Bundle extras){
22           }
23           @Override
24           public void onProviderEnabled(String provider){
25           }
26           @Override
27           public void onProviderDisabled(String provider){
28               updateView(null);
29           }
30           @Override
```

```
31              public void onLocationChanged(Location location){
32                      updateView(location);
33              }
34          };
35          @Override
36          public void onCreate(Bundle savedInstanceState){
37                  super.onCreate(savedInstanceState);
38                  setContentView(R.layout.activity_main);
39                  getLocationButton = (Button)findViewById(R.id.get_location);//获取 id 为 get_location 的按钮
40                  stopGetLocation = (Button)findViewById(R.id.stop_location);//获取 id 为 stop_location 的按钮
41                  showLocationTextView = (TextView)findViewById(R.id.show_location);//获取 id 为 show_location 的按钮
42                  getLocationButton.setOnClickListener(this);//绑定单击事件监听器
43                  stopGetLocation.setOnClickListener(this);
44                  locationManager = (LocationManager)getSystemService(Context.LOCATION_SERVICE);//获取位置管理器
45                  //指定由 GPS 定位来获取位置信息
46                  location = locationManager.getLastKnownLocation(LocationManager.GPS_PROVIDER);
47                  updateView(location); //更新位置信息
48          }
49          @Override
50          public void onClick(View v){
51                  int id = v.getId();
52                  if(R.id.get_location == id){//获取 id 为 get_location 的按钮
53                          getLocation();
54                  }else if(R.id.stop_location == id){//获取 id 为 stop_location 的按钮
55                          locationManager.removeUpdates(myLocationListener);//移除位置监听器
56                  }
57          }
58          private void getLocation(){
59                  //设置监听器,当距离超过 minInstance,且时间超过 minTime 时更新
60                  //因为位置信息实时更新,所以将第二个参数和第三个参数都设置成 0
61          locationManager.requestLocationUpdates(LocationManager.GPS_PROVIDER, 0, 0, myLocationListener);
62          }
63          /**
64          * 更新位置信息
65          * @param location
66          */
67          private void updateView(Location location){
68                  if(location == null){
```

```
69                showLocationTextView.setText("未定位到当前位置");
70                return;
71            }
72            double latitude = location.getLatitude();//纬度
73            double longitude = location.getLongitude();//经度
74            showLocationTextView.setText("纬度:" + latitude + "\n" + "经度:" + longitude);
75        }
76        @Override
77        publicbooleanonCreateOptionsMenu(Menu menu){
78            getMenuInflater().inflate(R.menu.activity_main, menu);
79            return true;
80        }
81    }
```

上述代码中通过调用 Context. getSystemService(Context. LOCATION_SERVICE)方法来获取位置管理器 LocationManager 对象，有了 LocationManager 对象之后，就可以开始监听位置的变化了。可以通过调用 LocationManager 对象的 requestLocationUpdates(String provider, long minTime, float minDistance, LocationListener listener)方法来监听位置的变化。

对于第一个参数，有两个可选值，分别为 LocationManager. NETWORK_PROVIDER 和 LocationManager. GPS_PROVIDER，前者用于移动网络获取位置，后者则是通过 GPS 定位。

（2）由于是通过 GPS 定位的方式去获取用户所处的地理位置信息，所以还需要在 AndroidManifest. xml 文件中添加相应的权限，具体代码如下：

```
1     <manifest xmlns:android="http://schemas.android.com/apk/res/android"
2         package="com.example.chapter12_1"
3         android:versionCode="1"
4         android:versionName="1.0">

5         <uses-sdk
6             android:minSdkVersion="8"
7             android:targetSdkVersion="15"/>
8     <uses-permissionandroid:name="android.permission.ACCESS_FINE_LOCATION"/>

9     <application
10        android:icon="@drawable/ic_launcher"
11        android:label="@string/app_name"
12        android:theme="@style/AppTheme">
13        <activity
14            android:name=".MainActivity"
15            android:label="@string/title_activity_main">
16            <intent-filter>
17            <action android:name="android.intent.action.MAIN"/>

18            <category android:name="android.intent.category.LAUNCHER"/>
19            </intent-filter>
20        </activity>
```

```
21     </application>
22     <uses-permissionandroid:name="android.permission.ACCESS_FINE_LOCATION"/>
23     </manifest>
```

（3）在布局文件中添加两个按钮，代码如下：

```
1      <LinearLayoutxmlns:android="http://schemas.android.com/apk/res/android"
2          xmlns:tools="http://schemas.android.com/tools"
3          android:layout_width="fill_parent"
4          android:layout_height="fill_parent"
5          android:orientation="vertical">
6          <!--获取地理信息位置-->
7          <Button
8          android:id="@+id/get_location"
9          android:layout_width="fill_parent"
10         android:layout_height="wrap_content"
11         android:text="GPS 定位"></Button>
12         <!--停止获取地理信息位置-->
13         <Button android:id="@+id/stop_location"
14         android:layout_width="fill_parent"
15         android:layout_height="wrap_content"
16         android:text="停止"></Button>
17         <!--显示地理信息位置-->
18         <TextView
19             android:id="@+id/show_location"
20             android:layout_width="wrap_content"
21             android:layout_height="wrap_content"></TextView>
22     </LinearLayout>
```

（4）开启模拟器，并在 DDMS 中设置经纬度，运行结果如图 9-1 所示。

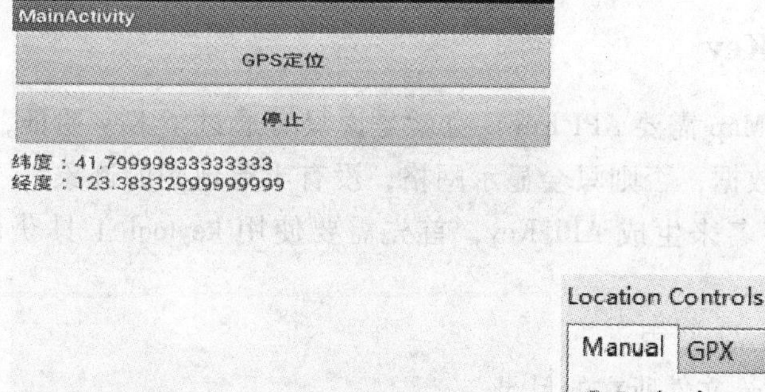

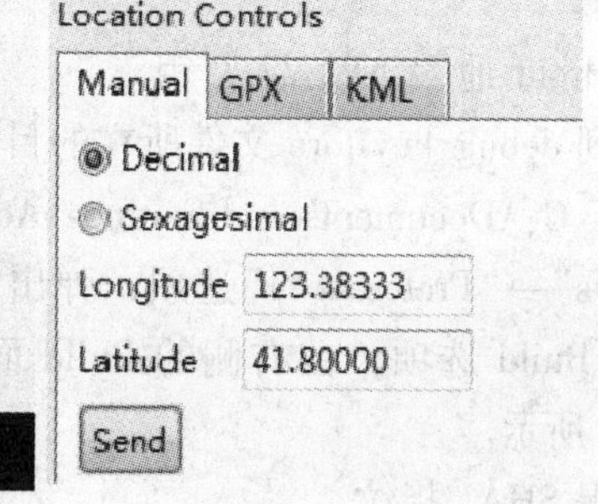

图 9-1 在模拟器中的显示在 DDMS 中的设置

9.2 Google 地图的使用

Google 地图（Google Map）是 Google 公司提供的电子地图服务。地图包含地标、线条、形状等信息，提供矢量地图、卫星照片、地形图等三种视图。Google 地图于 2005 年 2 月 8 日在谷歌博客上首次公布，并于 2005 年 6 月 20 日将覆盖范围从原先的美国、英国、加拿大扩大至全球。Android 作为 Google 公司下的产品，融合了 Google 地图的优秀功能。下面介绍 Android 系统中 Google 地图与应用项目整合的方法。

9.2.1 Google Maps 包

Android 提供了一组访问 Google Maps 的 API，借助 Google Maps 及定位 API，就可以在地图上显示出用户当前的地理位置。在 Android 中定义了一个名为 com. google. android. maps 的包，Google Maps 包位于 Android 系统 SDK 安装目录的“\add-ons\addon-google_apis-google-(API 版本号) \libs”下，其中 API 版本号为 API 版本所代表的数字。包中含有一系列用于在 Google Maps 上显示、控制和层叠信息的功能类，以下是该包中最重要的几个类。

MapActivity：这个类是用于显示 Google Map 的 Activity 类，它需要连接底层网络。MapActivity 是一个抽象类，任何想要显示 MapView 的 Activity 都需要派生自 MapActivity，并且在其派生类的 onCreate( ) 中，都要创建一个 MapView 实例。

MapView：MapView 是用于显示地图的 View 组件。它派生自 android. view. ViewGroup。它必须和 MapActivity 配合使用，而且只能被 MapActivity 创建，这是因为 MapView 需要通过后台的线程来连接网络或者文件系统，而这些线程需要有 MapActivity 来管理。

MapController：该类用于控制地图的移动、缩放等操作。

Overlay：这是一个可显示于地图之上的可绘制的对象。

GeoPoint：这是一个包含经纬度位置的对象。

9.2.2 获得 Map API Key

Android 上面的 Google Map 需要 API Key，也就是说只有通过了 Key 验证，用户编写的 Maps 应用才可以下载地图数据，否则只会显示网格，没有半点地图的迹象。一般根据应用程序的 keystore 的 SHA1 签名来生成 API Key，首先需要使用 keytool 工具获得 keystore 的 SHA1 签名。

获得 Android 地图 API V2 密钥

（1）找到 debug. keystore 文件所在的目录

通常位于 C:\Documents and Settings\Administrator\. android\目录中。可以通过 Eclipse 中打开“Windows”→“Preferences”选项，弹出“Preferences”对话框，在左侧的选项栏中选择 Android 下的 Build 选项，在右侧的 Build 面板中就能看到 debug. keystore 文件所在的位置，页面如图 9-2 所示。

（2）获得 SHA1 指纹

打开 cmd 命令行，使用 JDK 自带的 keytool 工具通过 keystore 生成 SHA1 指纹，在命令行

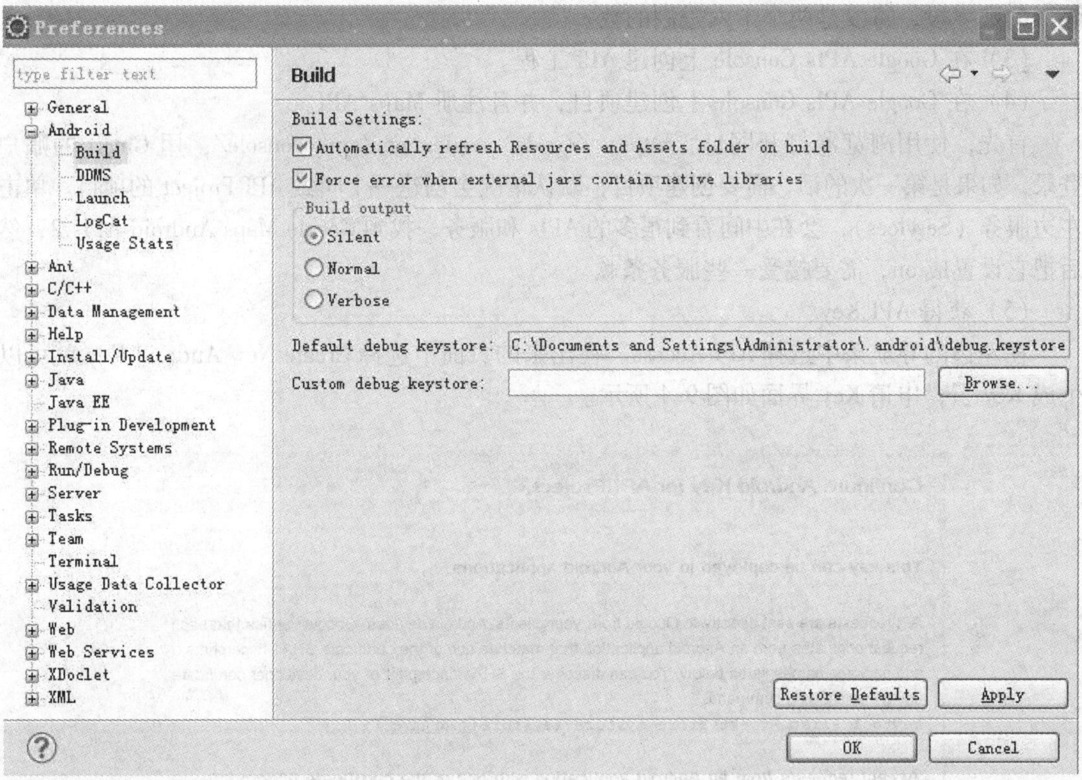

图 9-2　查看 debug. keystore 文件所在的位置

输入如下代码：

```
keytool - list - alias androddebugkey - keystore “c: \Documents and Settings \ Administrator \
. android \debug. keystore” - storepass android - keypass android
```

结果如图 9-3 所示，其中 SHA1 那一行包含了证书的 SHA-1 fingerprint，是二十段用冒号

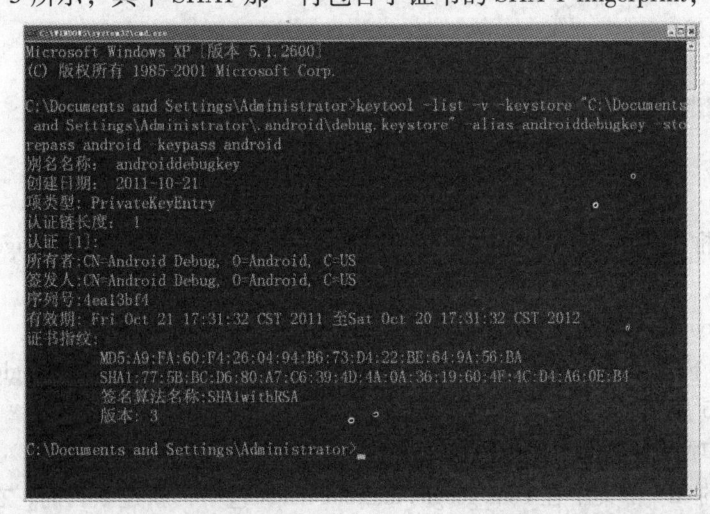

图 9-3　使用 keytool 工具生成 SHA1 认证指纹

隔开的数字段，每段是两个十六进制的数。

（3）在 Google APIs Console 上创建 API 工程。

（4）在 Google APIs Console 上创建项目，并且注册 Maps API。

首先，使用浏览器打开网址：https：//code. google. com/apis/console/，用 Gmail 的账户登录，如果是第一次的话，需要创建项目，默认情况会创建一个叫做 API Project 的项目。单击左边服务（Services），会在中间看到很多的 APIs 和服务，找到 Google Maps Android API v2，然后把它设置成 on，需要接受一些服务条款。

（5）获得 API Key

在左边的导航条中选择 API Access。在出来的页面中选择 Create New Android Key 就可以生成 Key 了，申请 Key 界面如图 9-4 所示。

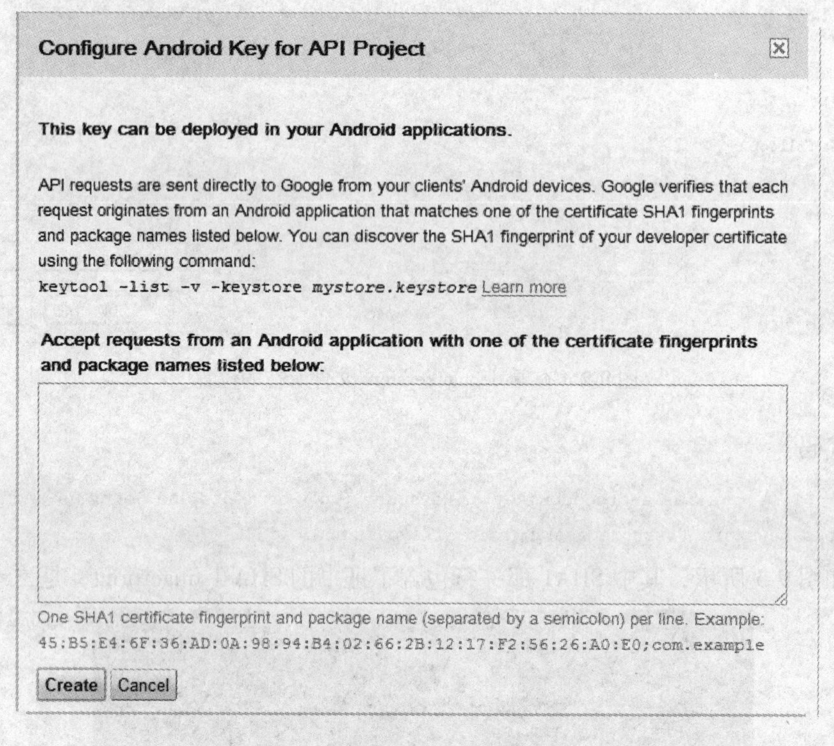

图 9-4　API 工程中申请 Key 界面

然后在对话框中填入 SHA-1 指纹，使用分号隔开，然后是应用的 package name。然后就会生成一个 Key。

（6）创建 Google API 的 AVD 模拟器

在创建 AVD 模拟器设备时，选择 Target 项时要选择 Google APIs（Google Inc.）的运行环境项，如图 9-5 所示。

（7）在新建项目时选用 Google API 版本

Google 地图的 API 都集中在 com. google. android. maps 包中，新建项目时，在“New Android Project”对话框中要选择 Google API 版本，这样在程序设计时才能导入 com. google. android. maps 包，如图 9-6 所示。

Edit Android Virtual Device (AVD)
AVD Name:
androidGoogleAPI
Device:
Nexus S (4.0", 480 × 800: hdpi)
Target:
Google APIs (Google Inc.) - API Level 7
CPU/ABI:
ARM (armeabi)
Keyboard:
Hardware keyboard present
Skin:
Display a skin with hardware controls
Front Camera:
None
Back Camera:
None
Memory Options:
RAM: 343
VM Heap: 32
Internal Storage:
200
MiB
SD Card:
Size:
MiB
File:
Browse...
Emulation Options:
Snapshot
Use Host GPU
Override the existing AVD with the same name
OK
Cancel

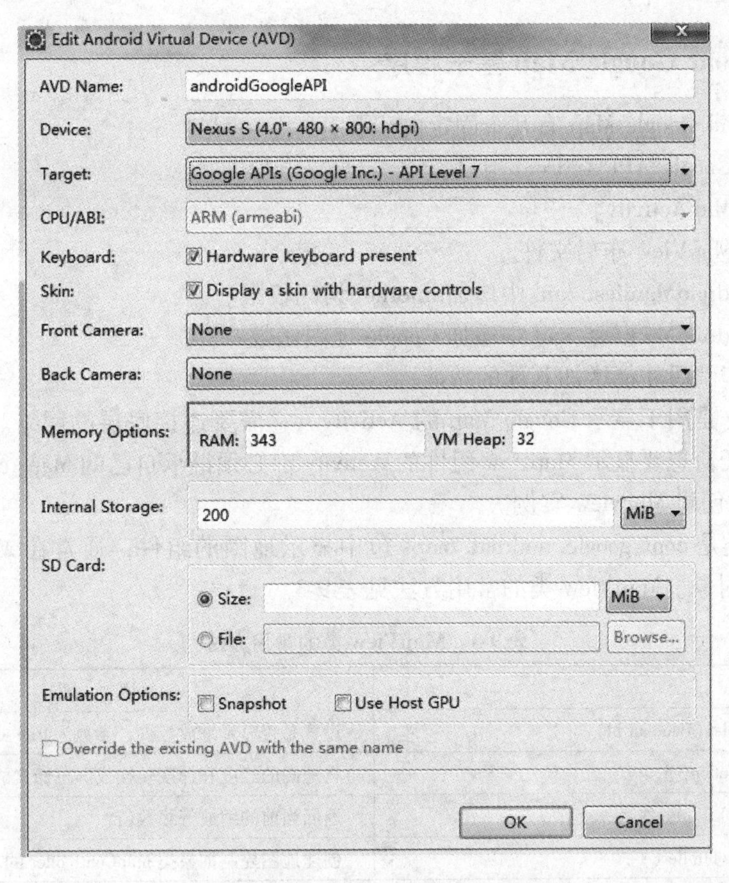

图 9-5 创建 Google API 的 AVD 模拟器

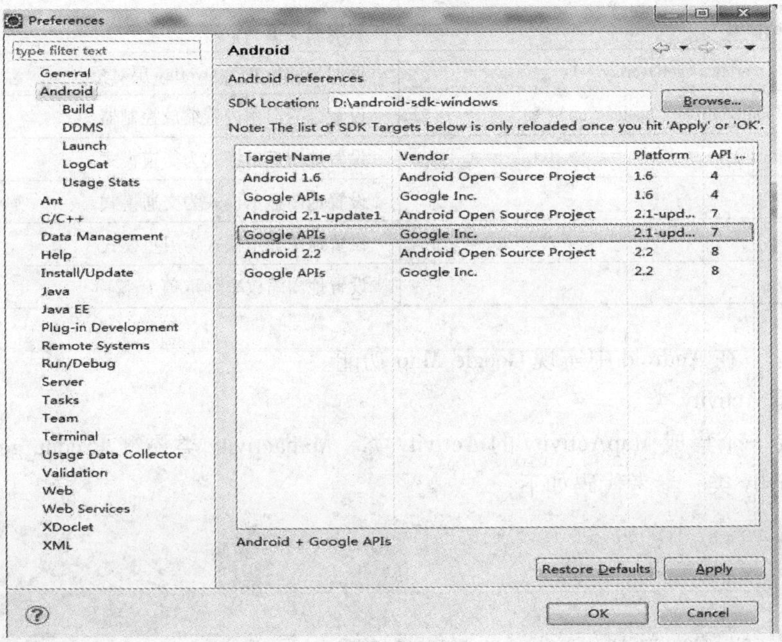

Preferences
type filter text
General
Android
Build
DDMS
Launch
LogCat
Usage Stats
Ant
C/C++
Data Management
Help
Install/Update
Java
Java EE
Plug-in Development
Remote Systems
Run/Debug
Server
Tasks
Team
Terminal
Usage Data Collector
Validation
Web
Web Services
XDoclet
XML
Android
Android Preferences
SDK Location: D:\android-sdk-windows
Browse...
Note: The list of SDK Targets below is only reloaded once you hit 'Apply' or 'OK'.
Target Name | Vendor | Platform | API ...
Android 1.6 | Android Open Source Project | 1.6 | 4
Google APIs | Google Inc. | 1.6 | 4
Android 2.1-update1 | Android Open Source Project | 2.1-upd... | 7
Google APIs | Google Inc. | 2.1-upd... | 7
Android 2.2 | Android Open Source Project | 2.2 | 8
Google APIs | Google Inc. | 2.2 | 8
Android + Google APIs
Restore Defaults
Apply
OK
Cancel

图 9-6 选择 Google API 版本

9.2.3 Android Google Map 基干程序

编写 Android Google Map 有几个固定的步骤：

（1）创建 Google API Android 项目。

（2）编写 MapActivity。

（3）编写 MapView 布局文件。

（4）在 AndroidManifest. xml 中增加 internet 访问权限。

（5）在 AndroidManifest. xml 中增加 Google 地图函数库。

完成上述步骤后才能增加其他的功能。

MapActivity 是用于显示 Google Map 的 Activity。它需要连接底层的网络。MapActivity 是一个抽象类，任何想要显示 Mapview 组件的 Activity 都必须继承自己的 MapActivity。并在 onCreate( ) 方法中创建 Mapview 实例。

MapView 类是 com. google. android. maps 包中显示地图的组件，通常在 MapActivity 中创建 MapView 的对象。MapView 类的常用方法见表 9-3。

表 9-3 MapView 类的常用方法

方　法	说　明
displayZoomControls（Boolean bl）	设置是否显示缩放控件，参数为 true 或 false
Projection getProjection( )	将地图的经度和纬度转换成屏幕像素的实际坐标
int getZoomLevel( )	返回当前的缩放等级数值
Mapcontroller getController( )	创建地图控制抽象类 MapControoller 的对象
GeoPoint getMapCenter( )	获得地图中心
int getLatitudeSpan( )	获取纬度值
int getLongitudeSpan( )	获取经度值
java. util. List < Overlay > getOverlays( )	返回当前所有的 Overlay 层对象
void setButilInZoomControls（boolean on）	设置是否启用内置缩放控制器
void setStreetView( )	设置地图显示模式为街道模式
void setTraffic( )	设置地图显示模式为交通模式
void setSatellite( )	设置地图显示模式为卫星模式
void setZoom( )	设置地图缩放率（取值 1 ~ 21）

【例 9-2】 在 Android 中实现 Google Map 功能。

（1）准备 Activity 类

首先需要一个集成 MapActivity 的 Activity 类，Mapactivity 类要继承 com. google. android. maps. MapActivity 类，参考代码如下：

```
1        Class MyGPSActivity extends MapActivity{
2            MapView mapview;
3            @Override
4            public void onCreate(Bundle savelInstanceState){
5                   super.onCreate(savedInstanceState);
```

```
6           setContentView(R.layout.main);
            ……
7           @Override
8       Protected BooleanisRouteDisplayed(){
9           Return false;
10      }
11  }
```

MapAcitivity 类中有一个抽象方法 isRouteDisplayed( )，该方法是告知 Google 服务器是否使用了显示“道路信息”有关服务，例如行车路线等。如果使用服务则这个方法就应该返回 true，否则返回 false。

若要成功引用 GoogleMaps API，则还必须先在 AndroidManifest.xml 中定义如下信息：

```
<user-libraryandroid:name="com.google.android.maps"/>
```

（2）使用 MapView

若要让地图显示，则需要将 Map View 加入到应用中来。例如在布局文件（main.xml）中加入如下代码：

```
1   <com.google.android.maps.MapView
2       android:id="@+id/map"
3       android:layout_width="wrap_content"
4       android:layout_height="wrap_content"
5       android:enabled="true"
6       android:clickable="true"
7       android:apiKey="API_Key_String"
8   />
```

另外，若要使用 Google Map 服务，则还需要一个 API key。需要在 Google 官网上申请，具体申请方式在此不赘述。

接下来补全 MyGPSActivity 类的代码，在此使用 MapView，代码如下：

```
1   class MyGPSActivity extends MapActivity{
2       @Override
3       public void onCreate(Bundle savedInstanceState){
4           //创建并初始化地图
5           gMapView=(MapView)findViewById(R.id.myGPS);
6           GeoPoint p=new GeoPoint((int)(lat* 10000000,(int)(long* 1000000));
7           gMapView.setSatellite(true);
8           mc=gMapView.getController();
9           mc.setCenter(p);
10          mc.setZoom(14);
11      }
12  }
```

（3）添加缩放控件

```
1   //将缩放控件添加到地图上
2   ZoomControls zoomControls=(ZoomControls)gMapVies.getZoomControls();
3   zoomControls.setLayoutParams(new ViewGroup.LayoutParams(LayoutParams.WRAP_CONTENT, Layout-
Params.WRAP_CONTENT));
```

```
4    gMapView.addView(zoomControls);
5    gMapView.displayZoomControls(true);
```

（4）添加 Map Overlay

例如，通过下面的代码可以定义一个 overlay。

```
1      Class MyLovationOverlay extends com.google.android.map.Overlay{
2          @Override
3          public Boolean draw(Canvas canvas,MapView,Booleanshadow,long when){
4                 super.draw(canvas,mapVies,shadow);
5                 Paint paint = new Paint();
6                 //将经纬度转换成实际屏幕坐标
7                 Point myScreenCoords = new Point();
8                 mapView.getProjection().toPixels(p,myScreenCoords);
9                 paint.setStrokeWidth(1);
10                paint.setARGB(255,255,255,255);
11                paint.setStyle(Paint.Style.STROKE);
12         Bitmap bmp = BitmapFactory.decodeResource(getResources(),R.drawable.marker);
13                canvas.drawBitmap(bmp,myScreenCoords.x,myScreenCoords.y,paint);
14         canvas.drawText("how are you…",myScreenCoords.x,myScreenCoords.y,paint);
15                return true;
16         }
17     }
```

通过上面的 overlay 会在地图上显示一段文本，接下来可以把这个 overlay 添加到地图中，具体代码如下所示，Google API Maps 显示效果如图 9-7 所示。

```
1    MyLocationOverlay myLocationOverlay = new MyLocationOverlay();
2    List<Overlay> list = gMapView.getOverlays();
3    List.add(myLocationOverlay);
```

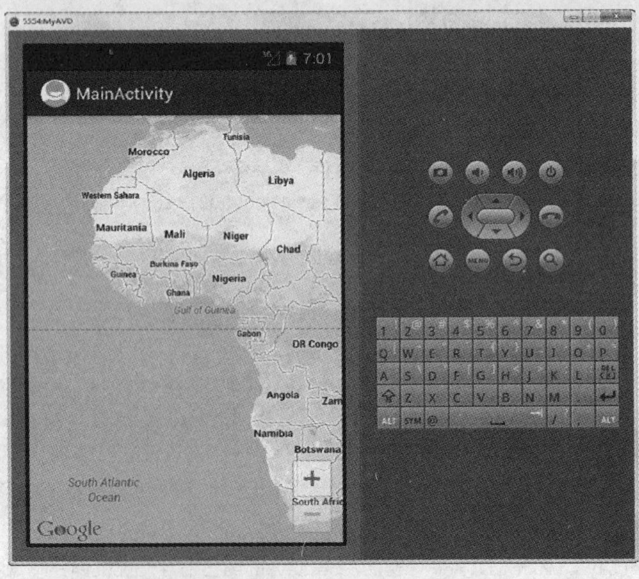

图 9-7　Google API Maps 显示

9.3 GPS 与 Google 地图结合

在地图应用程序中，随着坐标点的变化，GPS 发出通知，可是要把移动的位置标注在地图中需要在地图上建立叠加层。地图和地图叠加层分别位于两个不同的图层。

1. 叠加层

叠加层是地图上与纬度/经度坐标绑定的对象，会随拖动或缩放地图而移动。叠加层表示的是“添加”到地图中以标明点、线、区域或对象集合的对象。

Maps API 包含以下几种叠加层：

- 地图上的单个位置是使用标记显示的。标记有时可显示自定义的图标图片，这种情况下标记通常被称为“图标”。标记和图标是 Marker 类型的对象。
- 地图上的线是使用折线（表示一系列按顺序排列的位置）显示的。线是 Polyline 类型的对象。
- 地图上的不规则形状区域是使用多边形（类似于折线）显示的。与折线相同的是，多边形也是由一系列按顺序排列的位置构成的；不同的是，多边形定义的是封闭区域。
- 地图图层可使用叠加层地图类型显示。可以通过创建自定义地图类型来创建自己的图块集，自定义地图类型可取代基本地图图块集，或作为叠加层显示在现有基本地图图块集之上。
- 信息窗口也是特殊类型的叠加层，用于在指定地图位置上方的弹出式气泡框内显示内容（通常是文字或图片）。
- 还可以实现自己的自定义叠加层。这些自定义叠加层可实现 OverlayView 接口。

叠加层通常在构造时添加到地图中。所有叠加层都会定义构造中所用的 Options 对象，以指定应显示叠加层的地图。也可以使用叠加层的 setMap( )方法向其传递要添加叠加层的地图，从而直接在该地图上添加叠加层。

2. 经纬度位置类 GeoPoint

该类为不可变类，表示一对经纬度值，以微度的整数形式存储，GeoPoint 对象构造后不可再修改经纬度值但可返回该对象。其构造方法为 GeoPoint（int latitudeE6，int longitudeE6）。该类中的主要方法及说明如表 9-4 所示。

表 9-4 GeoPoint 类的主要方法及说明

方 法	说 明
boolean equals（java. lang. Object o）	指示其他某个 GeoPoint 对象是否与此 GeoPoint 对象“相等”
int getLatitudeE6( )	返回该 GeoPoint 对象的纬度，单位微度（度 $\times 10^{-6}$）
int getLongitudeE6( )	返回该 GeoPoint 对象的经度，单位微度（度 $\times 10^{-6}$）
int hashCode( )	返回该 GeoPoint 对象的哈希码值
java. lang. String toString( )	返回该 GeoPoint 对象的字符串表示

在 Android 地图类中，地图的位置标识为 GeoPoint 对象，在使用基于位置的返回服务 Location 对象之前，需要把位置的经纬度转换为以微度为单位的 GeoPoint 对象。其形式如下：

```
double lat =38.438892* 1E6;
double lng =-122.290863* 1E6;
GeoPoint point = new GeoPoint(lat.intValue(), lng.intValue());
```

3. 屏幕坐标

要根据物理位置添加注释，需要在地理位置和屏幕之间进行转换。而 Projection 接口正是用于屏幕像素点坐标（存储为 Point）系统和地球表面经纬度点坐标（存储为 GeoPoint）系统之间的变换。它所提供的主要方法如表 9-5 所示。

表 9-5　Projection 的主要方法

方　法	说　明
GeoPoint fromPixels（int x，int y）	该方法提供了这样的像素点转换器。用一对像素坐标创建一个新的 GeoPoint 对象，像素点坐标是相对于 MapView 的左上角的坐标
android. graphics. Point toPixels(GeoPoint in，android. graphics. Point out)	该方法提供了一个投影变换。将给定的 GeoPoint 对象转换到屏幕像素坐标，该坐标是相对于 MapView 左上角的坐标
float metersToEquatorPixels(float meters)	把一个以米为单位的距离（沿着赤道）转换到当前缩放级别的像素单位（水平）

在使用时需先通过 getProjection 来获得 MapView 的投影，使用方式如下所示：

```
Projection projection = mapView. getProjection();
projection. toPixles(Geopoint, point);
projection. fromPixles(point x, point y);
```

4. Overlay

要想在 MapView 中添加注释和单击处理，就必须创建和使用覆盖（Overlay），每一个覆盖都可以直接在画布上绘制 2D 基本图形。可以向一个地图中添加多个覆盖。所有覆盖层都在地图图层上面，每个覆盖层均可以对用户的单击事件作出响应。在这里创建的覆盖层继承了 Overlay 类的子类，并通过重载 draw( )方法来为指定位置添加注解，重载 ontap( )方法来处理用户的单击操作。Overlay 中常见的方法如表 9-6 所示。

表 9-6　贴图类 Overlay 中常见的方法

方　法	说　明
draw（Canvas canvas，Map View mapView，boolean shadow，long when）	在地图贴片图层上绘制标注
drawAt（android. graphics. Canvas，android. graphics. drawable. Drawable drawable，int x，int y，boolean shadow）	在指定坐标（x，y）处绘制标注
onKeyDown（int keyCode，android. view. KeyEvent event，MapView mapView）	处理按下某个按键事件
onKeyUp（int keyCode，android. view. KeyEvent event，MapView mapView）	处理抬起某个按键事件
onTouchEvent（android. view. MotionEvent e，MapView mapView）	处理触摸事件

创建 overlay 的大致方式如下所示：

```
1     public class TextOverlay extends Overlay{
2       @Override
3       public void draw(Canvas canvas, Map View mapView, boolean shadow){\
4             if(shadow = = false){

5             }else{
6             }
```

```
7            super.draw(canvas, mapView, shadow)
8        }
9        @Override
10       public boolean ontap(GeoPoint p,MapView mapView){
11           return false;
12       }
13   }
```

5. MyLocationOverlay

在 GPS 与地图的结合当中，势必会用到在地图中显示当前位置，从而进行接下来的一切活动。而 MyLocationOverlay 则是一个专门设计的本地覆盖，它用来在一个 MapView 中显示当前位置和方向。子类能覆盖方法 dispatchTap()去处理对当前位置的单击。

为了开启这个 overlay 的功能，需要去调用 enableMyLocation()方法或 enableCompass()方法，或调用 Activity 中的 Activity. onResume()方法。记住，当在后台时，要在 Activity 中的 Activity. onPause()方法中调用相应的 disableMyLocation()方法或 disableCompass()方法来关闭这个功能。由于构造函数也能携带一个 MapController，当在屏幕外时，它可以通过平移地图来保持“我的位置”这个点的可见性，同时也携带一个 View，当位置和朝向改变时，调用 View. postInvalidate()方法使其无效。Runnables 由 runOnFirstFix（java. lang. Runnable）方法提供，一旦有了一个位置，它将被运行。例如，定位地图中心和用缩放来显示位置。在这里常用的方法如表 9-7 所示。

表 9-7　MyLocationOverlay 中常用的方法

方　法	说　明
void disableMyLocation()	停止位置更新
protected void drawCompass（android. graphics. Canvas canvas, float bearing)	绘制指南针小图标
protected void drawMyLocation（android. graphics. Canvas canvas, MapView mapView, android. location. Location lastFix, GeoPoint myLocation, long when)	绘制“我的位置”点
booleanenableMyLocation()	尝试去启用 MyLocation，向 LocationManager. GPS_PROVIDER 和 LocationManager. NETWORK_PROVIDER 注册更新
GeoPoint getMyLocation()	返回一个对应于最新设置的用户位置的 GeoPoint
boolean isMyLocationEnabled()	检查位置感知功能是否被打开（通过 GPS 或网络）
boolean onSnapToItem（int x, int y, Point snapPoint, MapView mapView)	检查给定的（x，y）是否非常接近于一个导致当前单击动作的 item

要想使用 MyLocationOverlay，需要创建一个新的实例，并把它传入应用程序的上下文和目标 MapView，然后把它添加到 MapView 的覆盖列表中，其方式如下所示：

```
1    List<Overlay>overlays=mapView.getOverlays();
2    MyLocationOverlay myLocationOverlay=new MyLocationOverlay(this, mapView);
3    Overlays.add(myLocationOverlay);
```

6. ItemizedOverlay

在地图上做标记可以使用 Overlay 的子类 ItemizedOverlay 来简化实现，该类继承自 Overlay，它包含了一个 OverlayItems 列表。自定义覆盖物或标注管理类，通过 ItemizedOverlay 可以向地图添加一个或多个自定义覆盖物或标注。通常需要编写一个继承于 ItemizedOverlay 的子类，并重写其中的方法。该类中有一个或者多个 ItemizedOverlay，每个 ItemizedOverlay 代表一个标记。该类中常用的方法如表 9-8 所示。

表 9-8　ItemizedOverlay 类中常用的方法

方　法	说　明
static DrawableboundCenter(Drawable drawable)	调整 Drawable 对象边界，使（0，0）是这个 Drawable 对象的中心点
static DrawableboundCenterBottom(Drawable drawable)	调整 Drawable 对象的边界，使（0，0）是这个 Drawable 对象底部的中心点
protected abstract　ItemcreateItem(int i)	子类通过该方法创建实体 item
GeoPointgetCenter( )	获取标注集合的中心点
ItemgetFocus( )	返回当前焦点选中的 item
protected　intgetIndexToDraw(int drawOrder)	通过绘制顺序索引，获取加入集合的次序
booleanonLongPress(GeoPoint p, MotionEvent event, MapView mapView)	处理一个长按覆盖物事件
void setOnFocusChangeListener (ItemizedOverlay. OnFocusChangeListener listener)	设置焦点变换时的监听

【例 9-3】　实例实现 GPS 与 Google Map 的结合。

其中 MainActivity. java 中的主要代码如下：

```
1       /* *
2       * 图层类 ItemsOverlay
3       * /
4       private class LocationItemsOverlay extends ItemizedOverlay < OverlayItem > {
5           private List < OverlayItem > items = null;
6           private Drawable marker = null;

7           public LocationItemsOverlay(Drawable marker, List < OverlayItem > items) {
8                super(marker);
9                this. marker = marker;
10               this. items = items;
11               populate();
12          }

13          @ Override
14          protected OverlayItem createItem(int i) {
15               return items. get(i);
16          }

17          @ Override
18          public void draw(Canvas canvas, MapView mapView, boolean shadow) {
19               super. draw(canvas, mapView, shadow);
```

```
20                boundCenterBottom(marker);
21            }
22            /* *
23            * 单击坐标点时候触发的事件
24            * /
25            @Override
26            protected boolean onTap(int i){
27                Toast.makeText(MyMapActivity.this, items.get(i).getTitle() + "\n" + items.get(i)
.getSnippet(), Toast.LENGTH_LONG).show();
29                return true;
30            }

31            @Override
32            public int size(){
33                return items.size();
34            }
35        }

36        /* *
37        * 我的位置
38        * /
39        private void whereAmI(){
40            locationManager = (LocationManager)getSystemService(Context.LOCATION_SERVICE);
41            locationManager.requestLocationUpdates(LocationManager.GPS_PROVIDER, 1000, 0, mLocationLis-
tener);
42            me = new MyLocationOverlay(this, mapView);
43            me.enableMyLocation();

44            mapView.getOverlays().add(me);
45        }
46        /* *
47        * 开启 GPS 服务
48        * /
49        @Override
50        public void onResume(){
51            super.onResume();
52            whereAmI();
53        }
54        /* *
55        * 关闭 GPS 服务
56        * /
57        @Override
58        public void onPause(){
59            super.onPause();
60            if(locationManager! =null){
61                locationManager.removeUpdates(mLocationListener);
62            }
```

```
63          if(me! =null){
64               me.disableMyLocation();
65          }
66     }
```

运行效果如图 9-8 所示。

图 9-8　GPS 位置在 Google Map 的显示

本 章 小 结

本章主要介绍了 GPS 的概念、系统架构以及底层驱动的编写，并通过例子讲解了 GPS 在 Android 上的应用。

习　　题

9-1　编写一个地图应用程序，使其能在地图上显示任意提供的经纬度坐标。

9-2　利用 GPS 设计一个具有普通视图、交通视图和卫星视图三种模式并具有缩放功能的地图。

第 10 章　语音与短信服务

使用手机打电话及发送短信是手机用户最常用的功能，在 Android 设备上实现是非常简单的。本章将介绍如何使用 Android 电话服务以及短信服务。其中电话服务包括监视电话状态和电话呼叫，发起呼叫和监视来电信息；短信服务则包含发送和接收 SMS 消息。

10.1　电话服务的硬件支持

随着技术的发展，如今只支持 WiFi 的 Android 设备不断地涌入市场，因此并不是所有的硬件都支持电话服务。因此，为了指定应用程序需要设备支持电话服务，应在应用程序的 manifest 文件中添加一个 uses-feature 节点：

```
<uses-feature android:name="android.hardware.telephony"android:required="true"/>
```

如果一个应用程序试图使用电话服务 API，但是它们又不是必须的，那么可以在使用之前检查设备是否有电话硬件。为此，需要使用 PackageManager 的 hasSystemFeature( ) 方法，并指定 FEATURE_ TELEPHONY 功能来检测是否包含电话服务功能。其代码如下：

```
1    /** Check if this device has a telephony */
2    private Boolean checkTelephonyHardware(Context context) {
3        if(context.getPackageManager().hasSystemFeature(
            PackageManager.FEATURE_TELEPHONY)){
4            // this device has a telephony
5            return true;
6        } else {
7            // no telephony on this device
8            return false;
9        }
10   }
```

可以通过在应用程序生命周期的早期检查是否支持电话服务，来调整应用程序的 UI 和行为。

10.2　Android 系统电话服务框架

Android 系统的电话服务 API 提供了一种对基本电话信息进行监听的方式，不但可以获取网络类型和连接状态等基本信息，还可以对电话号码等字符串进行相应的操作。Android 系统电话服务业务涉及的框架结构如图 10-1 所示。

Android 电话服务业务框架采用了分层结构，共跨越了四层：

（1）电话业务应用（Telephony Apps）：包括了电话、MMS 和 STK 等应用程序。

（2）电话业务框架（Telephony Frameworks）：提供 TelephonyManager 类，包含数据连

接、通话、信息和SIM相关的API。

(3) 无线通信接口层(RIL),主要位于User Libraries层中的HAL,提供AP(Application Processor)和BP(Baseband Processor)之间的通信功能。

(4) 调制解调器(MODEM),位于BP,主要负责实际的无线通信能力处理。

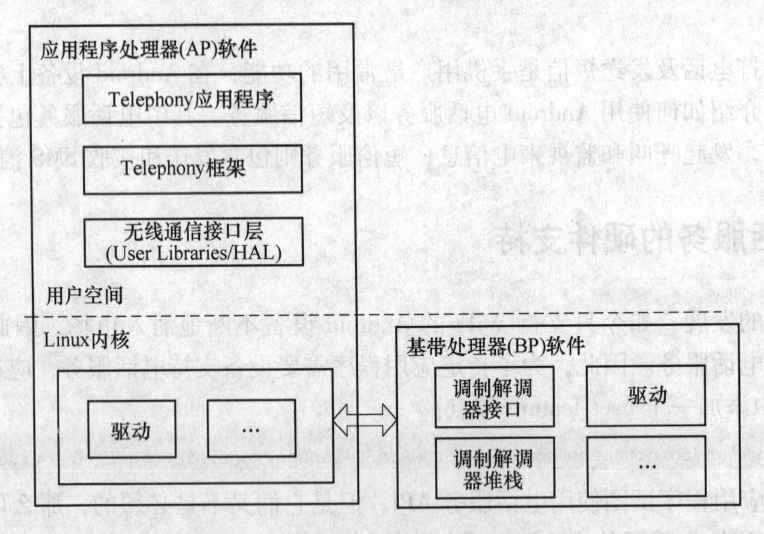

图10-1 Android Telephony分层结构

通过图10-1所示,Android电话业务跨越AP和BP,符合智能手机硬件的基本结构。Android系统在AP上运行,而电话业务运行在Linux内核之上的用户空间。Android电话业务采用了分层结构的设计,包括Java Applications、Java Frameworks和User Libraries层,与Android操作系统整体分层结构保持一致。Telephony框架层中应用层和框架层提供的Telephony服务如表10-1所示。

表10-1 Telephony服务

名　称	说　明
PhoneInterfaceManager	是ITelephony接口的实现
PhoneInterfaceManage	通过PhoneApp、CallManager、Phone对象实现相应功能
IccSmsInterfaceManager	短消息服务,是Isms接口的实现
IccPhoneBookInterfaceManager	电话本服务,是IIccPhoneBook接口的实现
PhoneSubInfo	提供用户信息读取服务,是IPhoneSubInfo接口的实现
TelephonyRegistry	提供应用层的消息登记服务,是ITelephonyRegistry接口的实现

其中,Telephony框架层是以Phone对象为核心的,Phone接口及其子类管理整个手机的Telephony功能。Telephony提供的功能包括语音、短信、SIM卡信息处理、数据连接和SIM卡应用开发工具箱(STK)等。

10.3 语音服务

Android电话服务API使应用程序能够访问底层的电话硬件栈,从而允许将呼叫处理程序和检查电话状态功能集成到自己的应用程序中。下面将介绍如何监视电话服务的属性状态、如何在应用程序中监视并控制电话。

10.3.1 TelephoneManager 类

TelephoneManager 类可用于访问有关设备上提供的电话服务信息。使用 TelephoneManager 类提供的方法可以确定电话服务和状态，以及访问某些类型的用户信息。同时应用程序还可以注册一个监听器以接收电话状态变化。TelephoneManager 对象也是通过 Context. getSystemService（Context. TELEPHONY_SERVICE）生成的。

TelephoneManager 类常用 API 如表 10-2 所示。

表 10-2 TelephoneManager 类常用 API

名　称	说　明
int getCallState( )	返回 int 值，表示在设备上的呼叫状态
CellLocation getCellLocation( )	返回移动设备当前的位置（CellLocation）
int getDataActivity( )	返回 int 值，表示活动的数据连接类型
int getDataState( )	返回 int 值，表示活动的数据连接状态
String getDeviceId( )	返回唯一的设备 ID，例如 IMEI GSM 和 MEID CDMA 手机
String getDeviceSoftwareVersion( )	返回设备的软件版本号，例如 IMEI / SV GSM 手机
String getLine1Number( )	返回 1 号线的电话号码，例如 MSISDN 用于 GSM 电话
int getNetworkType( )	返回一个 int 值，表示目前在设备上使用的无线电技术（网络类型）
int getPhoneType( )	返回设备的类型（手机制式）
int getSimState( )	返回一个 int 值，表示 SIM 卡设备的状态
boolean isNetworkRoaming( )	返回 true，如果该设备被认为是漫游当前网络上，支持 GSM 目的

Android 还为电话提供一个附加监听类 PhoneStateListener，用来监听电话服务状态、信号强度、消息等。PhoneStateListener 中提供了用于监听的 API 状态，以供开发者使用。

onCallStateChanged( )：设备通话状态改变时调用。

onServiceStateChanger( )：设备服务状态改变时调用。

onSignalStrengthsChanged( )：网络信号强度改变时调用。

10.3.2 访问电话服务的属性及状态

通过上述可知，对电话服务 API 的访问是由 Telephony Manager 类进行管理的，使用 getSystemService( )方法可以访问 Telephony Manager。

```
TelephonyManager telManager = (Telephony Manager) getSystemService(Context.TELEPHONY_SERVICE);
```

Telephony Manager 提供了对许多电话服务属性的直接访问，包括设备、网络、SIM 以及数据状态的详细信息。

使用 Telephony Manager 可以获得电话类型（GSM/CDMA/SIP）、唯一 ID、电话号码、软件版本信息；当连接网络时可以获得移动国家代码、网络代码、国家 ISO 代码、网络运行时的名称和连接网络类型，读取电话状态的方法如表 10-3 所示。

表 10-3　读取电话状态的方法

方　法	说　明
public String getDeviceId	读取 GSM 手机的 IMEI 或 CDMA 手机的 MEID
public String getLine1Nunber	读取电话号码
public String getDeviceISoftwareVersion	读取手机上软件版本
public String getNetworkCountryIso	读取网络运营商国家
public String getNetworkOperator	读取网络运营商代码
public String getNetworkOperatorName	读取网络运营商名称
public String getSimCountryIso	读取 SIM 卡运营商国家
public String getSimOperator	读取 SIM 运营商代码
public String getSimOperatorName	读取 SIM 卡运营商名称

上述属性中除了访问电话类型之外，访问其中的任意属性都需要在 manifest 文件中添加访问权限。

```
<uses-permissionandroid:name="android.permission.READ_PHONE_STATE"/>
```

【例 10-1】　如何获得 SIM 卡信息。

（1）新建名为 TelephonyManager 的工程，向 res/layout/activity_ main.xml 文件中添加如下代码：

```
1  <LinearLayoutxmlns:android="http://schemas.android.com/apk/res/android"
2     xmlns:tools="http://schemas.android.com/tools"
3     android:layout_width="fill_parent"
4     android:layout_height="fill_parent"
5     android:orientation="vertical">
6     <!--显示相关 Telephony 信息-->
7     <TextView
8         android:id="@+id/telephony_info"
9         android:layout_width="wrap_content"
10        android:layout_height="wrap_content"
11        android:textSize="18sp"  />
12 </LinearLayout>
```

（2）由于需添加访问权限，因此需要对 AndroidManifest.xml 文件进行编辑。详细信息如下：

```
1  <?xml version="1.0" encoding="utf-8"? >
2  <manifest xmlns:android="http://schemas.android.com/apk/res/android"
3     package="com.example.telephonymanager"
4     android:versionCode="1"
5     android:versionName="1.0" >
6      <!--读取相关 Telephony Services 权限-->
7      <uses-permission android:name="android.permission.READ_PHONE_STATE"></uses-permission
       >
```

```
8       <uses-sdk
9           android:minSdkVersion = "8"
10          android:targetSdkVersion = "18" />
11      <application
12          android:allowBackup = "true"
13          android:icon = "@drawable/ic_launcher"
14          android:label = "@string/app_name"
15          android:theme = "@style/AppTheme" >
16          <activity
17              android:name = "com.example.telephonymanager.MainActivity"
18              android:label = "@string/app_name" >
19              <intent-filter>
20                  <action android:name = "android.intent.action.MAIN" />
21                  <category android:name = "android.intent.category.LAUNCHER" />
22              </intent-filter>
23          </activity>
24      </application>
25  </manifest>
```

（3）向 MainActivity.java 中添加如下代码：

```
1   package com.example.telephonymanager;
2   import android.app.Activity;
3   import android.content.Context;
4   import android.os.Bundle;
5   import android.telephony.TelephonyManager;
6   import android.view.Menu;
7   import android.widget.TextView;
8   public classMainActivity extends Activity {
9       TextView showInfo;//声明显示相关信息的文本控件
10      TelephonyManager telManager;//声明电话管理器
11      @Override
12      public void onCreate(Bundle savedInstanceState) {
13          super.onCreate(savedInstanceState);
14          setContentView(R.layout.activity_main);
15          showInfo = (TextView)findViewById(R.id.telephony_info);//找到 id 为 telephony_info 的文本
控件
16          telManager = (TelephonyManager)getSystemService(Context.TELEPHONY_SERVICE);//实例化电话
管理器
17          StringBuilder sb = new StringBuilder();
18          sb.append("设备 ID:").append(telManager.getDeviceId()).append("\n");
19          sb.append("IMSI:").append(telManager.getSubscriberId()).append("\n");
20          sb.append("网络运营商国家:").append(telManager.getNetworkCountryIso()).append("\n");
21          sb.append("网络运营商代码:").append(telManager.getNetworkOperator()).append("\n");
22          sb.append("网络运营商名称:").append(telManager.getNetworkOperatorName()).append("\n");
```

```
23          sb.append("SIM卡运营商国家:").append(telManager.getSimCountryIso()).append("\n");
24          sb.append("SIM运营商代码:").append(telManager.getSimOperator()).append("\n");
25          sb.append("SIM卡运营商名称:").append(telManager.getSimOperatorName()).append("\n");
26          sb.append("IMEI:").append(telManager.getSimSerialNumber()).append("\n");
27          showInfo.setText(sb);
28      }
29      @Override
30      public boolean onCreateOptionsMenu(Menu menu) {
31          getMenuInflater().inflate(R.menu.main, menu);
32          return true;
33      }
34  }
```

（4）运行效果如图 10-2 所示。

图 10-2　读取 SIM 卡信息

10.3.3　监听来电信息

Android 电话服务 API 可以用来监视电话状态和相关信息（如来电）的变化。实现手机电话状态的监听，主要依靠两个类：TelephonyManger 和 PhoneStateListener。当然有一些状态变化也作为 Intent 广播出去。为了监视并管理电话状态应该获取其 READ_PHONE_STATE 权限。

```
<uses-permissionandroid:name="android.permission.READ_PHONE_STATE"/>
```

应用程序可以注册 listener 来监听电话状态的改变。主要静态成员常量（它们对应 PhoneStateListener.LISTEN_CALL_STATE 所监听到的内容）：

int CALL_STATE_IDLE 空闲状态，没有任何活动。

int CALL_STATE_OFFHOOK 摘机状态，至少有一个电话活动。该活动或是拨打（dialing）或是通话，或是 on hold，并且没有电话是 ringing or waiting。

int CALL_STATE_RINGING 来电状态，电话铃声响起的那段时间或正在通话又来新电，新来电话不得不等待的那段时间。

手机通话状态在广播中的对应值：

EXTRA_STATE_IDLE 用于表示 CALL_STATE_IDLE 状态

EXTRA_STATE_OFFHOOK 用于表示 CALL_STATE_OFFHOOK 状态

EXTRA_STATE_RINGING 用于表示 CALL_STATE_RINGING 状态

ACTION_PHONE_STATE_CHANGED 在广播中用 ACTION_PHONE_STATE_CHANGED 这个 Action 来标识通话状态改变的广播（Intent）。

实现对来电信息的监听应该在 PhoneStateListener 的实例中重写 onCallStateChanged 方法进行注册，以便在呼叫状态发生变化时接收通知。

```
1  class MyPhoneStateListener extends PhoneStateListener {
2      @Override
3      public void onCallStateChanged(int state, String incomingNumber) {
4          switch(state) {
5          case TelephonyManager.CALL_STATE_IDLE: //空闲
6              break;
7          case TelephonyManager.CALL_STATE_RINGING: //来电
8              break;
9          case TelephonyManager.CALL_STATE_OFFHOOK: //摘机（正在通话中）
10             break;
11         }
12     }
13 }
```

对传入呼叫的监听也可以使用 Intent Receiver 来实现，当电话状态改变时，TelephonyManager 会发出 TelephonyManager. ACTION_PHONE_STATE_CHANGED 的广播。通过在 manifest 文件中注册一个监听此广播的接收器就可以随时监听来电信息，注册信息如下：

```
1  <receiver android:name = "PhoneStateChangedReceiver" >
2      <intent-filter >
3          <actionandroid:name = "android.intent.action.PHONE_STATE"/ >
4      </intent-filter >
5  </receiver >
```

注册完之后则可以通过以下代码实现对电话状态的监听。

```
1  public class MyBroadcastReceiver extends BroadcastReceiver {
2      private static final String TAG = "MyBroadcastReceiver";
3      @Override
4      public void onReceive(Context context, Intent intent) {
5          String action = intent.getAction();
6          Log.i(TAG, "[Broadcast]" + action);

7          //呼入电话
8      if(action.equals(TelephonyManager.ACTION_PHONE_STATE_CHANGED)) {
9              Log.i(TAG, "[Broadcast]PHONE_STATE");
10             doReceivePhone(context,intent);
11         }
12     }
13     /* *
14     * 处理电话广播
```

```
15      * @param context
16      * @param intent
17      */
18      public void doReceivePhone(Context context, Intent intent) {
19          String phoneNumber = intent.getStringExtra(TelephonyManager.EXTRA_INCOMING_NUMBER);
20          TelephonyManager telephony = (TelephonyManager) context.getSystemService (Con-
text.TELEPHONY_SERVICE);
21          int state = telephony.getCallState();
22          switch (state) {
23          case TelephonyManager.CALL_STATE_RINGING:
24              Log.i(TAG, "[Broadcast]等待接电话 = " + phoneNumber);
25              break;
26          case TelephonyManager.CALL_STATE_IDLE:
27              Log.i(TAG, "[Broadcast]电话挂断 = " + phoneNumber);
28              break;
29          case TelephonyManager.CALL_STATE_OFFHOOK:
30              Log.i(TAG, "[Broadcast]通话中 = " + phoneNumber);
31              break;
32          }
33      }
34  }
```

10.3.4 监听去电信息

对于去电信息的监听大致可以分为三步：使用继承自 BroadcastReceiver 类的 Receiver；在 AndroidManifest.xml 中，配置好 Receiver，并拦截相应的 BroadCastAction；添加相应权限。与监听来电信息所不同的是这里需要添加：<uses-permission android：name = " android.permission.PROCESS_OUTGOING_CALLS" />权限。其添加的所有权限如下：

```
1  <receiver android:name = ".broadcase.PhoneStateBroadCastReceiver">
2      <intent-filter>
3          <actionandroid:name = "android.intent.action.PHONE_STATE"/>
4          <action android:name = "android.intent.action.NEW_OUTGOING_CALL" />
5      </intent-filter>
6  </receiver>
7  <uses-permission android:name = "android.permission.READ_PHONE_STATE" />
8  <uses-permission android:name = "android.permission.PROCESS_OUTGOING_CALLS" />
```

接下来只需在 mainactivity 中填写如下代码便可以实现对去电的监听。

```
1  public class MainActivity extends Activity {
2      private static final String TAG = "MainActivity";

3      private BroadcastReceiver outgoingCallReceiver = new BroadcastReceiver() {
4          @Override
5          public void onReceive(Context context, Intent intent) {
```

```
6            String action = intent.getAction();
7            Log.i(TAG, "[Broadcast]" + action);
8            //呼出电话
9            if(action.equals(Intent.ACTION_NEW_OUTGOING_CALL)){
10               String outPhoneNumber = intent.getStringExtra(Intent.EXTRA_PHONE_NUMBER);
11               Log.i(TAG, "[Broadcast]ACTION_NEW_OUTGOING_CALL:" + outPhoneNumber);
12               //this.setResultData(null);
13               //这里可以更改呼出电话号码。如果设置为 null,电话就永远不会拨出了
15           }
16       }
17   };
18   @Override
19   public void onCreate(Bundle savedInstanceState) {
20       super.onCreate(savedInstanceState);
21       setContentView(R.layout.activity_main);
22       IntentFilter intentFilter = new IntentFilter();
23       intentFilter.addAction(Intent.ACTION_NEW_OUTGOING_CALL);
24       intentFilter.setPriority(Integer.MAX_VALUE);
25       registerReceiver(outgoingCallReceiver, intentFilter);
26   }

27   @Override
28   public boolean onCreateOptionsMenu(Menu menu) {
29       getMenuInflater().inflate(R.menu.activity_main, menu);
30       return true;
31   }
32 }
```

上述代码在 onCreate( )方法中注册了一个对拨出电话广播进行监听的广播接收者 Receiver，并且在 MainActivity 中是以内部类的形式来定义这个广播接收者。当监听到用户拨出电话过后，广播接收者的 onReceive( )方法会自动被回调，可以通过获取 Intent 对象的相关参数来获取用户拨出的具体电话号码。

10.4 短消息服务

短消息服务从诞生到现如今大致经历了三个阶段：从 20 世纪 80 年代初的短消息业务 SMS 到之后的增强短信业务 EMS 再到如今的多媒体短信服务 MMS。其中最受欢迎的莫过于 SMS 和 MMS。SMS 以简单方便的使用功能受到大众的欢迎，不过始终是属于第一代的无线数据服务，在内容和应用方面存在单调和枯燥的限制。随着 3G 网络的普及，MMS 给手机用户提供了一种全新的手机沟通方式，使用户充分感受到通信的乐趣。

10.4.1 SMS 和 MMS 简介

最早的短消息业务（Short Messaging Service，SMS）是现在普及率最高的一种短消息业

务。这种短消息可以是文本的（包括文字和数字）短消息，也可以是二元非文本短消息（例如图片和铃声）。多媒体短信服务（Multimedia Messaging Service，MMS）最大的特色就是支持多媒体功能，可以在高速传输技术 EDGE（Enhanced Data rates for GSM Erolution）和 GPRS 的支持下，以无线应用协议（WAP）为载体传送视频片段、图片、声音和文字，传送方式除了在手机间传送外，还可以是手机与计算机之间的传送。EDGE 是一种提高数据速率的新技术，是“全球通”向第三代移动通信系统（IMT-2000）过渡的台阶。它也被称为“GSM 384”，因为这种技术能使“全球通”的数据速率由目前的 9. 6kbit/s 提高到 384kbit/s，这种速率可以支持语音、因特网浏览、电子邮件、会议电视等多种高速数据业务。

SMS 和 MMS 是成熟的移动技术，在下面的小节中将重点介绍在 Android 应用程序中如何发送、接收短消息及跟踪发送状态。

10. 4. 2　SMS 消息的发送与跟踪

在 Android 中可以通过 Intent 从应用程序中发送 SMS 和 MMS，也可以通过 SMSManager 类来处理 SMS。大多数情况下，最优的实现方案是利用 Intent 调用另一个程序来发送 SMS 和 MMS 消息。

需要使用 Intent 中的 ACTION_SENDTO 动作来调用 startActivity，并且在 sms_bodyextra 中包含想要发送的信息。

```
1  private void send1(String number, String message){
2      Uri uri = Uri.parse("smsto:" + number);
3      Intent sendIntent = new Intent(Intent.ACTION_VIEW, uri);
4      sendIntent.putExtra("sms_body", message);
5      startActivity(sendIntent);
6  }
```

上述代码的第 2 行可自动设置接收方的号码，第 4 行则是要发送的内容。还可以向其中添加文件类型，比如发送图片则可通过 sendIntent. setType（“image/jpeg”）；通过 intent. setType（" vnd. android-dir/mms-sms"）；则可以跳转到短信界面。

除此之外经常会用到 SMSManager 对 SMS 消息进行处理。Android API 中提供了 smsManager 类用于处理短信。其中 sendTextMessage（String destinationAddress，String scAddress，String text，PendingIntent sentIntent，PendingIntent deliveryIntent）函数就是发送短信的方法。第一个参数 destinationAddress 表示的就是需要发送对象的手机号码；第二个参数 scAddress 表示的是服务中心短信的地址，一般情况下写为 null；第三个参数 text 表示的是要发送的短信内容；第四个参数 sentIntent 是一个 PendingIntent 对象，它用来告诉用户发送短信是否成功；第五个参数 deliverIntent 也是一个 PendingIntent 对象，它用来确认对方是否成功接收。

一个应用程序要具备发送短信功能，需要在 androidManifest. xml 中加入 android. permission. SEND_SMS 权限。

一般在发短信时会看到“发送成功”或是“信息已送达”等字样，Android 提供了相应的监听机制。它主要监听 PendingIntent 中的动作即 sendTextMessage 中的第四个参数 sendintent 及第五个参数 deliveryintent。sendIntent 会在消息发送成功或失败时被触发，主要状态如表 10-4 所示。

表 10-4 sendIntent 状态

状 态	说 明
Activity. RESULT_OK	一次成功的发送
RESULT_ERROR_GENERIC_FAILURE	普通的发送失败
RESULT_ERROR_RADIO_OFF	电话无线信号被关闭
RESULT_ERROR_NULL_PDU	一次 PDU（协议数据单元）错误
RESULT_ERROR_NO_SERVICE	没有可用的手机服务

deliveryIntent 则是在接收人接收到 SMS 时被触发，下面的代码是典型的发送信息及监视其状态的例子。

```
1   String SENT_SMS_ACTION = "SENT_SMS_ACTION";
2   String DELIVERED_SMS_ACTION = "DELIVERED_SMS_ACTION";
3   // Create the sentIntent parameter
4   Intent sentIntent = new Intent(SENT_SMS_ACTION);
5   PendingIntent sentPI = PendingIntent.getBroadcast(getApplicationContext(),0,sentIntent,0);
6   // Create the deliveryIntent parameter
7   Intent deliveryIntent = new Intent(DELIVERED_SMS_ACTION);
8   PendingIntent deliverPI = PendingIntent.getBroadcast (getApplicationContext (), 0, deliveryIn-
tent,0);
9   // Register the Broadcast Receivers
10  registerReceiver(new BroadcastReceiver() {
11  @Override
12  public void onReceive(Context _context, Intent _intent) {
13      switch (getResultCode()) {
14      case Activity.RESULT_OK:
15          [… send success actions … ]; break;
16      case SmsManager.RESULT_ERROR_GENERIC_FAILURE:
17          [… generic failure actions … ]; break;
18      case SmsManager.RESULT_ERROR_RADIO_OFF:
19          [… radio off failure actions … ]; break;
20      case SmsManager.RESULT_ERROR_NULL_PDU:
21          [… null PDU failure actions … ]; break;}}},new IntentFilter(SENT_SMS_ACTION));
22          registerReceiver(new BroadcastReceiver() {
23          @Override
24          public void onReceive(Context _context, Intent _intent) {
25          [… SMS delivered actions … ]
26      }
27  },newIntentFilter(DELIVERED_SMS_ACTION));
28  // Send the message
29  smsManager.sendTextMessage(sendTo, null, myMessage, sentPI, deliverPI);
```

在 SMSManager 中除了采用典型的 sendTextMessage 方法外，还可以采用以下两种方式：

（1）sendDataMessage 方法将 Data 格式的 SMS 传送到特定程序的 Port。采用函数 send-

DataMessage (String destinationAddress, String scAddress, short destinationPort, byte [ ] data, PendingIntentsent Intent, PendingIntent deliveryIntent)，典型的发送程序如下：

```
1   public class SmsReceiver extends BroadcastReceiver {

2       @Override
3       public void onReceive(Context context, Intent intent) {
4           Bundle bundle = intent.getExtras();
5           SmsMessage[] msgs = null;
6           String phone;
7           String message;

8           if(bundle ! = null){
9               Object[] pdus = (Object[])bundle.get("pdus");
10              msgs = new SmsMessage[pdus.length];
11              for(inti = 0; i < msgs.length; i ++){
12                  msgs[i] = SmsMessage.createFromPdu((byte[])pdus[i]);
13                  phone = msgs[i].getOriginatingAddress();
14                  byte data[] = SmsMessage.createFromPdu((byte[])pdus[i]).getUserData();
15                  message = new String(data);
16              }
17          }
18      }
19  }
```

（2）sendMultipartTextMessage 方法发送超长文字短信，参数与 sendTextMessage 类似，无非是短信内容变成了用 divideMessage 拆成的 ArrayList，两个广播也是如此，在此不再举例说明。

10.4.3 SMS 消息的接收

短信的接收，需要实现 BroadcastReceiver 类。当一个 SMS 消息被接收时，就会广播一个包含了 android. provider. Telephony. SMS_RECEIVED 动作的 Intent。注意：这是一个字符串字面量（String literal），但是 SDK 当前并没有包括这个字符串的引用，因此当要在应用程序中使用它时必须显式地指定它。现在开始构建一个 SMS 接收程序。

（1）跟 SMS 发送程序类似，要在清单文件 AndroidManifest. xml 中指定权限允许接收。

```
SMS: <uses-permissionandroid:name = "android.permission.RECEIVER_SMS" />
```

为了能够回发短信，还应该加上发送的权限。

（2）应用程序监听 SMS 意图广播，SMS 广播 Intent 包含了到来的 SMS 细节。要从其中提取出 SmsMessage 对象，这样就要用到 pdu key 来提取一个 SMS PDUs（protocol description units—封装了一个 SMS 消息和它的元数据）数组，每个元素表示一个 SMS 消息。为了将每个 PDU byte 数组转化为一个 SMS 消息对象，需要调用 SmsMessage. createFromPdu。每个 SmsMessage 包含 SMS 消息的详细信息，包括起始地址（电话号码）、时间戳、消息体。下面编写一个接收短信的类 SmsReceiver，代码如下：

```
1   public class SmsReceiver extends BroadcastReceiver {
2       @Override
3       public void onReceive(Context _context, Intent _intent) {
```

```
4            if (_intent.getAction().equals(SMS_RECEIVER)) {
5                SmsManager sms = SmsManager.getDefault();
6                Bundle bundle = _intent.getExtras();
7                if (bundle ! = null) {
8                    Object[]pdus = (Object[]) bundle.get("pdus");
9                    SmsMessage[] messages = new SmsMessage[pdus.length];
10                   for (int i = 0; i < pdus.length; i ++)
11                       messages[i] = SmsMessage.createFromPdu((byte[]) pdus[i]);
12                   for (SmsMessagemessage : messages) {
13                       String msg = message.getMessageBody();
14                       String to = message.getOriginatingAddress();
15                       if (msg.toLowerCase().startsWith(queryString)) {
16                           String out = msg.substring(queryString.length());
17                           sms.sendTextMessage(to, null, out, null, null);
18                           Toast.makeText(_context, "success", Toast.LENGTH_LONG).show();
19                       }
20                   }
21               }
22           }
23       }
24       private static final String queryString = "@ echo";
25       private static final String SMS_RECEIVER = "android.provider.Telephony.SMS_RECEIVED";
26   }
```

上面代码的功能是从接收到的广播意图中提取来电号码、短信内容，然后将短信加上@echo 头部回发给来电号码，并在屏幕上显示一个 Toast 消息提示成功。

【例 10-2】 实例实现短信的收发，主要功能是从名为 5554 的模拟器发送信息到名为 5556 的模拟器。

（1）首先建立名为 SmsManager 的项目，并在 activity_main.xml 中添加布局文件，代码如下：

```
1   <? xml version = "1.0" encoding = "utf-8"? >
2       < AbsoluteLayout
3           android:layout_width = "fill_parent"
4           android:layout_height = "fill_parent"
5           xmlns:android = "http://schemas.android.com/apk/res/android"
6       >
7       < TextView
8           android:layout_width = "wrap_content"
9           android:layout_height = "wrap_content"
10          android:text = "收件人:"
11          android:textSize = "16sp"
12          android:layout_x = "0px"
13          android:layout_y = "12px"
14      >
```

```
15      </TextView>
16      <EditText
17         android:id="@+id/myEditText1"
18         android:layout_width="fill_parent"
19         android:layout_height="wrap_content"
20         android:text=""
21         android:textSize="18sp"
22         android:layout_x="100px"
23         android:layout_y="2px"
24      >
25      </EditText>
26      <EditText
27         android:id="@+id/myEditText2"
28         android:layout_width="fill_parent"
29         android:layout_height="223px"
30         android:text=""
31         android:textSize="18sp"
32         android:layout_x="0px"
33         android:layout_y="52px"
34          >
35      </EditText>
36      <Button
37         android:id="@+id/myButton1"
38         android:layout_width="242px"
39         android:layout_height="wrap_content"
40         android:text="发送短信"
41         android:layout_x="230px"
42         android:layout_y="302px"
43      >
44      </Button>
45  </AbsoluteLayout>
```

（2）下面进行主程序设计

```
1   package com.example.smsmanager;
2   import android.app.Activity;
3   import android.app.PendingIntent;
4   import android.content.Intent;
5   import android.os.Bundle;
6   import android.telephony.gsm.SmsManager;
7   import android.view.View;
8   import android.widget.Button;
9   import android.widget.EditText;
10  import android.widget.Toast;
11  public classMainActivity extends Activity {
12      private Button mButton1;
13      private EditText mEditText1;
```

```
14      private EditText mEditText2;
15      public void onCreate(Bundle savedInstanceState) {
16          super.onCreate(savedInstanceState);
17          setContentView(R.layout.activity_main);
18          //获取资源
19          mEditText1 = (EditText)findViewById(R.id.myEditText1);
20          mEditText2 = (EditText)findViewById(R.id.myEditText2);
21          mButton1 = (Button)findViewById(R.id.myButton1);
22          //发送短信的响应
23          mButton1.setOnClickListener(new Button.OnClickListener(){
24              public void onClick(View v) {
25                  //获取发送地址和发送内容
26                  String messageAddress = mEditText1.getText().toString();
27                  String messageContent = mEditText2.getText().toString();

28                  //构建一取得 default instance 的 SmsManager 对象
29                  SmsManager smsManager = SmsManager.getDefault();
30                  //检查输入内容是否为空,这里为了简单就没有判断是否是号码,短信内容长度的限制也没有做
31                  if(messageAddress.trim().length()! =0 &&messageContent.trim().length()! =0)
{
32                      try{
33                          PendingIntent pintent = PendingIntent.getBroadcast(MainActivity.
this, 0, new Intent(), 0);
34                          smsManager.sendTextMessage(messageAddress, null, messageContent, pin-
tent, null);
35                      }catch(Exception e){
36                          e.printStackTrace();
37                      }
38                      //提示发送成功
39                      Toast.makeText(MainActivity.this, "发送成功", Toast.LENGTH_LONG).show();
40                  }
41                  else{
42                      Toast.makeText(MainActivity.this, "发送地址或者内容不能为空", Toast.LENGTH_
SHORT).show();
43                  }
44              }
45          });
46      }
47  }
```

（3）在 AndroidManifest.xml 中注册权限，完整代码如下：

```
1   <?xml version="1.0" encoding="utf-8"?>
2       <manifest xmlns:android="http://schemas.android.com/apk/res/android"
3           package="com.example.smsmanager"
4           android:versionCode="1"
```

```
5           android:versionName = "1.0" >
6           <uses-sdk
7               android:minSdkVersion = "8"
8               android:targetSdkVersion = "18" />
9       <application
10          android:allowBackup = "true"
11          android:icon = "@drawable/ic_launcher"
12          android:label = "@string/app_name"
13          android:theme = "@style/AppTheme" >
14          <activity
15              android:name = "com.example.smsmanager.MainActivity"
16              android:label = "@string/app_name" >
17              <intent-filter>
18                  <action android:name = "android.intent.action.MAIN" />
19                  <category android:name = "android.intent.category.LAUNCHER" />
20              </intent-filter>
21          </activity>
22      </application>
23      <uses-permission android:name = "android.permission.SEND_SMS" > </uses-permission>
24  </manifest>
```

（4）运行结果如图 10-3 所示。

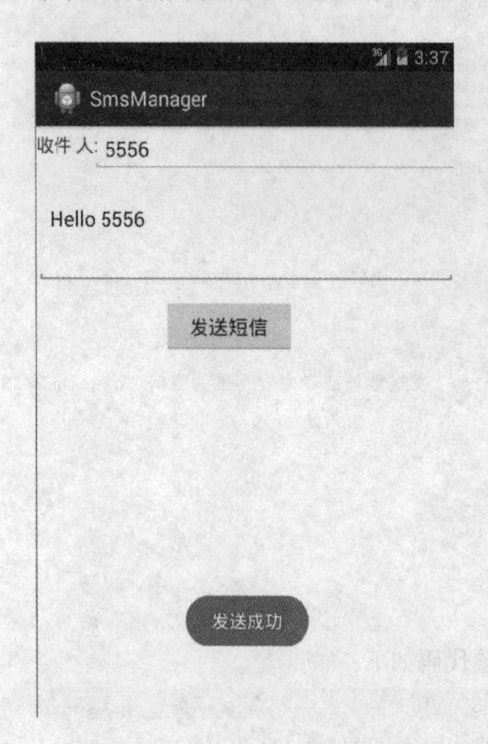

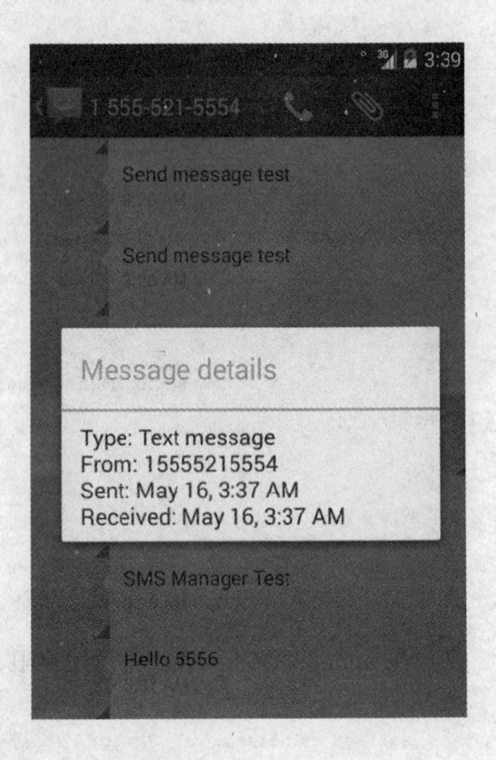

图 10-3　SmsManager 发送和接收短消息

本章小结

本章主要介绍了 Android 中对语音及短消息的访问，重点介绍了利用 Telephony 类来监听来电和去电信息；利用 SMSManager 来发送和接收短消息，并用 PendingIntent 对发送消息进行跟踪。

习 题

10-1 创建一个自定义的拨号程序，并且在模拟器间进行通信。

10-2 设计一个群发短信的应用程序，并显示发送状态。

第 11 章　Android 传感器应用

Android 是一个面向应用程序开发的平台，它拥有具有吸引力的用户界面元素、数据管理和网络应用等优秀的功能。Android 还提供了许多颇具特色的接口，比如传感器。传感器是让 Android 设备区别于其他计算机的重要功能。如果没有传感器，Android 设备只是一个动力不足、屏幕太小的 Web 浏览器，同时其输入机制也很笨拙。只要手机设备的硬件提供了这些传感器，Android 应用就可以通过传感器来获取设备的外界条件，包括设备的运行状态、当前摆放方向、外界的磁场、温度和压力等。Android 系统提供了驱动程序去管理这些传感器，当传感器感知到外部环境发生改变时，Android 系统负责管理这些传感器数据。

翻看手机中的应用，就能发现大多数的应用都已经使用了传感器。比如微信中的“摇一摇”、各种运动记录 APP、火爆的游戏“神庙逃亡”等，因此用好传感器已经是手机应用开发的必修课。

Android 1.5（API level 3）开始提供一套标准传感器以及相关的 API。到目前为止，系统已经内置了对多达 13 种传感器的支持，它们分别是：加速度传感器、环境温度传感器、重力传感器、陀螺仪、环境光照传感器、线性加速度传感器、磁力传感器、方向传感器、压力传感器、距离传感器、相对湿度传感器、旋转向量传感器和温度传感器。本章将介绍如何编写使用 Android 传感器的应用。

11.1　利用 Android 传感器

11.1.1　传感器的定义

传感器是一种检测装置，能感受到被测量的信息，并能将感受到的信息，按一定规律转换成电信号或其他所需形式的信息输出，以满足信息的传输、处理、存储、显示、记录和控制等要求。它是实现自动检测和自动控制的首要环节。在国家标准 GB 7665—1987 中对传感器下的定义是：“能感受规定的被测量件并按照一定的规律（数学函数法则）转换成可用信号的器件或装置，通常由敏感元件和转换元件组成”。

11.1.2　Android 中传感器关联类和接口

1. Sensor 类

Android 系统中内置了很多类型的传感器，这些传感器被封装在 Sensor 类中。Sensor 类是管理各种传感器共同属性（名字、供应商、类型、版本）的类。Sensor 类包括以下内容：

- 主要常量

在 Sensor 类，能使用的传感器的种类通过常量来定义（见表 11-1）。但是根据硬件的不同，传感器搭载是任意的。比如（HTC Desire 516t）有 ACCELEROMETER、PROXIMITY 和 LIGHT 三种传感器。

表 11-1　传感器的种类

常 量 名	说　明	返 回 值
String TYPE_ACCELEROMETER	加速度	1
String TYPE_GYROSCOPE	陀螺仪	4
String TYPE_LIGHT	光照	5
String TYPE_MAGNETIC_FIELD	磁力计	2
String TYPE_ORIENTATION	方位传感器	3
String TYPE_PRESSURE	压力传感器	6
String TYPE_PROXIMITY	距离传感器	8
String TYPE_TEMPRATURE	温度传感器	7
String TYPE_ALL	全部的传感器	-1

- 主要方法

该类中的主要方法是用来获取硬件传感器信息，其主要方法如表 11-2 所示。

表 11-2　Sensor 类的主要方法

方　法	返 回 值
public float getMaximumRange( )	返回传感器可测量的最大范围
public float getMinimumDelay( )	返回传感器的最小延迟
public String getName( )	返回传感器的名称
public float getPower( )	返回传感器的功率
public float getResolution( )	返回传感器的分辨率
public intgetType( )	返回传感器的类型
public String getVentor( )	返回传感器的供应商
public intgetVersion( )	返回传感器的版本

在定义了传感器中 Sensor 类的主要常量及方法后，就可通过如 sensor. getName( ) 方式来获取某一具体传感器的名称，其他具体信息的获取同该方法类似。

2. SensorManager 类

SensorManager 类就是所有传感器的一个综合管理类，包括了传感器的种类、采样率、精确度等，是 Android 为应用提供传感器硬件访问能力的系统服务。和其他系统服务一样，它允许用户注册或注销传感器相关事件。一旦注册成功，应用将会接收到从硬件传来的传感器数据。

- 主要常量

在 SensorManager 类中有很多个常量被定义，但是这些中最重要的是关于传感器反应速度的。一般用于注册监听器时为其指定延迟和测量速率。关于传感器的反应速度的常量如表 11-3所示。

表 11-3　关于传感器的反应速度的常量

常 量 名	说　明	返 回 值
int SENSOR_DELAY_FASTEST	在想取特别快的反应速度时使用	0
int SENSOR_DELAY_GAME	游戏用	1
int SENSOR_DELAY_UI	适用于用户界面功能，如旋转屏幕	2
int SENSOR_DELAY_NORMAL	默认值	3

- 主要方法

SensorManager 中常用的方法如表 11-4 所示，主要是用来获取传感器及注册和撤销传感器的监听器。

表 11-4　SensorManager 类中的常用方法

方　法	说　明
public boolean registerListener（SensorEventListener listener, Sensor sensor, int samplingPeriodUs）	为指定的传感器注册监听器
public void unregisterListener（SensorEventListener listener）	为所有传感器解除已注册的监听器
public void unregisterListener（SensorEventListener listener, Sensor sensor）	为指定的传感器解除已注册的监听器
public List < Sensor > getSensorList（int type）	获得可用传感器列表
pubilc SensorgetDefaultSensor（int type）	获得给定类型的默认传感器

SensorManager 不能直接生成 Instance 实例。SensorManager 的 Instance 是通过 Context 类定义的 getSystemService 方法取得的。其代码如下：

```
sensorManager = (SensorManager)this. getSystemService(SENSOR_SERVICE);
```

对于 Sensor 对象的访问提供了两种方法：getSensorList( ) 方法检索所有给定类型的传感器，getDefaultSensor( ) 返回指定类型的默认传感器。其实现代码如下：

```
获取某种传感器列表
List <Sensor>pressureSensors = sensorManager. getSensorList(Sensor. TYPE_PRESSURE);
获取某种默认的传感器
Sensor defaultAccelerometer = sensorManager. getDefaultSensor(Sensor. TYPE_ACCELEROMETER);
```

对于注册传感器，SensorManager 的常用方法如下：

```
public boolean registerListener(SensorEventListener listener, Sensor sensor, int samplingPeriodUs);
```

其中三个参数的含义为：

listener：监听传感器事件的监听器。该监听器需要实现 SensorEventListener 接口。

sensor：传感器对象。

samplingPeriodUs：指定获取传感器数据的频率。其具体的值已在表 11-3 中列出。在 Android 4. 0. 3 中，这些速率被硬编码为 0、20ms、67ms 和 200ms。也可以通过传入一个传感器速率值到注册器中来指定延迟。然而这些速率仅用于提示系统，因为接收事件的速率可能会快于或慢于指定的延迟。

【例 11-1】　实例实现如何获取真机设备上的各种传感器的信息。

获取手机中的传感器的步骤为：

（1）获取 SensorManager 对象。

（2）执行 SensorManager 对象的 getSensorList（）方法获取 Sensor 对象。

（3）获取 Sensor 对象中的各种属性。

1）新建工程 getSensorInfo，其布局文件源代码参见参考程序。

2）编辑 MainActivity. java 文件，具体代码如下：

```
1  package com.example.getsensorinfo;
2  import java.util.List;
3  import android.app.Activity;
4  import android.content.Context;
5  import android.hardware.Sensor;
6  import android.hardware.SensorManager;
7  import android.os.Bundle;
8  import android.view.Menu;
9  import android.widget.TextView;
10 public class MainActivity extends Activity {
11     TextView sensorInfoTextView;// 声明显示传感器信息文本
12     @Override
13     public void onCreate(Bundle savedInstanceState) {
14         super.onCreate(savedInstanceState);
15         setContentView(R.layout.activity_main);
16         sensorInfoTextView = (TextView) findViewById(R.id.sensor_info);// 找到 id 为 sensor_info
的 TextView 控件
17         // 声明并实例化传感器管理器
18         SensorManager sensorManager = (SensorManager) getSystemService(Context.SENSOR_SERVICE);
19         // 获得全部的传感器列表
20         List<Sensor> allSensors = sensorManager.getSensorList(Sensor.TYPE_ALL);
21         StringBuffer infoBuffer = new StringBuffer();//声明显示文本
22         infoBuffer.append("经检测该手机有" + allSensors.size() + "个传感器,它们分别是:\n");
23         //迭代传感器列表
24         for(Sensor sensor : allSensors){
25             switch(sensor.getType()){
26             caseSensor.TYPE_ACCELEROMETER:
27                 infoBuffer.append(sensor.getType() + " 加速度传感器 acceleromter");
28                 break;
29             caseSensor.TYPE_GYROSCOPE:
30                 infoBuffer.append(sensor.getType() + " 陀螺仪传感器 gyroscope");
31                 break;
32             caseSensor.TYPE_LIGHT:
33                 infoBuffer.append(sensor.getType() + " 环境光线传感器 light");
34                 break;
35             caseSensor.TYPE_MAGNETIC_FIELD:
36                 infoBuffer.append(sensor.getType() + " 电磁场传感器 magnetic field");
37                 break;
38             caseSensor.TYPE_ORIENTATION:
39                 infoBuffer.append(sensor.getType() + " 方向传感器 orientation");
40                 break;
41             caseSensor.TYPE_PRESSURE:
42                 infoBuffer.append(sensor.getType() + " 压力传感器 pressure");
43                 break;
44             caseSensor.TYPE_PROXIMITY:
```

```
45                  infoBuffer.append(sensor.getType() + " 距离传感器 proximity");
46                  break;
47              caseSensor.TYPE_TEMPERATURE:
48                  infoBuffer.append(sensor.getType() + " 温度传感器 temperature");
49                  break;
50              default:
51                  infoBuffer.append(sensor.getType() + " android 2.2 以下系统不支持触感器");
52                  break;
53              }
54              infoBuffer.append(" \n" + " 设备名称:" + sensor.getName() + " \n" + " 设备版本:" + sen-
sor.getVersion() + " \n" + " 供应商:" + sensor.getVendor() + " \n" + "最大量程:" + sensor.getMaximumRange()
 + " \n" + "传感器功率:" + sensor.getPower() + " \n" + "传感器分辨率:" + sensor.getResolution() + " \n" + "传感
器最小延迟:" + sensor.getMinDelay() + " \n" + "传感器类型:" + sensor.getType() + " \n");
55          }
56          sensorInfoTextView.setText(infoBuffer);
57      }
58      @Override
59      public boolean onCreateOptionsMenu(Menu menu) {
60          getMenuInflater().inflate(R.menu.main, menu);
61          return true;
62      }
63  }
```

最终的运行结果如图 11-1 所示。

3. SensorEvent 类

SensorEvent 类从本质上来说，它是一个数据结构，包含了硬件传感器输出到应用的信息。它是对从传感器事件上取得的信息进行整理管理的类。被管理的值全部用公用的 field 定义。其值如表 11-5 所示。

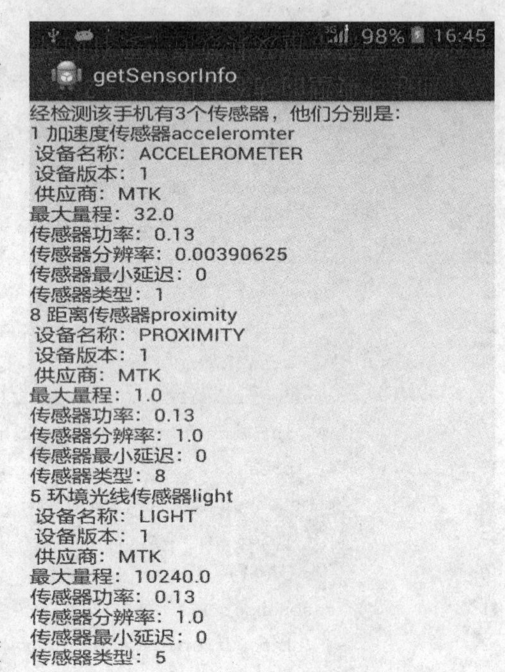

图 11-1 运行结果

表 11-5 SensorEvent 类的主要字段（field）

字　段	内　容
public int accuracy	传感器的精度
public Sensor sensor	Sensor 类中生成 SensorEvent 的实例
public long timestamp	SensorEvent 发生的时间，以毫秒为单位
public final float[ ] values	表示传感器数据的数组。数组的大小以及数组值的含义取决于产生数据的传感器

说明：传感器的精度指的是输出值的可靠度或者是“可信度”，而不是与物理值的接近程度。主要可分为以下几个等级：

- int SENSOR_STATUS_ACCURACY_HIGH
- int SENSOR_STATUS_ACCURACY_LOW
- int SENSOR_STATUS_ACCURACY_MEDIUM
- int SENSOR_STATUS_NO_CONTACT
- int SENSOR_STATUS_UNRELIABLE

4. SensorEventListener 类

SensorEventListener 类是提供回调以通知应用传感器相关事件的接口。为了能掌控这些事件，应用需要创建一个类实现 SensorEventListener 接口，并将其注册到 SensorManager。在这个封装的接口中可以获得传感器的值，其主要方法有以下两类：

- void onAccuracyChanged（Sensor sensor，int accuracy）：该方法在传感器的精准度发生改变时调用。其参数包括两个整数：一个表示传感器，另一个表示该传感器新的准确度。

- void onSensorChanged（SensorEvent event）：该方法在传感器值更改时调用。该方法只由受此应用程序监视的传感器调用。该方法的参数包括一个 SensorEvent 对象，该对象主要包括一组浮点数，表示传感器获得的方向、加速度等信息。

传感器值的取得需要通过 SensorManager 中的 registerListener 方法对加载 SensorEventListener 接口的对象进行登录处理。从登录开始到传感器的值取得的动作处理过程如图 11-2 所示。

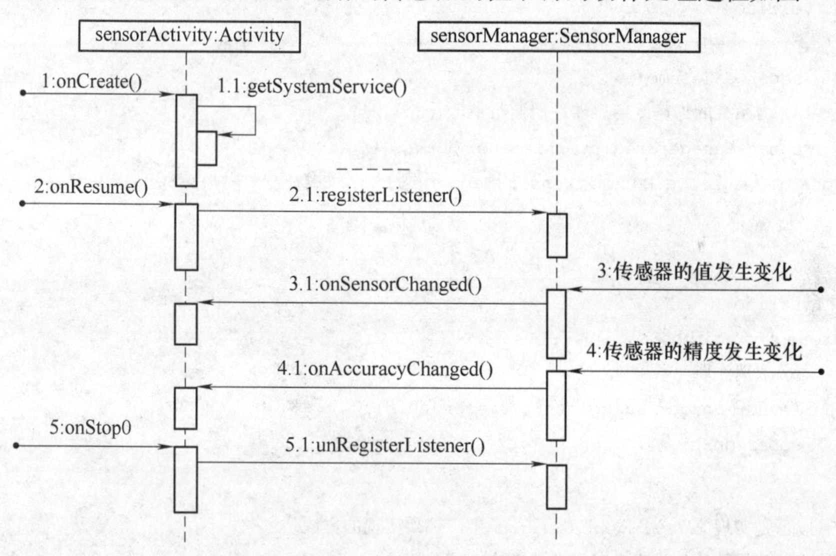

图 11-2 传感器取值过程

【例 11-2】 以加速度传感器为例来说明获取传感器值的具体步骤。

（1）新建工程 AccelerometerTest，其布局文件详见随书光盘中的代码

（2）向 MainActivity. java 文件中添加如下代码：

```
1   package com.example.accelerometertest;

2   import android.app.Activity;
3   import android.content.Context;
4   import android.hardware.Sensor;
5   import android.hardware.SensorEvent;
6   import android.hardware.SensorEventListener;
```

```
7   import android.hardware.SensorManager;
8   import android.os.Bundle;
9   import android.widget.EditText;
10  public class AccelerometerTest extends Activity
11      implements SensorEventListener{
12          //定义系统的 Sensor 管理器
13          SensorManager sensorManager;
14          EditText etTxt1;
15          @Override
16          public void onCreate(Bundle savedInstanceState){
17              super.onCreate(savedInstanceState);
18              setContentView(R.layout.main);
19              //获取程序界面上的文本框组件
20              etTxt1 = (EditText) findViewById(R.id.txt1);
21              //获取系统的传感器管理服务
22              sensorManager = (SensorManager) getSystemService(Context.SENSOR_SERVICE);
23          }
24          @Override
25              protected void onResume(){
26              super.onResume();
27              //为系统的加速度传感器注册监听器
28              sensorManager.registerListener(this,
29          sensorManager.getDefaultSensor(Sensor.TYPE_ACCELEROMETER), SensorManager.SENSOR_DELAY
_GAME);
30          }
31          @Override
32          protected void onStop(){
33              //取消注册
34              sensorManager.unregisterListener(this);
35              super.onStop();
36          }
37          //以下是实现 SensorEventListener 接口必须实现的方法
38          //当传感器的值发生改变时回调该方法
39          @Override
40          public void onSensorChanged(SensorEvent event){
41              float[] values = event.values;
42              StringBuilder sb = new StringBuilder();
43              sb.append("x 方向上的加速度:");
44              sb.append(values[0]);
45              sb.append("\ny 方向上的加速度:");
46              sb.append(values[1]);
47              sb.append("\nz 方向上的加速度:");
48              sb.append(values[2]);
49              etTxt1.setText(sb.toString());
```

```
50          }
51          // 当传感器精度改变时回调该方法。
52          @ Override
53          public void onAccuracyChanged(Sensor sensor, int accuracy){
54          }
55      }
```

本程序直接使用了 Activity 充当传感器监听器，因此该 Activity 实现了 SensorEventListener 接口，并实现了该接口中的 onSensorChanged( )方法。为了取得传感器的值需要加载 SensorEventListener，并通过 onResume( )方法进行注册，在该方法中实现 SensorEventListener 接口中的 onSensorChanged( )方法时是通过调用 SensorEvent 对象的 values( )方法来获取传感器的值，不同的传感器返回的值的个数是不等的。对于加速度传感器来说，它将返回三个值，分别代表手机设备在 x、y、z 三个方向上的加速度。最后在 onStop( )方法中解除所有传感器。其结果如图 11-3 所示。

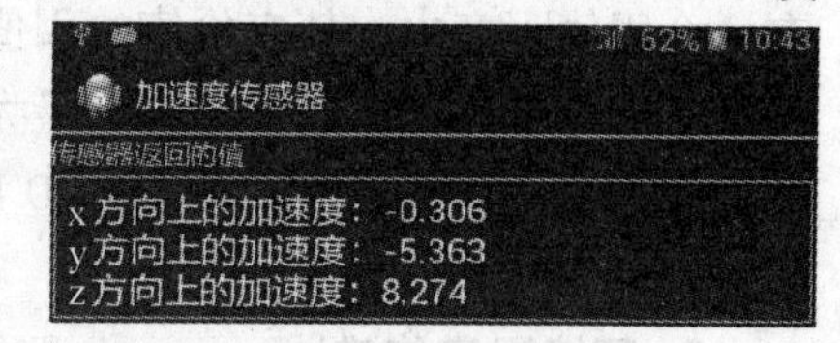

图 11-3 运行结果

11.2 Android 中常用的传感器

Android 中的传感器大致可以分为两大类：一类是感知环境的传感器；另一类则是感知设备方向和运动的传感器。在上一节中已经介绍了如何获得 Android 设备的加速度值，实际上 Android 系统对所有类型的传感器的处理完全一样，只不过是传感器的类型有所区别而已。接下来将具体介绍 Android 中常用的传感器。

11.2.1 感知环境

1. 光线传感器（Sensor. TYPE_LIGHT）

光线传感器主要是用来检测手机周围光的强度，位于一个小的黑色玻璃开口下面。它只是一个光敏二极管，工作方式和 LED 的物理原理相同，但是发光条件却正好相反。不是在施加电压时发光，而是在光入射时产生电压。与其他传感器不同的是，该传感器只读取一个数值即手机周围光的强度，单位为勒克斯（lx）。通常的动态范围为 1 ~ 30000 lx。在 Android 中这些范围被几个常量值所代替，常量值如表 11-6 所示。

表 11-6 光线传感器常量值

常 量	值
SensorManager. LIGHT_NO_MOON	0.001
SensorManager. LIGHT_FULLMOON	0.25
SensorManager. LIGHT_CLOUDY	100
SensorManager. LIGHT_SUNRISE	400
SensorManager. LIGHT_OVERCAST	10000
SensorManager. LIGHT_SHADE	20000
SensorManager. LIGHT_SUNLIGHT	110000
SensorManager. LIGHT_SUNLIGHT_MAX	120000

上面的八个常量只是临界值。读者在实际使用光线传感器时要根据实际情况确定一个范围。例如，当太阳逐渐升起时，它的值很可能会超过 LIGHT_SUNRISE，当它的值逐渐增大时，就会逐渐越过 LIGHT_OVERCAST，而达到 LIGHT_SHADE，当然，如果天特别好的话，也可能会达到 LIGHT_SUNLIGHT，甚至更高。关于其应用将在后续章节中介绍。

2. 接近传感器（Sensor. TYPE_PROXIMITY）

接近传感器包含一个在光电探测器边上的弱红外 LED。当有物体离传感器足够近时，光敏传感器会检测到反射的红外光。接近传感器分为两类：一类是用来检测物体与手机的距离，单位是 cm。而另一类则是现在大多数智能手机中的接近传感器用法——测量物体是否在一个阈值距离内，其有价值的阈值距离一般为 2 ~ 4cm。一些接近传感器只能返回远和近两个状态，因此，接近传感器将最大距离返回远状态，小于最大距离返回近状态。接近传感器可用于接听电话时自动关闭 LCD 屏幕以节省电量。一些芯片集成了接近传感器和光线传感器两者的功能。

3. 环境温度传感器（Sensor. TYPE_AMBIENT_TEMPERATURE）

环境温度传感器提供室内温度，单位为℃。这种传感器是为了取代已逐步淘汰的用于检测 CPU 温度的 Sensor. TYPE_TEMPERATURE。环境温度传感器会返回一个数据，该数据代表手机设备周围的温度。

11. 2. 2 感知设备方向和运动

1. 方向传感器（Sensor. TYPE_ORIENTATION）

方向传感器用于感应手机设备的摆放状态。方向传感器可返回三个角度，这三个角度即可确定手机的摆放状态。分别为：方向角、倾斜角以及旋转角。其坐标示意图如图 11-4 所示。

方向传感器比较特殊，因为它的数值是相对于绝对方向的。它得到的是手机设备的绝对姿态值。注意下面说的 x、y、z 轴均是手机自身的坐标轴。

第一个角度：方向角（Azimuth）——表示手机自身的 y 轴与地磁场北极方向的角度，即手机顶部朝向与正北方向的角度。当手机绕着自身的 z 轴旋转时，该角度值将发生改变。

例如该角度值为 0 时，表示手机顶部指向正北；该角度为 90°时，代表手机顶部指向正东；该角度为 180°时，代表手机顶部指向正南；该角度为 270°时，代表手机顶部指向正西。

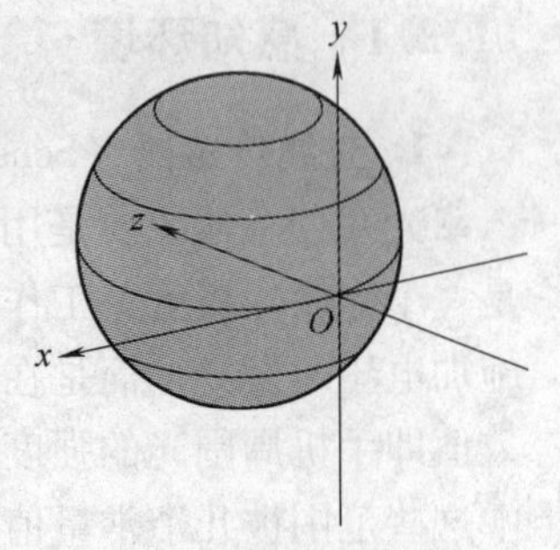

图 11-4　方向传感器坐标

第二个角度：倾斜角（Pitch）——表示手机顶部或尾部翘起的角度。当手机绕着自身的 x 轴旋转，该角度会发生变化，值的范围是 -180° ~ 180°。当 z 轴正向朝着 y 轴正向旋转时，该角度是正值；当 z 轴正向朝着 y 轴负向旋转时，该角度是负值。假设将手机屏幕朝上水平放在桌子上，如果桌子是完全水平的，该角度应该是 0°。假如从手机顶部抬起，直到将手机沿 x 轴旋转 180°（屏幕向下水平放在桌面上），这个过程中，该角度值会从 0 变化到 -180 如果从手机底部开始抬起，直到将手机沿 x 轴旋转 180°（屏幕向下水平放在桌面上），该角度的值会从 0 变化到 180。

第三个角度：旋转角（Roll）——表示手机左侧或右侧翘起的角度。当手机绕着自身 x

轴旋转时，该角度值将会发生变化，取值范围是 -90°~90°。当 z 轴正向朝着 x 轴正向旋转时，该角度是负值；当 z 轴正向朝着 x 轴负向旋转时，该角度是正值。利用方向传感器可开发出水平仪等有趣的应用。

2. 加速度传感器（Sensor. TYPE_ACCELEROMETER）

加速度传感器又叫 G-sensor，返回 x、y、z 三轴的加速度数值。该数值包含地心引力的影响，单位是 m/s^2。但其坐标与方向传感器不同。如图 11-5 所示，将手机平放在桌面上，x 轴方向默认值为 0，y 轴方向默认值为 0，z 轴方向默认为 $9.8m/s^2$。且对于这 3 个值有如下约定：

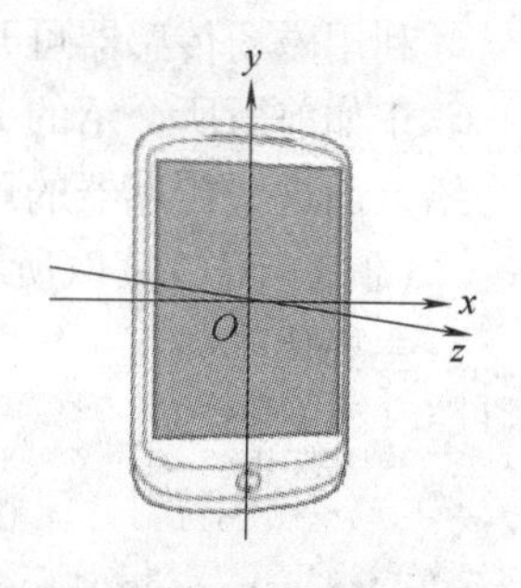

图 11-5　加速度传感器坐标

values［0］：x 轴上为负 gx；

values［1］：y 轴上为负 gy；

values［2］：z 轴上为负 yz。

加速计典型的动态范围为 0 ±2g 或者 ±4g，分辨率为 $0.1m/s^2$。在速度传感器中会经常用到的传感器常量及其数值主要有两种：

SensorManager. GRAVITY_EARTH：9. 80665

SensorManager. STANDARD_GRAVITY：9. 80665

加速度传感器可能是最为成熟的一种微机电传感器（MEMS）产品，市场上的加速度传感器种类很多。手机中常用的加速度传感器有 BOSCH（博世）的 BMA 系列，AMK 的 897X 系列，ST 的 LIS3X 系列等。这些传感器一般提供 ±2g 至 ±16g 的加速度测量范围，采用 I2C 或 SPI 接口和 MCU 相连，数据精度小于 16bit。

3. 磁场传感器（Sensor. TYPE_MAGNETIC_FIELD）

磁场传感器主要用于读取手机设备外部的磁场强度。即使周围没有任何的直接磁场，手机设备也始终会处于地球磁场中。随着手机设备摆放状态的改变，周围磁场在手机的 x、y、z 方向上的影响会发生改变。Android 输出的磁场以微特斯拉为单位。磁场传感器的分辨率为 0. 1μT，它的输出值绝对精确度较低，同时会受当地环境的影响。如果想获取高精度测量值，此时就需要用到 GeomagneticField 类，磁场传感器中常用的方法如表 11-7 所示。

表 11-7　磁场传感器中常用的方法

方　法	返 回 值
float getDeclination()	磁北和给定位置实际北方之间的角度
float getFieldStrength()	总的磁场强度以纳特斯拉为单位
float getHorizontalStrength()	水平方向的磁场强度以纳特斯拉为单位
float getInclination()	磁场向上或向下与水平线偏移的角度
float getX()	磁场向北分量以纳特斯拉为单位
float getY()	磁场向东分量以纳特斯拉为单位
float getZ()	磁场向下分量以纳特斯拉为单位

跟其他传感器一样在 SensorManager 中定义了相关的常量以及它们的值。具体如下：

SensorManager. MAGNETIC_FIELD_EARTH_MAX：60.0

SensorManager. MAGNETIC_FIELD_EARTH_MIN：30.0

利用磁场传感器可开发出指南针、罗盘等磁场应用。

正如前面所介绍的 Android 系统对所有的传感器处理方式都一样，下面通过一个实例来介绍上面这些传感器的用法。

【例 11-3】 读取所有传感器的值并显示。

主要代码如下：

```
1  public class MainActivity extends Activity implements SensorEventListener{
2      /* *  Called when the activity is first created. * /
3      private SensorManager sensorManager;
4      @Override
5      public void onCreate(Bundle savedInstanceState) {
6          super.onCreate(savedInstanceState);
7          //SensorManager 的接口取得
8          sensorManager = (SensorManager)this.getSystemService(SENSOR_SERVICE);
9          setContentView (R.layout.activity_main);
10     }

11     @Override
12     protected void onResume () {
13         super.onResume ();
14         List < Sensor > sensors = sensorManager.getSensorList (Sensor.TYPE_ALL);

15         //sensor1
16         for (Sensor s : sensors) {
17 sensorManager.registerListener (this, s, SensorManager.SENSOR_DELAY_NORMAL);
18     }
19     }

20     @Override
21     public void onAccuracyChanged (Sensor sensor, int accuracy) {
22         // TODO Auto-generated method stub
23     }
24     @Override
25     public void onSensorChanged (SensorEvent e) {
26         // TODO Auto-generated method stub
27         switch (e.sensor.getType ()) {
28         //加速度的值表示
29         caseSensor.TYPE_ACCELEROMETER: {
30             TextView x = (TextView) findViewById (R.id.x);
31             x.setText (" x:" + String.valueOf (e.values [SensorManager.DATA_X]));
32             TextView y = (TextView) findViewById (R.id.y);
33             y.setText (" y:" + String.valueOf (e.values [SensorManager.DATA_Y]));
```

```
34              TextView z = (TextView)findViewById(R.id.z);
35              z.setText("z:"+String.valueOf(e.values[SensorManager.DATA_Z]));
36              break;
37          }
38          //倾斜度的值表示
39          caseSensor.TYPE_ORIENTATION: {
40              TextView x= (TextView) findViewById (R.id.Azimuth);
41                  x.setText (" Azimuth" +String.valueOf (e.values [SensorManager.DATA_X]));
42                  TextView y= (TextView) findViewById (R.id.Pitch);
43                  y.setText (" Pitch:" +String.valueOf (e.values [SensorManager.DATA_Y]));
44                  TextView z= (TextView) findViewById (R.id.Roll);
45                  z.setText (" Roll:" +String.valueOf (e.values [SensorManager.DATA_Z]));
46                  break;
47          }
48          //磁力计的值表示
49          caseSensor.TYPE_MAGNETIC_FIELD: {
50              TextView x= (TextView) findViewById (R.id.magnetic_x);
51              x.setText (" x:" +String.valueOf (e.values [SensorManager.DATA_X]));
52              TextView y= (TextView) findViewById (R.id.magnetic_y);
53              y.setText (" y:" +String.valueOf (e.values [SensorManager.DATA_Y]));
54              TextView z= (TextView) findViewById (R.id.magnetic_z);
55              z.setText (" z:" +String.valueOf (e.values [SensorManager.DATA_Z]));
56              break;
57          }
58          //温度计值的取得
59          caseSensor.TYPE_TEMPERATURE: {
60              TextView x= (TextView) findViewById (R.id.degree);
61              x.setText (" Degree:" +String.valueOf (e.values [SensorManager.DATA_X]));
62          }
63          caseSensor.TYPE_PROXIMITY: {
64              TextView x= (TextView) findViewById (R.id.proximity);
65              x.setText (" proximity:" +String.valueOf (e.values [SensorManager.DATA_X]));
66          }
67          //近距离感应器值的取得
68          }
69      }
70      @Override
71      protected void onStop () {
72          super.onStop ();
73          sensorManager.unregisterListener (this);
74      }
75  }
```

最终的运行结果如图 11-6 所示。

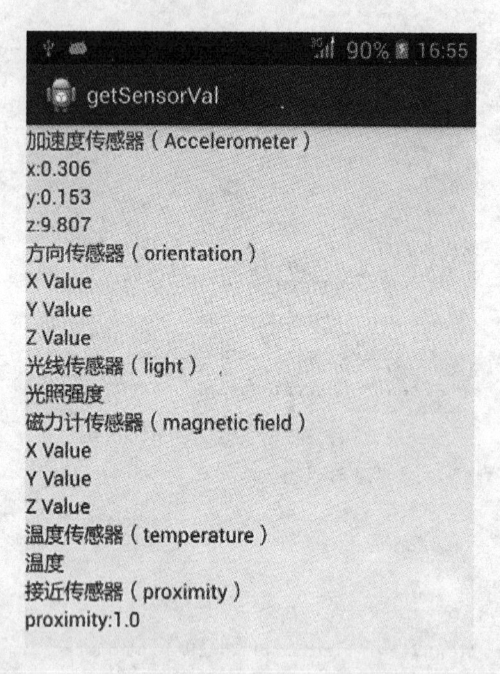

图 11-6　运行结果图

11.3　传感器应用案例

对传感器的支持是 Android 的特性之一，通过对传感器的使用可以开发出很多有趣的应用，下面就通过使用加速度传感器来开发两个有趣的应用。

11.3.1　Android 加速度传感器应用——实现手机摇一摇控制音乐播放

这个播放器的开始和停止是由晃动手机来实现的。其主要思路是：当手机晃动时，加速度的变化幅度超过设定的界限，视为决定播放或停止。与获取传感器值类似，首先通过 getSystemService 方法得到 SensorManager 对象，然后通过 RegisterListener 方法来对其进行注册，并在 onSensorChanged( )方法中设置监听器以实现该功能，最后注销传感器。

【例 11-4】　实现晃动手机控制播放器的应用程序。

其主要代码如下：(其他部分见随书代码)

```
1   public class MainActivity extends Activity implements SensorListener{
2       private MediaPlayer mp = null;//播放音乐
3       SensorManager sm;
4       private long currentTime,lastTime,duration;//记录传感器变化的时间变量
5       private float lastX,lastY,lastZ;//记录上一次振动位置
6       private float currentShake,totalShake;//记录振动幅度
7       private boolean isPlaying = false;//判断当前是否在播放音乐
8       @Override
9       public void onCreate(Bundle savedInstanceState) {
10          super.onCreate(savedInstanceState);
```

```
11          setContentView(R.layout.activity_main);
12          //载入音乐资源
13          mp=MediaPlayer.create (MainActivity.this, R.raw.jnstyle);
14          //获取系统服务
15          sm= (SensorManager) this.getSystemService (Context.SENSOR_SERVICE);
16          //注册传感器监听，监听加速器
17          sm.registerListener (this, SensorManager.SENSOR_ACCELEROMETER, SensorManager.SENSOR_
DELAY_NORMAL);
18      }
19      //覆写 SensorListener 的方法，传感器精度变化时调用
20      public void onAccuracyChanged (int sensor, int accuracy) {
21          // TODO Auto-generated method stub
22      }
23      //覆写 SensorListener 的方法，传感器值变化时调用
24      public void onSensorChanged (intsensor, float [] values) {
25          // TODO Auto-generated method stub
26          //获取当前振动位置
27          float x=values [0];
28          float y=values [1];
29          float z=values [2];
30          currentTime=System.currentTimeMillis (); //获取系统当期时间
31          if (lastX= =0&&lastY= =0&&lastZ= =0) //第一次振动
32              lastTime=currentTime;
33              if (currentTime-lastTime>200) {//当两次振动时间间隔大于200ms时捕捉振动
34                  duration=currentTime-lastTime;
35 currentShake= (Math.abs (x-lastX) + Math.abs (y-lastY) + Math.abs (z-lastZ)) /duration*
200; //获取振动幅度
36              }
37              totalShake+ =currentShake;
38              if (totalShake>100) {//两次总振动幅度大于100时触发传感器，控制音乐播放或暂停
39                  //数据恢复未振动状态
40                  totalShake=0;
41                  lastX=0;
42                  lastY=0;
43                  lastZ=0;
44                  lastTime=0;
45                  currentTime=0;
46                  if (! isPlaying) {
47                      mp.start ();
48                      isPlaying=true;
49                      ( (TextView) findViewById (R.id.sensor_hint)) .setText (" 当前状态：播放中
.....");  //提示信息
50                  } else {
51                      mp.stop ();
52                      isPlaying=false;
```

```
53                    ((TextView) findViewById (R.id.sensor_hint)) .setText (" 当前状态: 暂停中
....."); //提示信息
54                }
55            }
56            lastX = x; lastY = y; lastZ = z;
57            lastTime = currentTime;
58        }
59        public void stop () {//关闭 Activity 时注销监听
60        sm.unregisterListener (this);
61    }
62 }
```

运行结果如图 11-7 所示。

图 11-7　运行结果

11.3.2　Android 加速度传感器应用二——重力小球

该案例主要是通过小球的滚动来形象化地体现重力加速度的变化。其主要思路是：当晃动手机时在 x、y、z 方向上的加速度值会产生变化，这种变化通过小球的运动轨迹来体现。与案例相同的是都在 onSensorChanged()方法中对加速度值的变化进行监听并进行相应的处理。

【例 11-5】　实现一个小球滚动来体现重力加速度的变化例程。

其主要代码如下：（其余部分见随书代码）

```
1  public class AccelerometerActivity extends Activity {
2      private static final float MAX_ACCELEROMETER = 9.81f;
3      private SensorManager sensorManager;
4      private BallView ball;
5      private boolean success = false;
```

```
6       private boolean init = false;
7       private int container_width = 0;
8       private int container_height = 0;
9       private int ball_width = 0;
10      private int ball_height = 0;
11      private TextView prompt;
12      @Override
13      public void onCreate (Bundle savedInstanceState) {
14          super. onCreate (savedInstanceState);
15          setContentView (R. layout. main);
16          //获取感应器管理器
17          sensorManager = (SensorManager) getSystemService (SENSOR_SERVICE);
18          prompt = (TextView) findViewById (R. id. ball_prompt);
19      }
20      @Override
21      public void onWindowFocusChanged (booleanhasFocus) {//ball_container 控件显示出来后才能获取
其宽和高，所以在此方法得到其宽高
22          super. onWindowFocusChanged (hasFocus);
23          if (hasFocus&& ! init) {
24              View container = findViewById (R. id. ball_container);
25              container_width = container. getWidth ();
26              container_height = container. getHeight ();
27              ball = (BallView) findViewById (R. id. ball);
28              ball_width = ball. getWidth ();
29              ball_height = ball. getHeight ();
30              moveTo (0f, 0f);
31              init = true;
32          }
33      }
34      @Override
35      protected void onResume () {
36          Sensor sensor = sensorManager. getDefaultSensor (Sensor. TYPE_ACCELEROMETER); //获取重力
加速度感应器
37          success = sensorManager. registerListener (listener, sensor, SensorManager. SENSOR_DELAY_
GAME); //注册 listener，第三个参数是检测的精确度
38          super. onResume ();
39      }
40      @Override
41      protected void onPause () {
42          if (success) sensorManager. unregisterListener (listener);
43              super. onPause ();
44      }
45      privateSensorEventListener listener = new SensorEventListener () {
46          @Override
```

```
47              public void onSensorChanged(SensorEvent event) {
48              if (! init) return ;
49              float x = event. values[SensorManager. DATA_X];
50              float y = event. values  [SensorManager. DATA_Y];
51              float z = event. values  [SensorManager. DATA_Z];
52              prompt. setText  (" Sensor: "  + x + ", "  + y + ", "  + z);
53              //当重力 x, y 为 0 时，球处于中心位置，以 y 为轴心（固定不动），转动手机，x 会在（0 ~ 9.81）之间变化，负号代表方向
54              moveTo  (-x, y); //x 方向取反
55          }
56          @ Override
57          public void onAccuracyChanged  (Sensor sensor, int accuracy) {
58          }
59      };
60      private void moveTo  (float x, float y) {
61          int max_x =  (container_width- ball_width) / 2; //在 x 轴可移动的最大值
62          int max_y =  (container_height - ball_height) / 2; //在 y 轴可移动的最大值
63          //手机沿 x、y 轴垂直摆放时，自由落体加速度最大为 9.81，当手机沿 x、y 轴成某个角度摆放时，变量 x 和 y 即为该角度的加速度
64          float percentageX = x / MAX_ACCELEROMETER; //得到当前加速度的比率，如果手机沿 x 轴垂直摆放，比率为 100%，即球在 x 轴上移动到最大值
65          float percentageY = y / MAX_ACCELEROMETER;
66          int pixel_x =  (int) (max_x *  percentageX); //得到 x 轴偏移量
67          int pixel_y =  (int) (max_y *  percentageY); //得到 y 轴偏移量
68          //以球在中心位置的坐标为参考点，加上偏移量，得到球的对应位置，然后移动球到该位置
69          ball. moveTo  (max_x + pixel_x, max_y + pixel_y);
70      }
71  }
```

最终的运行效果如图 11-8 所示。

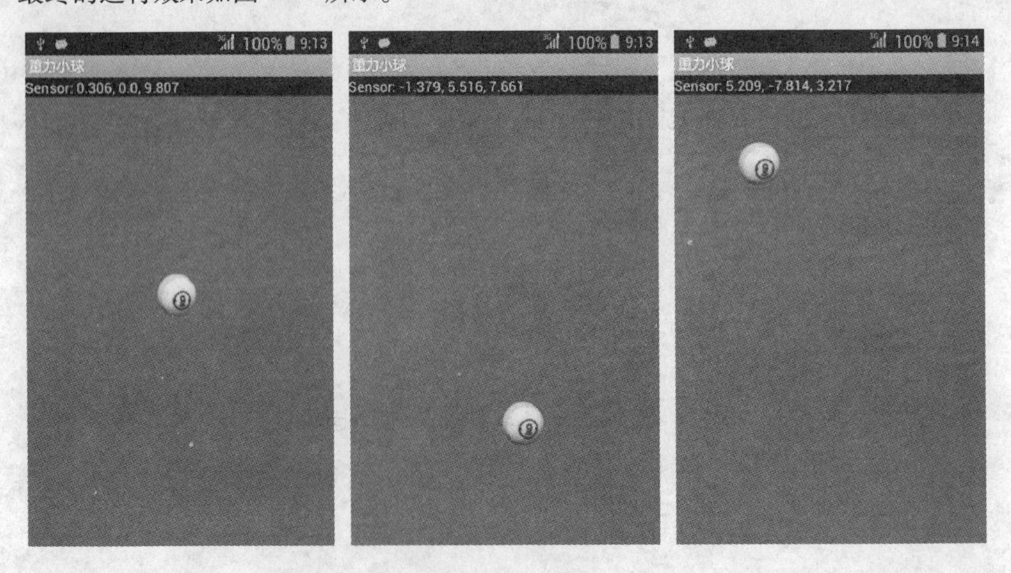

图 11-8　运行结果

本 章 小 结

Android 系统的特色之一就是对传感器的支持。本章详细介绍了 Android 系统所支持的传感器类型、如何使用传感器 API 来获取传感器数据、如何通过 SensorManager 来注册传感器监听器、如何在 SensorEventListener 中对传感器进行监听、如何使用几种常用的传感器等。最后通过使用最常用的加速度传感器开发有趣的应用来进一步介绍传感器开发的流程。

习　　题

11-1　利用 Android 中的传感器设计一个水平仪。

11-2　利用 Android 中的传感器设计一个应用程序实现微信中的“摇一摇”功能。

11-3　利用 Android 中的传感器设计一个计步器。

参考文献

[1] Zigurd Mednieks, Laird Dornin, G Blake Meike，等. Android 程序设计［M］. 祝洪凯，李妹芳，译. 2 版. 北京：机械工业出版社，2014.

[2] Onur Cinar. Android C ++高级编程——使用 NDK［M］. 于红，佘建伟，冯艳红，译. 北京：清华大学出版社，2014.

[3] 邓文渊. Android 开发基础教程［M］. 北京：人民邮电出版社，2014.

[4] 罗雷，韩建文，汪杰. Android 系统应用开发实战详解［M］. 北京：人民邮电出版社，2014.

[5] Marko Gargenta. Learning Android：Develop Mobile Apps Using Java and Eclipse［M］. 2nd，USA：O'Reilly Media，Inc，2014.

[6] 弗里森. Android 开发范例代码大全［M］. 赵凯，陶冶，译. 2 版. 北京：清华大学出版社，2014.

[7] Wei-Meng Lee. Android application development cookbook：93 recipes for building winning apps［M］. Indianapolis：John Wiley & Sons，2013.

[8] 王英强，等. Android 应用程序设计［M］. 北京：清华大学出版社，2013.

[9] 张思民. Android 应用程序设计［M］. 北京：清华大学出版社，2013.

[10] 秦建平. Android 编程宝典［M］. 北京：北京航天航空大学出版社，2013.

[11] 李刚. 疯狂 Android 讲义［M］. 3 版. 北京：电子工业出版社，2013.

[12] 迈耶. Android 4 高级编程［M］. 佘建伟，赵凯，译. 3 版. 北京：清华大学出版社，2013.

[13] Michael Burton，Donn Felker. Android application development for dummies［M］. New Jersey：Wiley，2012.

[14] Rob Huddleston. Android fully loaded / Rob Huddleston［M］. New Jersey：Wiley，2012.

[15] Jason Ostrander. Android UI Fundamentals：Develop & Design［M］. San Francisco：Peachpit Press，2012.

[16] Frank Ableson，Robi Sen，Chris King，等. Android in action［M］. Shelter Island，NY：Manning Publications Co.，2012.

[17] 吴亚锋，于复兴. Android 应用开发完全自学手册——核心技术、传感器、2D/3D、多媒体与典型案例［M］. 北京：人民邮电出版社，2012.

[18] 明日科技. Android 从入门到精通［M］. 北京：清华大学出版社，2012.

[19] 王石磊，吴峥. Android 多媒体应用开发实战详解：图像、音频、视频、2D 和 3D［M］. 北京：人民邮电出版社，2012.

[20] 关东升，等. Android 开发案例驱动教程［M］. 北京：机械工业出版社，2011.

[21] 林城. Google Android 2. X 应用开发实战［M］. 北京：清华大学出版社，2011.

[22] 杨丰盛. Android 技术内幕：系统卷［M］. 北京：机械工业出版社，2011.

[23] Jeff Friesen，DaveSmith. Android recipes：a problem - solution approach［M］. New York：APress，2011.

[24] Mark L，Murphy. Beginning Android 2［M］. New York：APress，2010.

[25] 吴亚峰，索依娜. Android 核心技术与实例详解［M］. 北京：电子工业出版社，2010.

[26] 杨丰盛. Android 应用开发揭秘［M］. 北京：机械工业出版社，2010.

[27] 余志龙，等. Google Android SDK 开发范例大全［M］. 2 版. 北京：人民邮电出版社，2010.

[28] 郭宏志. Android 应用开发详解［M］. 北京：电子工业出版社，2010.

[29] 靳岩，姚尚朗. Google Android 开发入门与实战［M］. 北京：人民邮电出版社，2009.

[30] 盖索林. Google Android 开发入门指南［M］. 北京：人民邮电出版社，2009.

[31] 申康. 蓝牙协议体系结构［J］. 微电子技术，2001（6）：55-57.